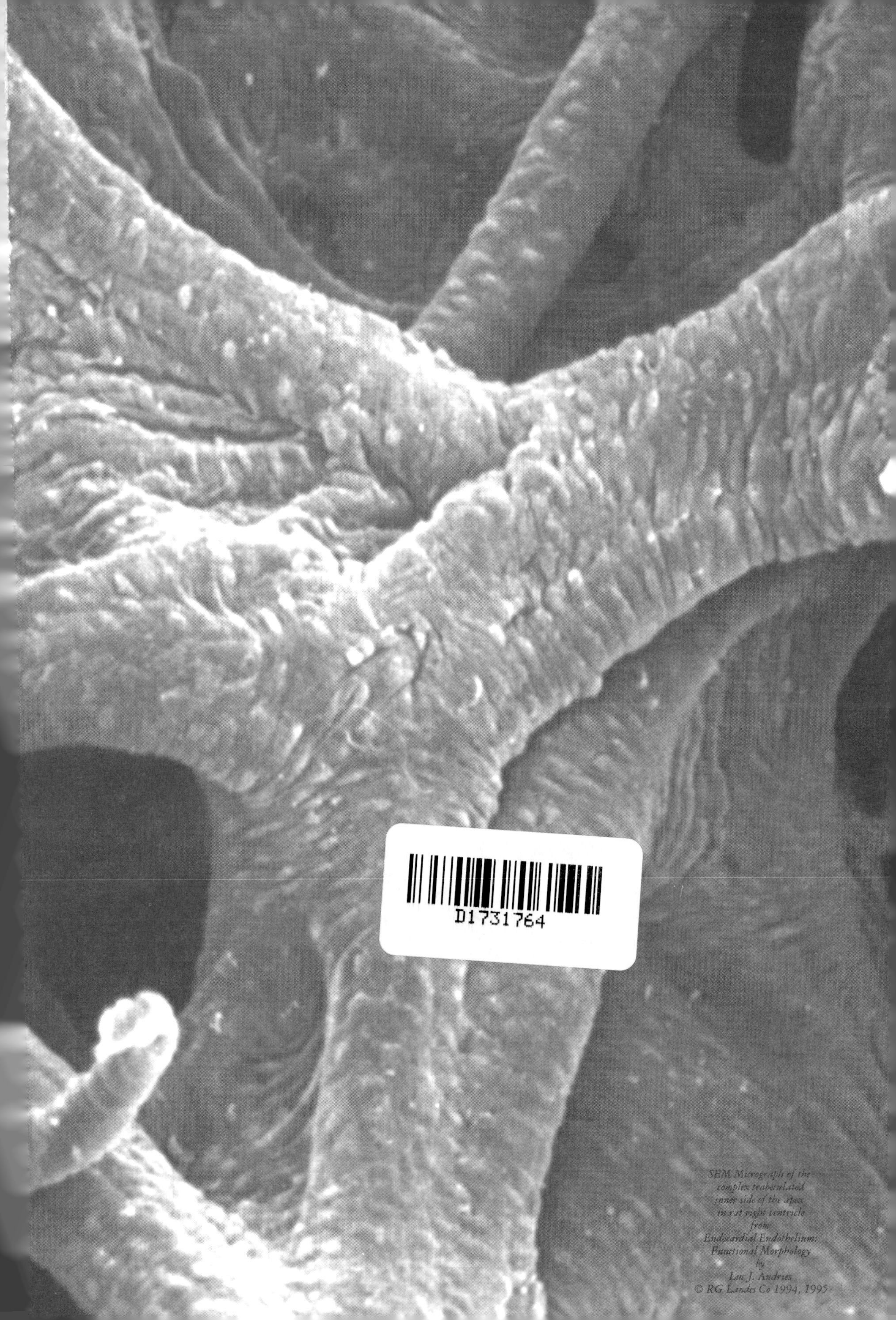

SEM Micrographs of the complex trabeculated inner side of the apex in rat right ventricle from Endocardial Endothelium: Functional Morphology by Luc J. Andries

MEDICAL INTELLIGENCE UNIT

MYOCARDIAL PRECONDITIONING

Cherry L. Wainwright
James R. Parratt

University of Strathclyde
Glasgow, Scotland

Springer
New York Berlin Heidelberg London Paris
Tokyo Hong Kong Barcelona Budapest

R.G. LANDES COMPANY
AUSTIN

MEDICAL INTELLIGENCE UNIT
MYOCARDIAL PRECONDITIONING

R.G. LANDES COMPANY
Austin, Texas, U.S.A.

Printed in the U.S.A.

Please address all inquiries to the Publishers:
R.G. Landes Company, 909 Pine Street, Georgetown, Texas, U.S.A. 78626
Phone: 512/ 863 7762; FAX: 512/ 863 0081

International distributor (except North America):
Springer-Verlag GmbH & Co. KG
Tiergartenstrasse 17, D-69121 Heidelberg, Germany

International ISBN: 3-540-60790-0

While the authors, editors and publisher believe that drug selection and dosage and the specifications and usage of equipment and devices, as set forth in this book, are in accord with current recommendations and practice at the time of publication, they make no warranty, expressed or implied, with respect to material described in this book. In view of the ongoing research, equipment development, changes in governmental regulations and the rapid accumulation of information relating to the biomedical sciences, the reader is urged to carefully review and evaluate the information provided herein.

Library of Congress Cataloging-in-Publication Data

Myocardial preconditioning / [edited by] Cherry L. Wainwright, James R. Parratt.
p. cm. — (Medical intelligence unit)
Includes bibliographical references and index.
ISBN 1-57059-333-7 (alk. paper)
1. Myocardial infarction--Prevention. 2. Coronary heart disease--Pathophysiology.
3. Heart--Adaptation. I. Wainwright, Cherry L., 1960- . II. Parratt, James R. III. Series.
[DNLM: 1. Myocardial Ischemia--physiopathology. WG 300 M997667 1996]
RC685.I6M96 1996
616.1'23705--dc20
DLM/DLC
for Library of Congress

96-4883
CIP

Publisher's Note

R.G. Landes Company publishes six book series: *Medical Intelligence Unit, Molecular Biology Intelligence Unit, Neuroscience Intelligence Unit, Tissue Engineering Intelligence Unit, Environmental Intelligence Unit* and *Biotechnology Intelligence Unit.* The authors of our books are acknowledged leaders in their fields and the topics are unique. Almost without exception, no other similar books exist on these topics.

Our goal is to publish books in important and rapidly changing areas of bioscience for sophisticated researchers and clinicians. To achieve this goal, we have accelerated our publishing program to conform to the fast pace in which information grows in bioscience. Most of our books are published within 90 to 120 days of receipt of the manuscript. We would like to thank our readers for their continuing interest and welcome any comments or suggestions they may have for future books.

Deborah Muir Molsberry
Publications Director
R.G. Landes Company

CONTENTS

1. Preconditioning Against Myocardial Infarction—Its Features and Adenosine-Mediated Mechanism 1

Tetsuji Miura

1.1. Introduction 1
1.2. Phenomenological Features of Preconditioning 2
1.3. Preconditioning and Concurrent Stunning 3
1.4. Adenosine Receptor Activation in Preconditioning 4
1.5. Possible Mechanisms Downstream to Adenosine Receptor Activation 7
1.6. Conclusions 12

2. Does Preconditioning Reduce Lethal Mechanical Reperfusion Injury? 19

David García-Dorado, Marisol Ruiz-Meana and José A. Barrabés

2.1. Introduction 19
2.2. Mechanical Reperfusion Injury 20
2.3. Effects of Preconditioning on Mechanical Injury 23
2.4. Conclusion 29

3. Ischemic Preconditioning Markedly Reduces the Severity of Ischemia and Reperfusion-Induced Arrhythmias; Role of Endogenous Myocardial Protective Substances 35

Agnes Vegh and James R. Parratt

3.1. Introduction 35
3.3. The Antiarrhythmic Effects of Ischemic Preconditioning—The Phenomenon 38
3.4. The Antiarrhythmic Effect of Ischemic Preconditioning—Possible Mechanisms 44
3.5. Conclusion 55

4. The Protective Effects of Preconditioning on Postischemic Contractile Dysfunction 61

Alison C. Cave

4.1. Introduction 61
4.2. The Preconditioning Protocol 61
4.3. Functional and Metabolic Changes During the Sustained Ischemic Period 64
4.4. Does Preconditioning Protect Against Reperfusion Injury? 68
4.5. Mechanism of Preconditioning-Induced Protection Against Contractile Dysfunction 69
4.6. Conclusions 74

5. Reduction of Infarct Size—"Preconditioning at a Distance" 79
Karin Przyklenk, Peter Whittaker, Michel Ovize and Robert A. Kloner
5.1. Introduction 79
5.2. Preconditioning and Infarct Size 80
5.3. Protection via Nonischemic Tissue 83
5.4. Mechanism(s) of Global Protection by Brief Regional Ischemia 85
5.5. Corroborating Evidence 88
5.6. Conclusion 92

6. Novel Approaches to Myocardial Preconditioning in Pigs 97
Ben C.G. Gho, Monique M.G. Koning, René L.J. Opstal, Eric van Klaarwater, Dirk J. Duncker and Pieter D. Verdouw
6.1. Introduction 97
6.2. Relation Between Infarct Size and Area at Risk in Control and Preconditioned Pigs 99
6.3. Transmural Distribution of Infarct Size in Control and Preconditioned Pigs 101
6.4. The Duration of Protection Afforded by Ischemic Preconditioning 101
6.5. Ischemic Preconditioning with a Partial Coronary Artery Occlusion Without Intervening Reperfusion 104
6.6. Myocardial Protection by a Period of Rapid Ventricular Pacing 108
6.7. Conclusions 111

7. Preconditioning in the Human Heart: Fact or Fantasy? 115
Clive S. Lawson
7.1. Introduction 115
7.2. Insights from Studies of Isolated Human Myocardium 117
7.3. Does Angina Protect Against Myocardial Infarction? 118
7.4. Does Preconditioning Occur During Balloon Angioplasty? 120
7.5. The 'Warm-Up' Phenomenon and Preconditioning 121
7.6. Preconditioning and Cardiac Surgery 122
7.7. Where Might Preconditioning Find a Role in Clinical Medicine? 123
7.8. Conclusions 124

8. Role of ATP-Sensitive Potassium Channels in Myocardial Preconditioning 129
Gary J. Grover
8.1. Introduction 129
8.2. K_{ATP}: General Considerations 130
8.3. Pharmacology of the Cardioprotective Effects of K_{ATP} Openers 130

8.4. Effect of K_{ATP} Blockers on Preconditioning ... 135
8.5. How does K_{ATP} Fit into the Cascade of Events in Preconditioning? ... 138
8.6. Conclusions ... 140

9. The Role of G Proteins in Myocardial Preconditioning ... 147
Lucia Piacentini and Nigel J. Pyne
9.1. Introduction ... 147
9.2. Modification of G Protein Function as a Consequence of Myocardial Ischemia ... 149
9.3. Modification of G Protein Function as a Consequence of Preconditioning ... 151
9.4. Activation of G Protein Coupled Receptors as a Mechanism of Preconditioning ... 152
9.5. Possible Interaction with Protein Kinase C ... 157
9.6. Conclusions and Future Directions for Research ... 158

10. Mimicking Preconditioning with Catecholamines ... 167
Tanya Ravingerová
10.1. Introduction ... 167
10.2. Myocardial Ischemia and the Release of Catecholamines ... 168
10.3. Catecholamines and Myocardial Injury ... 169
10.4. Mimicking Preconditioning with Catecholamines ... 171
10.5. Conclusion ... 180

11. Activation of Protein Kinase C is Critical to the Protection of Preconditioning ... 185
Michael V. Cohen, Yongge Liu and James M. Downey
11.1. Introduction ... 185
11.2. Protein Kinase C and Its Activation Pathways ... 185
11.3. Adenosine Triggers Ischemic Preconditioning ... 186
11.4. Adenosine Receptors Trigger as well as Mediate Protection ... 188
11.5. The Role of Protein Kinase C in Preconditioning ... 188
11.6. What is Preconditioning's Memory? ... 189
11.7. The Protein Kinase C Translocation Theory of Preconditioning ... 193
11.8. Does Translocation of PKC Account for the Upregulation? ... 194
11.9. The 5'-Nucleotidase Theory of Preconditioning ... 196
11.10. Other Receptors Can Precondition the Heart ... 196
11.11. Multiple Receptors Contribute to Ischemic Preconditioning ... 197
11.12. Bradykinin's Anti-Infarct Effect Depends upon PKC ... 198
11.13. The Signaling Pathway for Ischemic Preconditioning is Highly Redundant ... 198

11.14. Tolerance May Be an Obstacle to Therapeutic Application ... 200
11.15. The Elusive End-Effector ... 201
11.16. Conclusions ... 201

13. The Mechanism of Preconditioning—What Have We Learned from the Different Animal Species? ... 207
Cherry L. Wainwright and Wei Sun
12.1. Introduction ... 207
12.2. Endogenous Labile Mediators ... 208
12.3. Transduction/Signaling Mechanisms ... 216
12.4. Metabolic Changes ... 221
12.5. Conclusions ... 222

13. Myocardial Stress Response, Cytoprotective Proteins and the Second Window of Protection Against Infarction ... 233
Gary F. Baxter, Michael S. Marber and Derek M. Yellon
13.1. Introduction ... 233
13.2. The Stress Response, Protection and Cross-Tolerance ... 234
13.3. The Thermal Stress Response and Myocardial Protection ... 235
13.4. Evidence for the "Second Window of Protection" After Preconditioning ... 236
13.5. Timecourse of the Second Window of Protection ... 239
13.6. Signaling and Mediation of Delayed Protection ... 240
13.7. Clinical Implications of SWOP and Future Directions ... 244
13.8. Conclusion ... 245

14. Delayed Ischemic Preconditioning Induced by Drugs and by Cardiac Pacing ... 251
Agnes Vegh and James R. Parratt
14.1. Introduction ... 251
14.2. Delayed Myocardial Protection by Drugs ... 251
14.3. Delayed Myocardial Protection by Cardiac Pacing ... 254
14.4. Conclusions ... 256

15. Cardioprotective Effects of Chronic Hypoxia: Relation to Preconditioning ... 261
František Kolář
15.1. Introduction ... 261
15.2. Protective Effects of Chronic High Altitude Hypoxia (HAH) ... 262
15.3. Proposed Mechanisms of Protection by HAH ... 264
15.4. Chronic Hypoxia vs. Preconditioning—Conclusions ... 270

Index ... 277

EDITORS

Cherry L. Wainwright, BSc, PhD, FESC
Department of Physiology and Pharmacology
University of Strathclyde
Royal College
Glasgow, Scotland
Chapter 12

James R. Parratt, PhD, DScMDhc, FESC, FRSE, FRCPath
Department of Physiology and Pharmacology
University of Strathclyde
Royal College
Glasgow, Scotland
Chapters 3, 14

CONTRIBUTORS

José A. Barrabés, MD
Servicio de Cardiología
Hospital General Universitario Vall D'Hebron 119-129
Barcelona, Spain
Chapter 2

Gary F. Baxter, PhD, MIBiol, MRPharmS
The Hatter Institute for Cardiovascular Studies
Department of Academic and Clinical Cardiology
University College London Hospital
London, United Kingdom
Chapter 13

Alison C. Cave, BSc, PhD
Department of Radiological Sciences
Guy's Hospital
London, United Kingdom
Chapter 4

Michael V. Cohen, MD
Department of Physiology
University of South Alabama
College of Medicine
Mobile, Alabama, U.S.A.
Chapter 11

James M. Downey, PhD
Department of Physiology
University of South Alabama
College of Medicine
Mobile, Alabama, U.S.A.
Chapter 11

Dirk J. Duncker, MD, PhD
Department of Cardiology, Thoraxcenter
Rotterdam, The Netherlands
Chapter 6

CONTRIBUTORS

David García-Dorado, MD, FESC, FACC
Servicio de Cardiología
Hospital General Universitario Vall D'Hebron 119-129
Barcelona, Spain
Chapter 2

Ben C.G. Gho, MD
Department of Cardiology, Thoraxcenter
Rotterdam, The Netherlands
Chapter 6

Gary J. Grover, PhD
Department of Pharmacology
Bristol-Meyers Squibb Pharmaceutical Research Institute
Princeton, New Jersey, U.S.A.
Chapter 8

Robert A. Kloner, MD, PhD
Heart Institute
Good Samaritan Hospital
Department of Medicine
Section of Cardiology
University of Southern California
Los Angeles, California, U.S.A.
Chapter 5

František Kolář, PhD
Department of Developmental Cardiology
Institute of Physiology
Academy of Sciences of the Czech Republic
Prague, Czech Republic
Chapter 15

Monique M.G. Koning, MD, PhD
Department of Cardiology, Thoraxcenter
Rotterdam, The Netherlands
Chapter 6

Clive S. Lawson, MD, MRCP
Consultant Cardiologist
St. Thomas' Hospital, London and
Kent and Sussex Hospital
Tunbridge Wells, United Kingdom
Chapter 7

Yongge Liu, PhD
Department of Physiology
University of South Alabama
College of Medicine
Mobile, Alabama, U.S.A.
Chapter 11

Michael S. Marber, PhD, MRCP
The Hatter Institute for Cardiovascular Studies
Department of Academic and Clinical Cardiology
University College London Hospital
London, United Kingdom
Chapter 13

Tetsuji Miura, MD, PhD
Second Department of Internal Medicine
Sapporo Medical University School of Medicine
Sapporo, Japan
Chapter 1

CONTRIBUTORS

René L.J. Opstal, BS
Department of Cardiology,
Thoraxcenter
Rotterdam, The Netherlands
Chapter 6

Michel Ovize, MD, PhD
Hôpital Cardiologique
et Pneumologique Louis Pradel
University of Claude Bernard
Cedex, Lyon, France
Chapter 5

Lucia Piacentini, BSc, PhD
Universität Heidelberg
Innere Medezin III
Heidelberg, Germany
Chapter 9

Karen Przyklenk, PhD
Heart Institute
Good Samaritan Hospital
Department of Medicine
Section of Cardiology
University of Southern California
Los Angeles, California, U.S.A.
Chapter 5

Nigel J. Pyne, BSc, PhD
Department of Physiology
and Pharmacology
University of Strathclyde
Royal College
Glasgow, Scotland
Chapter 9

Tanya Ravingerová, MD, PhD
Institute for Heart Research
Slovak Academy of Sciences
Bratislava, Dúbravská cesta 9,
Republic of Slovakia
Chapter 10

Marisol Ruiz-Meana, Vet D
Servicio de Cardiología
Hospital General Universitario Vall
D'Hebron 119-129
Barcelona, Spain
Chapter 2

Wei Sun, MD, PhD
Department of Physiology
and Pharmacology
University of Strathclyde
Royal College
Glasgow, Scotland
Chapter 12

Eric van Klaarwater, MD
Department of Cardiology,
Thoraxcenter
Rotterdam, The Netherlands
Chapter 6

Agnes Vegh, PhD
Department of Pharmacology
Albert Szent-Gyorgyi Medical
University
Szeged, Hungary
Chapters 3, 14

CONTRIBUTORS

Pieter D. Verdouw, PhD
Professor of Experimental Cardiology
Department of Cardiology, Thoraxcenter
Rotterdam, The Netherlands
Chapter 6

Peter Whittaker, PhD
Heart Institute
Good Samaritan Hospital
Department of Medicine
Section of Cardiology
University of Southern California
Los Angeles, California, U.S.A.
Chapter 5

Derek M. Yellon, DSc, FACC, FESC
The Hatter Institute for Cardiovascular Studies
Department of Academic and Clinical Cardiology
University College London Hospital
London, United Kingdom
Chapter 13

PREFACE

Probably no other aspect of basic heart research has aroused so much attention over the past few years as the phenomenon of ischemic preconditioning. This is the ability of the heart to "adapt" to fairly long periods of ischemia, an adaptation triggered by brief periods of ischemia initiated by complete or brief periods of coronary artery occlusion or by rapid cardiac pacing. The importance of preconditioning is that the protection afforded is *powerful* (more powerful, for example, than drug intervention), *comprehensive* (protection against the development of myocardial necrosis, against life-threatening ventricular arrhythmias and against contractile failure resulting from long periods of ischemia and reperfusion) and, as more recently demonstrated, is *prolonged.* The potential pharmacological and clinical importance of this phenomenon is that if we understood the cellular mechanisms involved in this endogenous form of protection, many believe that such mechanisms could be exploited for therapeutic gain.

The present volume summarizes the considerable advances made in elucidating the mechanisms involved in this protection, with chapters written by "key players" in the field. Many of these are involved in a European-wide Scientific Network on "Endogenous Mechanisms of Cardioprotection" which is coordinated from the University of Strathclyde in Glasgow and involves 11 different laboratories in nine different European countries. This volume is thus, in part, a review of the activities of these different groups in the field of cardioprotection which has been made possible by financial support from the European Economic Commission (Grant No. ERB CT 924009). This support is especially acknowledged.

Most of the work summarized in this volume was triggered initially by the reports from two laboratories, one in the United States (that of Professor Robert Jennings at Duke University, North Carolina) and one in Europe (that of Professor Wolfgang Schaper at the Max Planck Institute in Bad Nauheim, Germany) and the Editors would like to acknowledge the outstanding contributions of these two fine scientists.

The Editors are grateful to each of the Authors for their willingness to contribute to this volume and trust that it will provide a reasonably up-to-date summary of current progress in this exciting area of basic heart research.

Cherry L. Wainwright
James R. Parratt

CHAPTER 1

Preconditioning Against Myocardial Infarction—Its Features and Adenosine-Mediated Mechanism

Tetsuji Miura

1.1. INTRODUCTION

Myocardial infarction is the leading cause of mortality and morbidity in industrialized countries. Prognosis of the patients is determined primarily by the extent of infarct in the heart (i.e., infarct size).[1] Because of this fact, for more than 20 years basic and clinical investigators have devoted much effort to limiting myocardial infarct size. There have been two approaches to this objective: early reperfusion and enhancement of myocardial resistance to ischemia. Several methodologies have been developed over a decade to recanalize the occluded coronary artery and it has been established that reperfusion improves hemodynamics, decreases infarct size, and improves prognosis.[2] In contrast, attempts to enhance myocardial resistance against infarction by pharmacological agents[3-7] has achieved only limited success in animal experiments. However, in 1986 Murry and co-workers[8] found that exposing the myocardium to brief ischemia markedly limits the infarct size due to the subsequent 40 minutes of coronary occlusion in canine hearts. This cardioprotective effect, termed "preconditioning," was not accompanied by alterations of coronary collateral flow, indicating that myocardial resistance to infarction was directly enhanced. In order to understand this novel form of cardioprotection against infarction, we

Myocardial Preconditioning, edited by Cherry L. Wainwright and James R. Parratt.

undertook a series of studies to clarify the features and mechanisms of preconditioning.

1.2. PHENOMENOLOGICAL FEATURES OF PRECONDITIONING

In order to design experiments to analyze the mechanism of preconditioning it is important to understand the relation between various preconditioning protocols and their effects. Accordingly, we examined if the duration of preconditioning ischemia, or the number of preconditioning episodes, would determine the myocardial resistance to necrosis. We employed a rabbit model of infarction, which has the advantage that infarct size variation due to collateral blood flow[9] need not be considered in this collateral deficient species.[10] In our series of experiments, myocardial infarction was induced by a 30 minute coronary occlusion, and both infarct and area at risk (i.e., the area of occluded coronary bed) were estimated at 3 or 72 hours after reperfusion.

Figure 1.1a illustrates the relationship between the duration of ischemia employed to precondition the heart and the infarct size after preconditioning. There was no sharp threshold for preconditioning ischemia, but the extent of infarct size limitation correlates with the duration of the preconditioning ischemia in the range of 2-5 minutes.[11,12] Because 10 minutes of ischemia results in focal necrosis in rabbit hearts,[10] preconditioning with ischemia longer than 5 minutes is unlikely to be further protective. Indeed, Yamasaki et al[13] recently showed that preconditioning with 15 minute ischemia does not protect rabbit hearts at all. Preconditioning with 2 minute ischemia afforded very slight protection. However, when it was repeated two times, infarct size was limited to the same level as by a single episode of 5 minutes preconditioning[11,14] (Fig. 1.1b), though such potentiation of the protective effect by repetition was not observed when the duration of the preconditioning ischemia was 5 minutes.[11,14] Similar observations have been reported in canine hearts.[17]

To examine how long the ischemic tolerance afforded by 5 minute preconditioning persists, we assessed the myocardial response to a 30 minute ischemic insult at various times during a 35 minute recovery period of reperfusion following the preconditioning. As shown in Figure 1.1c, 5 and 15 minutes after preconditioning, myocardial infarct size was significantly limited compared with nonpreconditioned controls.[11] However, 25 minutes after preconditioning, myocardial resistance against infarction does not reach statistical significance. On the other hand, the decay of the preconditioning effect appears to be longer-lived in canine[16] and swine[17] hearts. Significant tolerance against infarction was observed at 2 hours after preconditioning in canine hearts, though the effect was substantially attenuated compared with that at 5 minutes after preconditioning. Nevertheless, this persistence of increased ischemic tolerance after brief transient ischemia, the so-called "memory

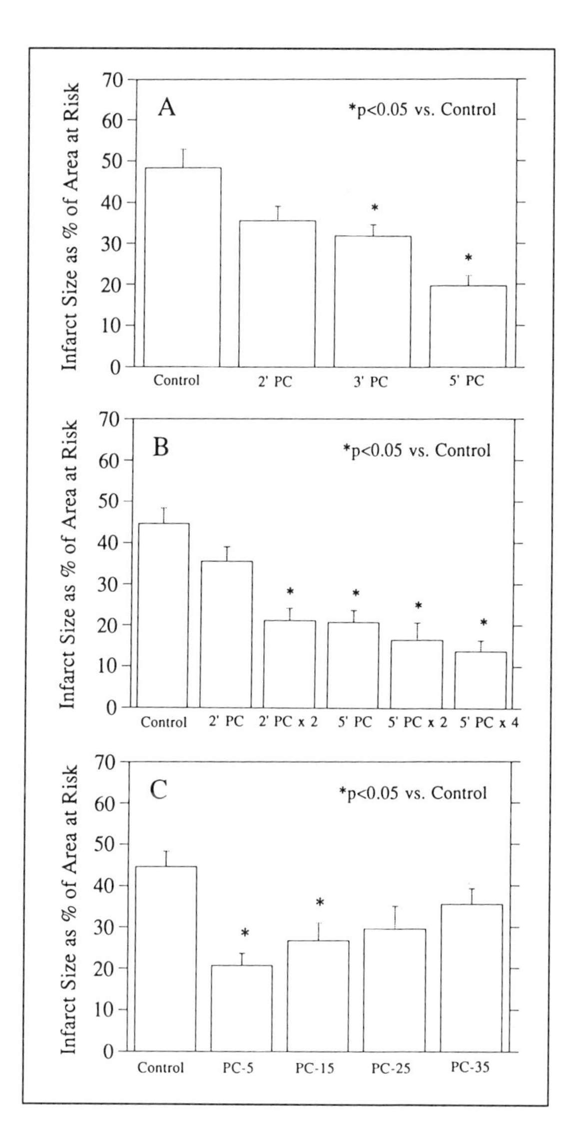

Fig. 1.1. Determinants of the infarct size limiting effect of preconditioning in rabbits. Infarct was induced by a 30 minute coronary occlusion, which was followed by reperfusion. Infarct size and area at risk were determined by histology and fluorescent particles respectively, 72 hours after reperfusion. (A) Duration of preconditioning ischemia. Control, nonpreconditioned controls; 2' PC, preconditioning with 2 minute ischemia; 3' PC, preconditioning with 3 minute ischemia; 5' PC, preconditioning with 5 minute ischemia. (B) Number of preconditioning episodes. 2' PC x 2, two cycles of preconditioning with 2 minute ischemia/5 minute reperfusion; 5' PC x 2, two cycles of preconditioning with 5 minute ischemia/5 minute reperfusion; 5' PC x 4, four cycles of preconditioning with 5 minute ischemia/5 minute reperfusion. (C) Decay of the infarct size limiting effect of 5 minute-preconditioning. Interval between the preconditioning ischemia and the sustained 30 minute ischemia was 5 minutes in PC-5, 15 minutes in PC-15, 25 minutes in PC-25, and 35 minutes in PC-35. Mean ± SEM. n = 9-12.

of preconditioning," is an intriguing phenomenon, although its mechanism is as yet undefined.

1.3. PRECONDITIONING AND CONCURRENT STUNNING

To precondition the heart in most of the studies, 5 minutes of ischemia have been used in rabbits and 5-15 minutes of ischemia in dogs. These periods of ischemia do not cause myocardial necrosis but are long enough to stun the myocardium. Although oxygen consumption is not necessarily reduced in the stunned myocardium,[19-21] the stunning might cause a reduction of myofibrillar ATP utilization and thus spare ATP. Furthermore, it could account for the reduced ATP utilization and proton accumulation in the preconditioned myocardium.[22,23] Accordingly, we tested the hypothesis that myocardial stunning contributes to preconditioning. First, we examined if the degree of stunning in the preconditioned myocardium correlates with the resistance to infarction. Rabbit hearts were preconditioned with two cycles of 2 minute ischemia/5 minute reperfusion, or one or two cycles of 5 minute ischemia/5 minute reperfusion. The regional systolic thickening fraction was 76.8 ± 7.2% of the baseline after the two cycles of 2 minute preconditioning, and 31.4 ± 9.2% and 34.3 ± 9.7% after one and two cycles of 5 minute preconditioning.[14] However, all three preconditioning protocols limited infarct size to the same extent, i.e., approximately 45% of untreated controls.[14] These results suggest that myocardial stunning is not sufficient to achieve the protection of preconditioning. Moreover, the preconditioning effect was shown to decay earlier than the stunning both in rabbits[11] and in dog hearts,[16] which also supports the concept that preconditioning cannot be explained by stunning alone.

However, a possibility remained that some degree of reduction in contractility may have a permissive role for cellular protection. To test this possibility, we[24] recently assessed whether preconditioning can protect the heart when contractile dysfunction due to stunning is prevented by infusion of an inotropic agent. In this series of experiments, preconditioning with two cycles of 2 minute ischemia/5 minute reperfusion reduced regional thickening fraction in rabbit hearts to 72.8 ± 4.7% of the pre-ischemic baseline value. This reduction of the thickening fraction was completely prevented by dobutamine infused at 10 μg/kg/min. However, cardioprotection afforded by preconditioning was not attenuated by the dobutamine treatment.[24] These results argue against the possibility that stunning plays a permissive role in the infarct size-limiting effect of preconditioning. Matsuda et al[25] independently conducted similar experiments in canine hearts to test the effect of dobutamine on preconditioning and reported comparable results. It should be noted that dobutamine improved contractility of the stunned myocardium, which does not necessarily mean that the cellular derangement underlying stunning was improved. However, observations from this laboratory[11,14,24] and others[16,25] indicate that reduced contractile function per se is not sufficient, nor necessary, for preconditioning against infarction.

1.4. ADENOSINE RECEPTOR ACTIVATION IN PRECONDITIONING

A number of biologically active substances are known to be produced during ischemia and reperfusion. Of these, adenosine was the first to claim the attention of investigators as playing a possible role in preconditioning. Liu et al[26] found that preconditioning in vivo was blocked by nonselective adenosine receptor blockers (8-sulphophenyl-theophylline and PD115,199) and that preconditioning was mimicked in vitro by transient exposure of the heart to adenosine and to the A_1 receptor agonist R(-)N_6-2-phenylisopropyl adenosine (R-PIA). However, they determined infarct size by tetrazolium staining, the accuracy of which is influenced by a number of conditions, including pharmacological agents.[27,28] Therefore, we aimed to confirm their results using a chronic rabbit model, which permits infarct sizing by "gold standard" histological methods. In this model, preconditioning was significantly attenuated by 8-phenyltheophylline and conversely, R-PIA was able to protect the myocardium against infarction (Fig. 1.2).[29] These results indicated that the adenosine receptor-mediated cardioprotection is not artefactual. These effects of adenosine receptor blockers and A_1 receptor agonists have now been confirmed in other species[30-32] except for the rat.[33,34] On the other hand, the A_2 agonist (CGS21680) was shown not to mimic preconditioning.[35] Taken together, these results

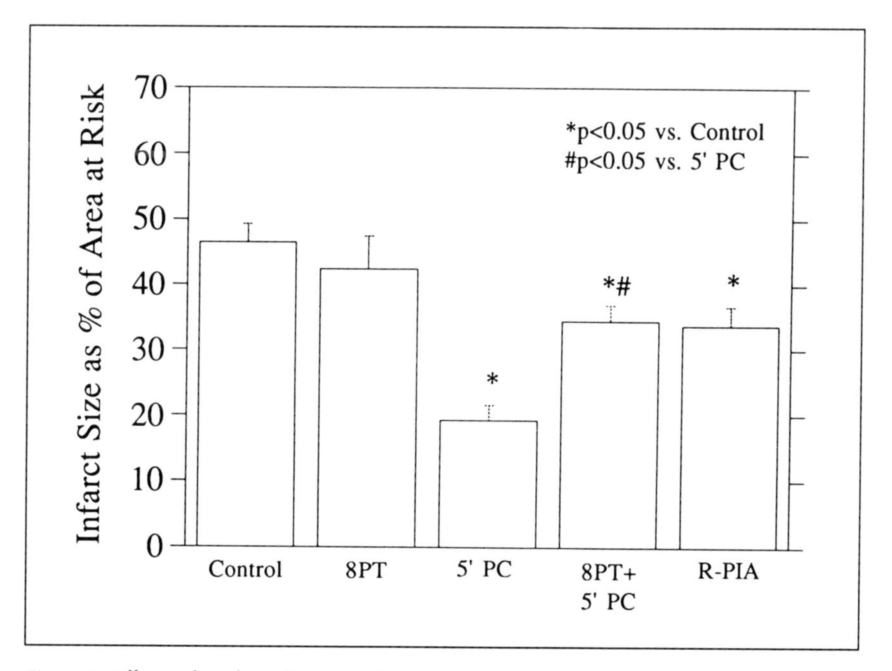

Fig. 1.2. Effect of 8-phenyltheophylline on preconditioning in rabbits with 5 minute ischemia and the effect of R-PIA on infarct size. Control, 30 minute coronary occlusion without pretreatment; 8PT, 8-phenyltheophylline (10 mg/kg) given intravenously 25 minutes before 30 minute coronary occlusion; 5' PC, preconditioning with 5 minute ischemia/5 minute reperfusion; 8PT + 5' PC, 8-phenyltheophylline was given 15 minutes before the 5 minute-preconditioning; R-PIA, 1 mg/kg of R-PIA was injected 15 minutes before the 30 ischemia, with atrial pacing at 240/min to avoid bradycardia by A_1 receptor stimulation. n = 8-12.

suggest that the subtype of the adenosine receptor involved is not A_2, but probably A_1.

Yet another form of support for the adenosine hypothesis comes from our observation of the effect of dipyridamole on preconditioning.[36] Dipyridamole is an inhibitor of nucleoside transport, and a recent study using microdialysis showed that this agent augments, by four-fold, the elevation of interstitial adenosine during ischemia.[37] As shown in Figure 1.3, 0.25 mg/kg of dipyridamole alone had no effect on infarct size, but markedly potentiated infarct size-limitation by preconditioning. Furthermore, this potentiation was attenuated by 8-phenyltheophylline, suggesting an involvement of adenosine receptors. It is possible that the effect of dipyridamole on phosphodiesterase[38] and prostacyclin[39] might also contribute to the potentiation of preconditioning. However, we believe that is not the case, since similar results were subsequently obtained with both dilazep and a specific nucleoside transport inhibitor, R75231.[40] Tsuchida et al[41] found that an adenosine 'modulator,' acadesine, also lowered the temporal threshold for preconditioning. Further, we recently observed that dipyridamole

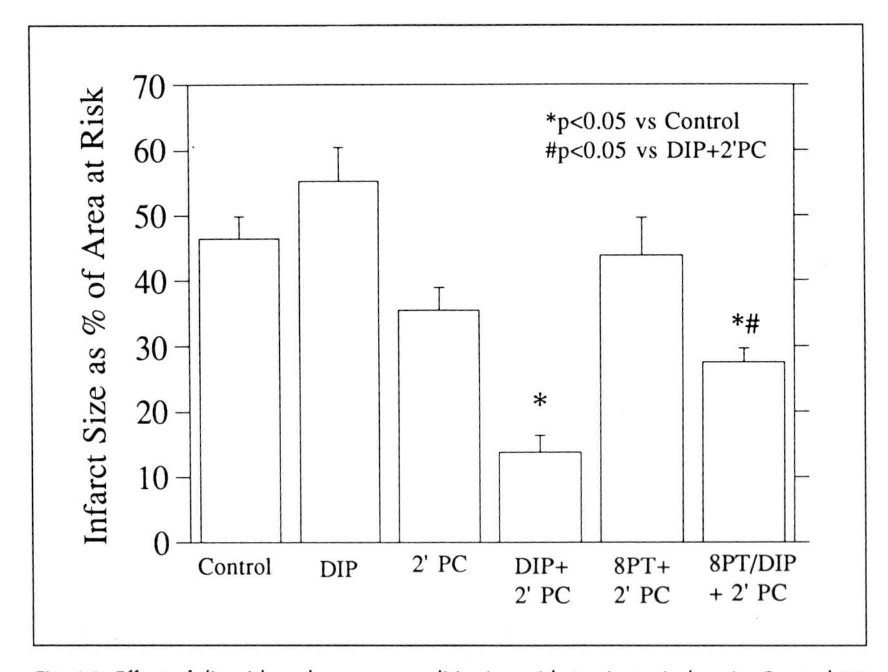

Fig. 1.3. Effect of dipyridamole on preconditioning with 2 minute ischemia. Control, 30 minute coronary occlusion without pretreatment; 2' PC, preconditioning with 2 minute ischemia/5 minute reperfusion; DIP, dipyridamole (0.25 mg/kg) was injected intravenously 22 minutes before the 30 minute ischemia; DIP + 2' PC, dipyridamole was given 15 minutes before preconditioning with 2 minute ischemia; 8PT + 2' PC, 8-phenyltheophylline (10 mg/kg) was given 15 minutes before preconditioning with 2 minute ischemia; 8PT/DIP + 2' PC, 8-phenyltheophylline was given 5 minutes before dipyridamole injection, which was followed by 2 minute-preconditioning. n = 7-13.

could potentiate only modestly preconditioning with **3 minute** ischemia, in contrast with its marked effect on preconditioning with a **2 minute** period of ischemia.[42] It appears then that the interstitial adenosine level achieved during preconditioning ischemia is a major determinant of the myocardial resistance afforded by preconditioning.

Another important contributor to the adenosine debate concerns the role of ecto-5'-nucleotidase. When there is no shortage of oxygen this enzyme does not appear to contribute to adenosine production. Adenosine is generated from hydrolysis of S-adenosylhomocysteine (SAH) by SAH-hydrolase and dephosphorylation of AMP by cytosolic 5'-nucleotidase.[43,44] Adenosine production may also be substrate-regulated; a slight decrease in ATP potential can substantially increase free AMP by the myokinase reaction.[43] However, once the myocardium is rendered ischemic, ecto 5'-nucleotidase may also contribute to adenosine production[45,46] by dephosphorylation of adenine nucleotides leaked from the cytosol,[47] and of ATP cotransmitted with noradrenaline from sympathetic nerves.[48,49] Kitakaze et al[50,51] proposed that activation of this ecto-5'-nucleotidase by preconditioning is the key to cardioprotection. Their proposal is primarily based on two lines of evidence in canine hearts. Firstly, the cellular protection by preconditioning occurred with parallel changes of ecto-5'-nucleotidase and levels of adenosine released after reperfusion.[50] Secondly, an inhibitor of this enzyme, α,β-methylene adenosine 5'-diphosphate (AOPCP), abolished the infarct size limitation by preconditioning in dog hearts.[51] However, the ecto-5'-nucleotidase hypothesis contradicts a finding by Van Wylen[52] that the elevation of interstitial adenosine during ischemia is not enhanced, but rather significantly *reduced* after preconditioning. This observation is also consistent with an earlier finding that ATP utilization is reduced in preconditioned myocardium.[22,23] A recent study[53] from this laboratory also raises doubts about the role of ecto-5'-nucleotidase, at least in rabbit hearts. In this species the infusion of AOPCP (0.17-1.5 mg/kg/min) into the left atrium markedly inhibited the hemodynamic responses (i.e., reduction of left ventricular dP/dt and blood pressure fall) to ATP and AMP challenge, demonstrating the efficacy of AOPCP to inhibit ecto-5'-nucleotidase in vivo. However, these doses of AOPCP failed to block preconditioning against infarction, suggesting that the adenosine level needed to trigger preconditioning in rabbit hearts may be achieved by pathways other than through ecto-5'-nucleotidase.

1.5. POSSIBLE MECHANISMS DOWNSTREAM TO ADENOSINE RECEPTOR ACTIVATION

The mechanisms through which adenosine (probably A_1) receptor activation enhances myocardial tolerance against infarction remain unclear. However, of the intracellular effectors of adenosine receptors,[43] ATP sensitive potassium channels and phospholipase C may play crucial roles in the cellular protection of preconditioning.

1.5.1. ATP Sensitive Potassium Channel (K_{ATP})

K_{ATP} is a subtype of the potassium channel modulated primarily by intracellular ATP levels, and Kirsch et al[54] found that this channel is linked to A_1 receptors via G_i protein. Gross and co-workers[30,55,56] proposed that activation of this channel contributes to preconditioning (reviewed by Grover in chapter 8). Evidence for this proposal has been obtained from dogs in which myocardial infarcts were produced by 60 minutes of ischemia instituted 10 minutes after a 5 minute preconditioning episode. In this study,[55] the administration of a specific K_{ATP} blocker, glibenclamide, abolished the protective effect of preconditioning. Conversely, a K_{ATP} opener, RP52891, administered before ischemia in nonpreconditioned hearts, was able to mimic preconditioning in limiting infarct size. Furthermore, Grover et al[30] found that cardioprotection by an A_1 receptor agonist, R-PIA, was abolished by glibenclamide. These effects of K_{ATP} blockers (glibenclamide, 5-hydroxydecanoate) on preconditioning and A_1-receptor mediated protection were subsequently confirmed in other canine[30,56] and swine[32,57] models of infarction.

However, conflicting results have been observed in rabbits. In a study by Toombs et al,[58] 0.3 mg/kg of glibenclamide prevented infarct size limitation by preconditioning, whereas Thornton et al[59] failed to detect any inhibition of preconditioning by 0.15-3.0 mg/kg of glibenclamide. The only methodological difference between these two studies was the anesthetic used, i.e., ketamine/xylazine vs. pentobarbital. Because of this difference in anesthesia, heart rates were about 40% lower in the rabbits used by Toombs et al[58] compared with those in the study by Thornton et al.[59] It is also interesting to note that the level of the rate-pressure product in rabbits anesthetized with ketamine/xylazine (about 16,000) was very similar to that in the canine[55,56] and swine[32,57] models, in which inhibition of preconditioning by sulfonylureas has been observed.

Accordingly, we suggested two possibilities to explain the difference in results. Firstly, the negative results in the study by Thornton et al[59] may be attributable to the large amount of adenosine produced in the animals with tachycardia, which may have overwhelmed any inhibitory effect of glibenclamide. Secondly, xylazine, an adjunct drug used by Toombs et al,[58] might be responsible for the contradictory findings in rabbits. To test these possibilities, we[12] assessed whether the effect of glibenclamide on preconditioning is modified (1) when rabbits were pretreated with prazosin and metoprolol to reduced myocardial oxygen consumption, and (2) when xylazine was added to pentobarbital anesthesia (Table 1.1). Under pentobarbital anesthesia alone, neither modest protection by preconditioning with 3 minute ischemia, nor the potent effect of a 5 minute preconditioning, were attenuated by glibenclamide. Glibenclamide also failed to block preconditioning even when the rate-pressure product was reduced to approximately 60%

by prazosin/metoprolol. However, when xylazine was added to the anesthetic regimen, glibenclamide did inhibit preconditioning in rabbit hearts. These results suggest that K_{ATP} mediates preconditioning in rabbit hearts, as in canine and swine hearts, and also that the role of K_{ATP} in preconditioning may be modified or obscured by conditions of anesthesia. We cannot specify a mechanism through which xylazine altered the response of preconditioning to glibenclamide in rabbits. However, it is probably not through reducing myocardial oxygen consumption by its α_2-agonistic action,[60,61] since the rate-pressure product was similar after both xylazine and metoprolol/prazosin injection. Obviously, some other factors, including species difference, must explain why glibenclamide prevents preconditioning in dogs anesthetized with pentobarbital.[55,56]

The mechanism whereby K_{ATP} opening protects cardiac myocytes from infarction remains poorly understood. However, two lines of evidence suggest that this protection is probably not through alterations

Table 1.1. Effect of glibenclamide on preconditioning in rabbits

Treatment	n	RPP(/1000)	%IS/AAR
Pentobarbital anesthesia			
Untreated	12	27.5 ± 1.1	49.2 ± 3.3
Glib	8	25.9 ± 1.7	58.5 ± 5.3
3-min PC	9	27.4 ± 1.9	31.7 ± 2.8^{a,b}
Glib/3-min PC	6	26.0 ± 2.0	37.3 ± 7.9^{b}
5-min PC	14	26.7 ± 1.3	19.6 ± 2.5^{a,b}
Glib/5-min PC	6	26.8 ± 1.8	25.7 ± 5.9^{a,b}
Pentobarbital plus autonomic blockade			
Praz/Met	8	15.7 ± 1.1	46.2 ± 2.2
Praz/Met/Glib	7	14.6 ± 0.9	46.6 ± 3.0
Praz/Met/PC	7	17.1 ± 1.2	12.2 ± 3.7^{c}
Praz/Met/Glib/PC	6	17.1 ± 1.8	14.9 ± 4.3^{c}
Pentobarbital plus xylazine			
Xylazine	8	11.8 ± 0.9	49.9 ± 5.5
Xylazine/Glib	6	12.5 ± 0.7	44.4 ± 7.3
Xylazine/PC	6	11.2 ± 1.1	20.0 ± 5.6^{d}
Xylazine/Glib/PC	7	12.3 ± 0.7	33.5 ± 2.6

RPP, rate-pressure product immediately before the coronary occlusion.
%IS/AAR-infarct size after 30 minutes ischemia expressed as % of area at risk.
Glib, 0.3 mg/kg glibenclamide; PC, preconditioning;
Praz, 0.15 mg/kg prazosin; Met, 0.15 mg/kg metoprolol;
Xylazine, 6 mg/kg xylazine.
Values are mean ± SEM.
^{a}P<0.05 vs. Untreated; ^{b}P<0.05 vs. Glib;
^{c}P<0.05 vs. Praz/Met and Praz/Met/Glib;
^{d}P<0.05 vs. Xylazine and Xylazine/Glib.

in membrane potential by opening this channel. First, in a study by Jennings et al,[62] preconditioning was conducted in canine myocardium in vivo, after which the hearts were arrested with saturated KCl. The hearts were then excised from the dogs and incubated in vitro at 37°C without coronary perfusion. The rate of ATP depletion and anaerobic glycolysis during this total ischemia was much slower in the preconditioned myocardium than in the nonpreconditioned region. These findings are essentially the same as those observed in situ, although in the total ischemia model the electrical activity had been abolished by the saturated KCl. Secondly, Sleph and Grover[63] recently reported that administration of K_{ATP} openers before potassium cardioplegia provides additional protection against LDH release and contractile dysfunction after 30 minutes global ischemia in isolated rat hearts. Thus, shortening of the action potential and the concomitant shortening of the time window for calcium influx are unlikely to explain K_{ATP}-mediated protection in preconditioning.

1.5.2. Phospholipase C-Protein Kinase C

Phospholipase C is known as one of the effectors of A_1 receptors, although this link has been studied mainly in noncardiac tissues.[43] Activation of this enzyme degrades the membrane phospholipid to produce two second messengers, diacylglycerol and inositol trisphosphate (IP_3). Diacylglycerol, as a cofactor, activates protein kinase C (PKC). The importance of the activity of this enzyme activity in preconditioning was suggested by Ytrehus et al[64] and is reviewed further in chapter 11. They found that preconditioning was abolished by two PKC inhibitors, staurosporine and polymyxin B. Conversely, PKC-activating phorbol esters, 4β-phorbol 12-myristate 13-acetate (PMA) and 1-oleyl-2-acetyl glycerol, administered before ischemia, limit infarct size in isolated nonpreconditioned rabbit hearts. However, it has not been determined whether PKC activation necessary for cardioprotection is provoked by stimulation of A_1-receptors, or of other types of receptors. To test this, we assessed the effects of PKC inhibitors on infarct size-limitation by an A_1 receptor agonist.[65] If PKC is indeed activated by A_1-receptor stimulation during preconditioning, then PKC inhibitors such as staurosporine and polymixin B should abolish the infarct size limitation by R-PIA. Prior to the infarct size experiment, we examined the effects of staurosporine and polymyxin B on the inotropic response to low dose PMA[66] to confirm the efficacy of the PKC inhibitors in vivo. In open chest rabbits, 0.02 μg/kg of PMA was injected into the left atrium, and 3 minutes later, a second dose of 0.05 μg/kg was administered. The hemodynamic response peaked at approximately 30 seconds after the bolus injection of PMA. Left ventricular dP/dt_{max} after 0.02 and 0.05 μg/kg of PMA was significantly increased by 5% and 10%, respectively. In contrast, an equimolar dose of 4α-phorbol 12,13-didecanote, a non-PKC activating phorbol ester, had no effect on LV dP/dt_{max}. When the rabbits were pretreated with 50 µg/kg of staurosporine and 2.5 mg/kg of polymyxin B, LV dP/dt_{max} was not increased by the PMA challenge. These results suggest that these doses of the PKC inhibitors are appropriate to block PKC in rabbit hearts in situ.

As shown in Figure 1.4, neither 50 µg/kg of staurosporine nor 2.5 mg/kg of polymyxin B modified infarct size after 30 minutes of ischemia in rabbits. However, infarct size limitation by 1.0 mg/kg of R-PIA was completely abolished when the rabbits were given staurosporine or polymyxin B prior to the onset of ischemia. Although these PKC inhibitors reduce blood pressure by 5~15 mmHg, their inhibitory effect was not correlated with the rate-pressure product or with systemic blood pressure. Taken together, these results suggest that infarct size limitation by A_1 receptor stimulation requires PKC activity during an ischemic insult. Although it has not yet been proved that A_1 receptor stimulation does indeed activate PKC in the cardiac myocytes, Kohl et al[67] recently showed that R-PIA is capable of increasing IP_3 in

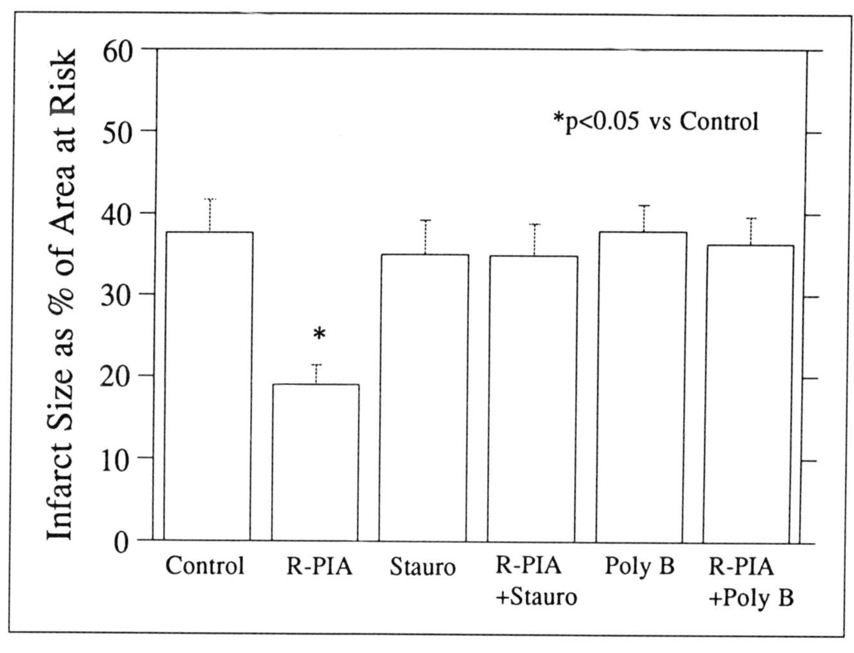

Fig. 1.4. Effects of staurosporine and polymyxin B on infarct size limitation by R-PIA in rabbits. Infarct was induced by 30 minute coronary occlusion and the infarcted area was determined by tetrazolium staining. R-PIA, administration of R-PIA (1 mg/kg i.v.) 15 minutes before the ischemia, with atrial pacing at 240 beats/min; Stauro, injection of staurosporine (50 µg/kg, i.v.) 5 minutes before the coronary occlusion; Poly B, injection of polymyxin B (2.5 mg/kg i.v.) 5 minutes before coronary occlusion. Staurosporine and polymyxin B were injected after the R-PIA administration in R-PIA + Stauro and R-PIA + Poly B groups, respectively. n = 8-12.

guinea-pig papillary muscle, and that this is inhibited by 8-cyclopentyl-1,3-dipropylxanthine, a selective A_1-receptor blocker. Thus, PKC activation by ischemic preconditioning may be a consequence of A_1 receptor stimulation, although other types of receptors linked to this enzyme could also contribute.

However, it should be noted that recent studies using canine and swine models of infarction argue against a major the role for PKC. Przyklenk et al[68] reported that preconditioning does not activate protein kinase and Vogt et al[69] failed to detect any inhibition of preconditioning by PKC blockers. Furthermore, there are some studies reporting deleterious effects of PKC activation on ischemia/reperfusion injury in cardiac[70] and noncardiac tissues.[71] Possible explanations for the conflicting results include the difference in the doses of PKC blockers used, the isoform of PKC activated, and species differences in the roles of PKC. Further investigations are necessary to determine whether PKC is indeed causally related to the cardioprotection afforded by preconditioning.

1.6. CONCLUSIONS

The dependence of the preconditioning effect on the duration and the number of preconditioning occlusions is probably through interstitial adenosine levels (and thus the level of adenosine receptor activation) achieved by each preconditioning protocol. However, the persistence of the protective effect remains an unexplained feature. Our studies support the theory that adenosine A_1-receptor activation and subsequent PKC activation enhance myocardial ischemic tolerance. K_{ATP} may be involved in preconditioning in rabbits, but its relative importance in canine and swine models of infarction is controversial. The mechanism of preconditioning, and possible species differences (reviewed in chapter 12) remains to be further elucidated before it can be adapted for therapeutic advantage in the clinic.

REFERENCES

1. van der Laarse AA, van Leeuwen FT, Krul R et al. The size of infarction as judged enzymatically in 1974 patients with acute myocardial infarction. Relation with symptomatology, infarct localization and type of infarction. Int J Cardiol 1988; 19:191-207.
2. Topol EJ. Thrombolytic intervention. In: Topol EJ, ed. Textbook of interventional cardiology. 2nd ed. Philadelphia: WB Saunders Company, 1994:68-111.
3. Reimer KA, Rasmussen MM, Jennings RB. Reduction by propranolol of myocardial necrosis following temporary coronary artery occlusion in dogs. Circ Res 1973; 33:353-363.
4. Hearse DJ, Yellon DM, Downey JM. Can beta blockers limit myocardial infarct size? Eur Heart J 1986; 7:925-930.

5. Kloner RA, Braunwald E. Effects of calcium antagonists on infarcting myocardium. Am J Cardiol 1987; 59:84B-94B.
6. Adachi T, Miura T, Noto T et al. Does verapamil limit myocardial infarct size in a heart deficient in xanthine oxidase? Clin Exp Pharmacol Physiol 1990; 17:769-779.
7. Downey JM, Yellon DM. Do free radicals contribute to myocardial cell death during ischemia-reperfusion? In: Yellon DM, Jennings RB, eds. Myocardial protection:the pathophysiology of reperfusion and reperfusion injury. New York: Raven Press, 1992:35-57.
8. Murry CE, Jennings RB, Reimer KA. Preconditioning with ischemia: A delay in lethal injury in ischemic myocardium. Circulation 1986; 74:1124-1136.
9. Miura T, Yellon DM, Hearse DJ et al. Determinants of infarct size during permanent occlusion of a coronary artery in the closed chest dog. J Am Coll Cardiol 1987; 9:647-654.
10. Miura T, Downey JM, Ooiwa H et al. Progression of myocardial infarction in a collateral flow deficient species. Jpn Heart J 1989; 30:695-708.
11. Miura T, Adachi T, Ogawa T et al. Myocardial infarct size-limiting effect of ischemic preconditioning: Its natural decay and the effect of repetitive preconditioning. Cardiovasc Pathol 1992; 1:147-154.
12. Miura T, Goto M, Miki T et al. Glibenclamide, a blocker of ATP-sensitive potassium channels, abolishes infarct size-limitation by preconditioning in rabbits anesthetised with xylazine/pentobarbital, but not with pentobarbital alone. J Cardiovasc Pharmacol 1995; 25:531-538.
13. Yamasaki K, Tanaka M, Yokota R et al. Preconditioning with 15 min ischemia extends myocardial infarct size after subsequent 30 min ischemia in rabbits. J Mol Cell Cardiol 1994; 26:CCXIV (Abstract).
14. Miura T, Goto M, Urabe K et al. Does myocardial stunning contribute to infarct size limitation by ischemic preconditioning? Circulation 1991; 84:2504-2512.
15. Li CG, Vasquez JA, Gallagher KP et al. Myocardial protection with preconditioning. Circulation 1990; 82:609-619.
16. Murry CE, Richard VJ, Jennings RB et al. Myocardial protection is lost before contractile function recovers from ischemic preconditioning. Am J Physiol 1991; 260:H796-H804.
17. Schwartz ER, Mohri M, Sacck S et al. Duration of infarct size limiting effect of ischemic preconditioning in the pig. Circulation 1991; 84 (suppl II): II-432 (Abstract).
18. Schaper W, Ito BR. The energetics of "stunned" myocardium. In: de Jong JW, ed. Myocardial Energy Metabolism. Dordrecht: Martinus Nijhoff Publishing 1988:203-213.
19. Dean EN, Schlafer M, Nicklas JM. The oxygen consumption paradox of "stunned myocardium" in dogs. Basic Res Cardiol 1990; 85:120-131.
20. Laster SB, Becker LC, Ambrosio G et al. Reduced aerobic metabolic efficiency in globally "stunned" myocardium. J Mol Cell Cardiol 1989; 21:419-426.

21. Ohgoshi Y, Goto Y, Futaki S et al. Increased oxygen cost of contractility in stunned myocardium of dog. Circ Res 1991; 69:975-988.
22. Murry CE, Richard VJ, Reimer KA et al. Ischemic preconditioning slows energy metabolism and delays ultrastructural damage during a sustained ischemic episode. Circ Res 1990; 66:913-931.
23. Kida M, Fujiwara H, Ishida M et al. Ischemic preconditioning preserves creatine phosphate and intracellular pH. Circulation 1991; 84:2495-2503.
24. Goto M, Miura T, Itoya M et al. Reduction of regional contractile function by preconditioning ischemia does not play a permissive role in the infarct size-limitation by the preconditioning. Basic Res Cardiol 1993; 88:594-606.
25. Matsuda M, Catena TG, Vander Heide RS et al. Cardiac protection by ischaemic preconditioning is not mediated by myocardial stunning. Cardiovasc Res 1993; 27:585-592.
26. Liu GS, Thornton J, Van Winkle D et al. Protection against infarction afforded by preconditioning is mediated by A_1 receptors in rabbit heart. Circulation 1991; 84:350-356.
27. Yellon DM, Maxwell MP, Hearse DJ et al. Infarct size limitation—real or artifactual. Studies with flurbiprofen using a reperfusion model. In: Dhalla NS, Hearse DJ eds. Advances in myocardiology. London: Plenum Press, 1985:619-627.
28. Shirato C, Miura T, Ooiwa H et al. Tetrazolium artefactually indicates superoxide dismutase-induced salvage in reperfused rabbit heart. J Mol Cell Cardiol 1989; 21:1187-1193.
29. Tsuchida A, Miura T, Miki T et al. Role of adenosine receptor activation in infarct size limitation by preconditioning in the heart. Cardiovasc Res 1992; 26:456-461.
30. Grover GJ, Sleph PG, Dzwonczyk S. Role of ATP-sensitive potassium channels in mediating preconditioning in the dog heart and their possible interaction with adenosine A_1-receptors. Circulation 1992; 86:1310-1316.
31. Yao Z, Gross GJ. A comparison of adenosine-induced cardioprotection and ischemic preconditioning in dogs. Comparison of adenosine-induced cardioprotection and ischemic preconditioning in dogs. Efficacy, time-course, and role of K_{ATP} channels. Circulation 1994; 89:1229-1236.
32. Van Winkle DM, Chien GL, Wolff RA et al. Cardioprotection provided by adenosine receptor activation is abolished by blockade of the K_{ATP} channel. Am J Physiol 1994; 266:H829-H839.
33. Li Y, Kloner RA. The cardioprotective effects of ischemic preconditioning are not mediated by adenosine receptors in rat hearts. Circulation 1993; 87:1642-1648.
34. Liu Y, Downey JM. Ischemic preconditioning protects rat heart against infarction. Am J Physiol 1993; 263:H1107-H1112.
35. Thornton JD, Liu GS, Olsson RA et al. Intravenous pretreatment with A_1-selective adenosine analogs protects the heart against infarction. Circulation 1992; 85:659-665.

36. Miura T, Ogawa T, Iwamoto T et al. Dipyridamole potentiates the myocardial infarct size-limiting effect of ischemic preconditioning. Circulation 1992; 86:979-985.
37. Numazawa K, Sakuma I, Kobayashi T et al. Dipyridamole protects canine heart against ischemia/reperfusion by raising cardiac interstitial adenosine via inhibition of its deamination as well as reuptake. Circulation 1993: (suppl I); I-431 (Abstract).
38. Asano T, Ochiai Y, Hidaka H. Selective inhibition of separated forms of human platelet cyclic nucleotide phosphodiesterase by platelet aggregation inhibitors. Mol Pharmacol 1977; 13:400-406.
39. Blass KE, Block HU, Forster W et al. Dipyridamole: a potent stimulator of prostacyclin (PGI_2) biosynthesis. Br J Pharmacol 1980; 68:71-73.
40. Itoya M, Miura T, Sakamoto J et al. Nucleoside transport inhibitors enhance the infarct size-limiting effect of ischemic preconditioning. J Cardiovasc Pharmacol 1994; 24:846-852.
41. Tsuchida A, Liu GS, Mullane K et al. Acadesine lowers temporal threshold for the myocardial infarct size-limiting effect of preconditioning. Cardiovasc Res 1993; 27:116-120.
42. Miki T, Miura T, Sakamoto J et al. Does preconditioning (PC) require myocardial adenosine level elevation during both PC ischemic and subsequent ischemic insult? J Mol Cell Cardiol 1994; 26: CCXVI (Abstract).
43. Olsson RA, Pearson JD. Cardiovascular purinoceptors. Physiol Rev 1990; 70:761-845.
44. Lloyd HGE, Schrader J. Adenosine metabolism in the guinea pig heart: the role of cytosolic S-adenosyl-L-homocysteine hydrolase, 5'-nucleotidase and adenosine kinase. Eur Heart J 1993; 14 (Suppl I):27-33.
45. Headrick JP, Matherne G, Berne RM. Myocardial adenosine formation during hypoxia: effects of ecto-5'-nucleotidase inhibition. J Mol Cell Cardiol 1992; 24:295-303.
46. Raatikainen MJP, Peuhkurinen KJ, Hassinen IE. Contribution of endothelium and cardiomyocytes to hypoxia-induced adenosine release. J Mol Cell Cardiol 1994; 26:1069-1080.
47. Chaudry IH. Does ATP cross the cell plasma membrane? Yale J Biol Med 1982; 55:1-10.
48. Johnson RG. Accumulation of biological amines into chromaffin granules: A model for hormone and neurotransmitter transport. Physiol Rev 1988; 68:232-307.
49. Fredholm EB, Hedqvist P, Lindstrom K et al. Release of nucleosides and nucleotides from the rabbit heart by sympathetic nerve stimulation. Acta Physiol Scand 1982; 116:285-295.
50. Kitakaze M, Hori M, Takashima S et al. Ischemic preconditioning increases adenosine release and 5'-nucleotidase activity during myocardial ischemic and reperfusion in dogs: implication for myocardial salvage. Circulation 1993; 87:208-215.

51. Kitakaze M, Hori M, Morioka T et al. Infarct size-limiting effect of ischemic preconditioning is blunted by inhibition of 5'-nucleotidase activity and attenuation of adenosine release. Circulation 1994; 89:1237-1246.
52. Van Wylen DGL. Effect of ischemic preconditioning on intersitial purine metabolite and lactate accumulation during myocardial ischemia. Circulation 1994; 89:2283-2289.
53. Miki T, Miura T, Suzuki K et al. Does ectosolic 5'-nucleotidase contribute to preconditioning in the rabbit heart? Circulation 1994; 90 (suppl I):I-477 (Abstract).
54. Kirsch GE, Codiana J, Birnbaumer L et al. Coupling of ATP-sensitive K channels to A_1 receptors by G_i proteins in rat ventricular myocytes. Am J Physiol 1990; 259:H820-H826.
55. Gross GJ, Auchampach JA. Blockade of ATP-sensitive potassium channels prevents myocardial preconditioning in dogs. Circ Res 1992; 70:223-233.
56. Auchampach JA, Gross GJ. Adenosine A_1 receptors, K_{ATP} channels, and ischemic preconditioning in dogs. Am J Physiol 1993; 264:H1327-H1336.
57. Schultz R, Rose J, Heusch G. Involvement of activation of ATP-dependent potassium channels in ischemic preconditioning in swine. Am J Physiol 1994; 267:H1341-H1352.
58. Toombs CF, Moore TL, Shebuski RJ. Limitation of infarct size in the rabbit by ischemic preconditioning is reversible with glibenclamide. Cardiovasc Res 1993; 27:617-622.
59. Thornton JD, Thornton CS, Sterling DL et al. Blockade of ATP-sensitive potassium channels increases infarct size but does not prevent preconditioning in rabbit hearts. Circ Res 1993; 72:44-49.
60. Wyatt JD, Scott AW, Richardson ME. The effects of prolonged ketamine-xylazine intravenous infusion on arterial blood pH, blood gases, mean arterial blood pressure, heart and respiratory rates, rectal temperature and reflexes in the rabbit. Lab Anim Sci 1989; 39:411-416.
61. Sylviana TJ, Bergman NG, Fox JG. Effects of yohimbine on bradycardia and duration of recumbency in ketamine/xylazine anesthetized ferrets. Lab Anim Sci 1990; 40:178-182.
62. Jennings RB, Murry CE, Reimer KA. Energy metabolism in preconditioning and control myocardium; effect of total ischemia. J Mol Cell Cardiol 1991; 23:1449-1458.
63. Sleph PG, Grover GJ. Protective effects of cromakalim and BMS-180448 in ischemic rat hearts treated with potassium cardioplegia. J Mol Cell Cardiol 1994; 26:CLXVII (Abstract).
64. Ytrehus K, Liu Y, Downey JM. Preconditioning protects ischaemic rabbit heart by protein kinase C activation. Am J Physiol. 1994; 266: H1145-H1152.
65. Sakamoto J, Miura T, Goto M et al. Limitation of myocardial infarct size by adenosine A_1-receptor activation is abolished by protein kinase C inhibitors in the rabbit. Cardiovasc Res 1995; 29:682-688.

66. Ward CA, Moffat MP. Positive and negative inotropic effects of phorbol 12-myristate 13-acetate: relationship to PKC-dependence and changes in $[Ca^{2+}]_i$. J Mol Cell Cardiol 1992; 24:937-948.
67. Kohl C, Linck B, Schmitz W et al. Effects of carbachol and R(-)-N6-phenylisopropyladenosine on myocardial inositol phosphate content and force of contraction. Br J Pharmacol 1990; 101:829-34.
68. Przyklenk K, Sussman MA, Kloner RA. Fluorescence microscopy reveals no evidence of protein kinase C activation in preconditioned canine myocardium. Circulation 1994; 90 (suppl. I):I-647 (Abstract).
69. Vogt A, Barancik M, Weihrauch D et al. Protein kinase C inhibitors reduce infarct size in pig heart in vivo. Circulation 1994; 90 (suppl. I):I-647 (Abstract).
70. Black SC, Fagbemi SO, Chi L et al. Phorbol ester-induced ventricular fibrillation in the Langendorff-perfused rabbit heart: antagonism by staurosporine and glibenclamide. J Mol Cell Cardiol 1993; 25:1427-38.
71. Hara H, Onodera H, Yoshidomi M et al. Staurosporine, a novel protein kinase C inhibitor, prevents postischemic neuronal damage in the gerbil and rat. J Cereb Blood Flow Metab 1990; 10:646-53.

CHAPTER 2

Does Preconditioning Reduce Lethal Mechanical Reperfusion Injury?

David García-Dorado, Marisol Ruiz-Meana and José A. Barrabés

2.1. INTRODUCTION

Ischemic heart disease is the leading cause of death in Europe. The extent of myocardial necrosis occurring during acute coronary syndrome is the main determinant of survival and quality of life in patients with ischemic heart disease. Coronary occlusion is usually transient during acute coronary syndromes due to either spontaneous or therapeutic recanalization. The tolerance of myocardium to ischemia-reperfusion is very poor as compared to other tissues, and short periods of severe ischemia (less than one hour) may result in massive cell necrosis. There is increasing evidence that this poor tolerance is only partly due to ischemic conditions themselves and that the specific conditions of reperfusion may aggravate the extent of injury. During the acute phase of reperfusion the additional injury seems due primarily to mechanical causes. Myocytes have a potent contractile machinery and are tightly interconnected to ensure the necessary electrical and mechanical coupling during normal function. These characteristics make cardiomyocytes particularly susceptible to mechanical injury during reperfusion.

Brief episodes of transient coronary occlusion have a powerful protective effect against myocardial necrosis secondary to a subsequent prolonged period of ischemia.[1] This intriguing phenomenon has been termed ischemic preconditioning, and has prompted a great amount of work to elucidate its mechanisms. Several mechanisms have been

Myocardial Preconditioning, edited by Cherry L. Wainwright and James R. Parratt.

suggested, but their relative contribution to the beneficial effects of ischemic preconditioning have not been elucidated. Indeed they may vary depending on conditions such as duration of preconditioning ischemia and reperfusion, duration of prolonged coronary occlusion, collateral flow and inter-species differences. It is fair to say that, despite the great advance in our understanding of the phenomenon of ischemic preconditioning, its mechanisms are far from being completely understood.

It is clear that preconditioning slows the progression of ischemic injury, as expressed by electrophysiologic, mechanical and biochemical derangements.[2-5] Preconditioning slows the progression of acidosis[2,5] and the fall in high energy phosphates,[6] the occurrence of monophasic potentials[3] and the appearance of ischemic rigor during coronary occlusion.[4] However, the effect of retarding the appearance of the manifestations of ischemic injury is relatively brief, ranging from 3-5 minutes in the curve of ATP depletion to 10-15 minutes in pH fall. After 30 minutes of acute coronary occlusion, the situation in preconditioned myocardium can hardly be distinguished from that in nonpreconditioned myocardium.[2,5] After 45-120 minutes of coronary occlusion the situation regarding various indicators of ischemia in preconditioned and in nonpreconditioned myocytes is virtually identical. Moreover, the reduction in the duration of the period of time during which myocytes suffer unfavorable conditions, such as severe ATP depletion or acidosis, afforded by preconditioning is short in relation to the total duration of exposure to these unfavorable conditions. This relatively short delay in the progression of ischemic injury is in contrast to the dramatic reduction in infarct size observed in preconditioned hearts after reperfusion.

This chapter will analyze some of the evidence supporting the hypothesis that part of the protection afforded by preconditioning is due to reduced mechanical reperfusion injury. The mechanisms leading to lethal mechanical damage during reperfusion will be briefly reviewed and the potential effects of ischemic preconditioning on each of these mechanisms will be discussed.

2.2. MECHANICAL REPERFUSION INJURY

Mechanical damage to cardiomyocytes during the early phase of reperfusion is the result of the combination of two circumstances: (1) reduced tolerance of the sarcolemma-cytoskeleton to withstand mechanical stress (i.e., increased fragility); and (2) generation of forces imposing an important mechanical stress on these structures.

2.2.1. Mechanical Fragility

Anoxia, ischemia and metabolic inhibition induce changes resulting in reduced mechanical resistance.[7-10] Myocytes may present a marked structural fragility at the time of reperfusion. This structural fragility has two main aspects: reduced resistance of the sarcolemma to stress, which can contribute to its rupture and directly cause cell death, and cytoskeletal fragility, which may render the myocyte more susceptible to collapse within itself under contractile forces. This second phenomenon, known as hypercontracture, does not result in sarcolemmal disruption and cell death in isolated cells.[8,11] However, in reperfused myocardium, hypercontracture of myocytes invariably results in sarcolemmal disruption due to the combined actions of cell-to-cell interaction and swelling.[8,12,13]

Sarcolemmal weakness can be the result of abnormalities in the lipid structure, in sarcolemmal proteins or both. It may also result from partial decoupling of the sarcolemma from its underlying cytoskeletal scaffold.[14] A weakened sarcolemma may become more easily disrupted when exposed to the mechanical forces produced by developing hypercontracture and osmotic swelling during reperfusion. Mechanical fragility secondary to anoxia may be modified by treatments acting during energy deprivation. It has been shown that increased sarcolemmal fragility can be reduced or prevented by interventions that reduce protein dephosphorylation.[15] Recent data suggest that fragility may be aggravated during reoxygenation. It has been recently demonstrated that the increased fragility of the reoxygenated cardiomyocytes can be prevented if inhibitors of lipid peroxidation (diphenylphenylenediamine) or nitric oxide (NO) donors in high concentration (SIN-1) are added upon reoxygenation.[16,17] Thirty minutes of this treatment are sufficient to avoid increased fragility.

Cytoskeletal fragility may contribute to development of hypercontracture. Hypercontracture, i.e., the extreme shortening of the contractile machinery, is not a result of force development alone but requires the collapse of the cytoskeleton. This collapse is the result of breaks in the connections between contractile proteins and the surrounding cytoskeletal network.[18] Recent observations from the group of Piper in Giessen suggest that cytoskeletal changes occurring during the anoxic period render the reoxygenated cell more susceptible to hypercontracture than could be expected for normal cells with the same elevation of cytosolic calcium.[19] These authors suggest that the cytoskeletal damage involved in this pathomechanism may be the result of protein hydrolysis, or delocalization of constitutive proteins, or due to changes in their phosphorylation state. Reduced elasticity caused by rigor-bond formation may contribute to this phenomenon.[20]

2.2.2. Mechanical Stress

a. Hypercontracture

In the reoxygenated myocardial cell a number of unusual mechanical forces are generated. The myofibrils may develop hypercontracture when they become reactivated upon resupply of oxidative energy in the presence of still elevated calcium concentrations.[21,23-25] Recent studies from Piper's group have shown that in reoxygenated cardiomyocytes calcium may be elevated during two early phases of recovery.[24,26] At the very onset of re-energization of the cell, calcium is gradually removed from the cytosol by sequestration into the sarcoplasmic reticulum. As this proceeds at limited speed, calcium concentrations persist for some time beyond the normal resting level.[24] The second phase where increased calcium concentrations are observed begins when most of the calcium has been taken up into the sarcoplasmic reticulum. This load begins to exceed the capacity of the organelle and as a result, calcium is repeatedly released from, and re-accumulated by, the sarcoplasmic reticulum. These oscillatory high transients of cytosolic calcium again represent a cause of hypercontracture. In ischemia-reperfused myocardium the first phase is unlikely to be the cause for hypercontracture because during this time myofibrils are largely paralyzed by the still acidotic cytosolic pH.[27] It is there due to the oscillatory release of calcium from the sarcoplasmic reticulum that the cells develop hypercontracture. Recent studies indicate that upon reoxygenation the myofibrils may exhibit an increased susceptibility to develop hypercontracture at a given rise of cytosolic calcium, probably due to degradation of some of the constitutive elements of the myofibrillar cytoskeletal network.[19] The combination of uncontrolled force and weakness of the cytoskeletal connections leads to hypercontracture. Studies using intramyocardial piezoelectric crystals have detected a reduction in end-diastolic length of reperfused myocardial segments by up to 65-70% of basal value. This myocardial shrinkage has been documented to occur during the first minutes of in situ coronary reperfusion, and its magnitude has been shown to closely correlate with the extent of myocardial necrosis and contraction band necrosis.[28] Transient contractile blockade with BDM during the first minutes of reoxygenation[29] or reperfusion[30,31] prevents hypercontracture and limits necrosis,[30,31] demonstrating that excessive and uncoordinated contractile activity may kill viable myocytes during reperfusion.

b. Osmotic swelling

During ischemia, ischemic by-products diffuse to the extracellular space and accumulate in both intracellular and extracellular compartments. This produces an important increase in osmotic pressure, which is of similar magnitude both inside and outside the ischemic myocyte.[9,32] During reperfusion, cell swelling occurs as a consequence of the rapid removal of ischemic metabolites from the extracellular space, creating a transarcolemmal osmotic gradient.[9,32-35] Cell swelling stretches the sarcolemma, which may rupture if its mechanical support by the underlying cytoskeleton is weakened. Even if swelling does not disrupt the sarcolemma, it may have effects on the electrophysiologic and mechanical behavior of myocytes. Swelling of myocytes and of endothelial cells may compress the microvascular bed and contribute to the "no-reflow" phenomenon.[9,34,36]

The importance of osmotic cell swelling as a cause of lethal mechanical reperfusion injury has been previously investigated indirectly by analyzing the effect of hyperosmotic reperfusion on infarct size.[9,34] Although the results are controversial, the discrepancies could be explained by differences in the study protocols. Remarkably, all studies using highly hyperosmotic reperfusion after periods of ischemia not far beyond the irreversibility threshold have yielded positive results.[9]

In a recent study, Ruiz-Meana et al have shown that osmotic cell swelling during reoxygenation induced by exposure to hypoosmotic medium, may kill freshly isolated adult rat cardiomyocytes.[8] In this model, reoxygenation-induced hypercontracture does not result in sarcolemmal rupture in the absence of swelling.[8,37] Interestingly, osmotic stress was unable to cause sarcolemmal disruption during reoxygenation when hypercontracture was prevented by transient contractile blockade. Furthermore, when hypercontracture and swelling were induced by exposure to hypoosmotic, high calcium media in myocytes not previously submitted to anoxia, it did not result in rupture. This suggests that swelling, hypercontracture, and anoxia-induced mechanical fragility may have additive effects to induce sarcolemmal rupture and cell death (Fig. 2.1).

c. Cell-to-cell mechanical interaction

In tissue the exchange of force between adjacent myocardial cells causes massive disruption of hypercontracting myocytes. Firm intercellular connections may prevent the sarcolemma following the extremely violent movement of underlying myofibrils during reoxygenation, so causing its disruption.[38,39] Hypercontracture may also impose mechanical stress on adjacent myocytes and the transmission of hypercontracture forces to adjacent cells may cause spreading of necrosis.[40] The importance of external mechanical forces in the tolerance of cardiomyocytes to ischemia/reperfusion is also suggested by the distribution of myocardial necrosis, secondary to short periods of coronary occlusion, along the interface between control and reperfused myocardium, an area of maximal mechanical stress.[41]

2.3. EFFECTS OF PRECONDITIONING ON MECHANICAL INJURY

Ischemic preconditioning modifies the response of myocytes to ischemia in several ways which could interfere with pathomechanisms involved in mechanical reperfusion injury.

2.3.1. Fragility

The pathophysiological mechanisms by which ischemic preconditioning may contribute to the prevention of cell fragility during ischemia

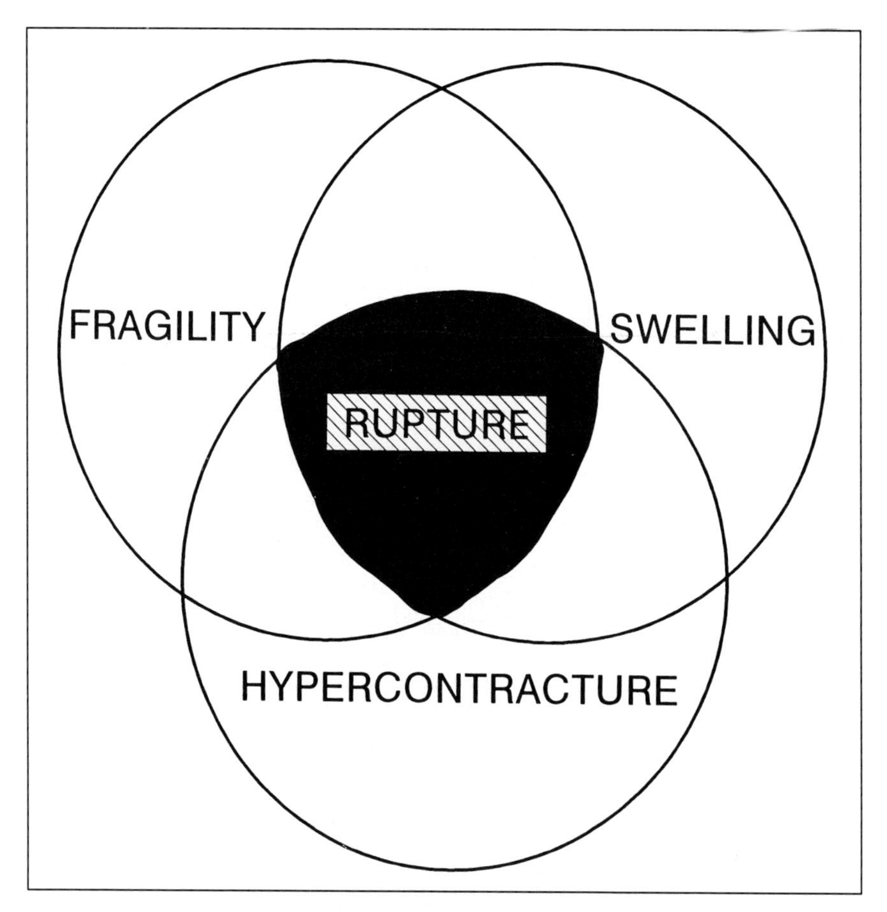

Fig. 2.1. Diagram showing the interrelationship between hypercontracture, osmotic cell swelling and anoxia-induced mechanical fragility in the genesis of sarcolemmal disruption and cell death during reoxygenation of isolated myocytes. None of these mechanisms alone is able to produce sarcolemmal disruption after one hour of metabolic inhibition; all three are required. Any intervention, such as ischemic preconditioning, which limits any of these mechanisms should have a protective effect against rupture. Adapted from ref. 8.

are summarized in Figure 2.2. A reduced rise in the cytoplasmic calcium concentrations could play a key role in the genesis of these mechanisms.

The effects of ischemic preconditioning on the rise in cytoplasmic calcium induced by ischemia have been well documented.[5,42] Measurements using 19F NMR of hearts loaded with 5FBAPT have shown that preconditioning slows and delays the rise induced by a subsequent period of prolonged ischemia.[5] This effect seems to be due to reduced H^+ accumulation, and reduced stimulation of Na^+/H^+ and Na^+/Ca^{2+} exchange (Fig. 2.2). The reduced H^+ accumulation during ischemia in preconditioned myocardium is mainly due to limited lactate

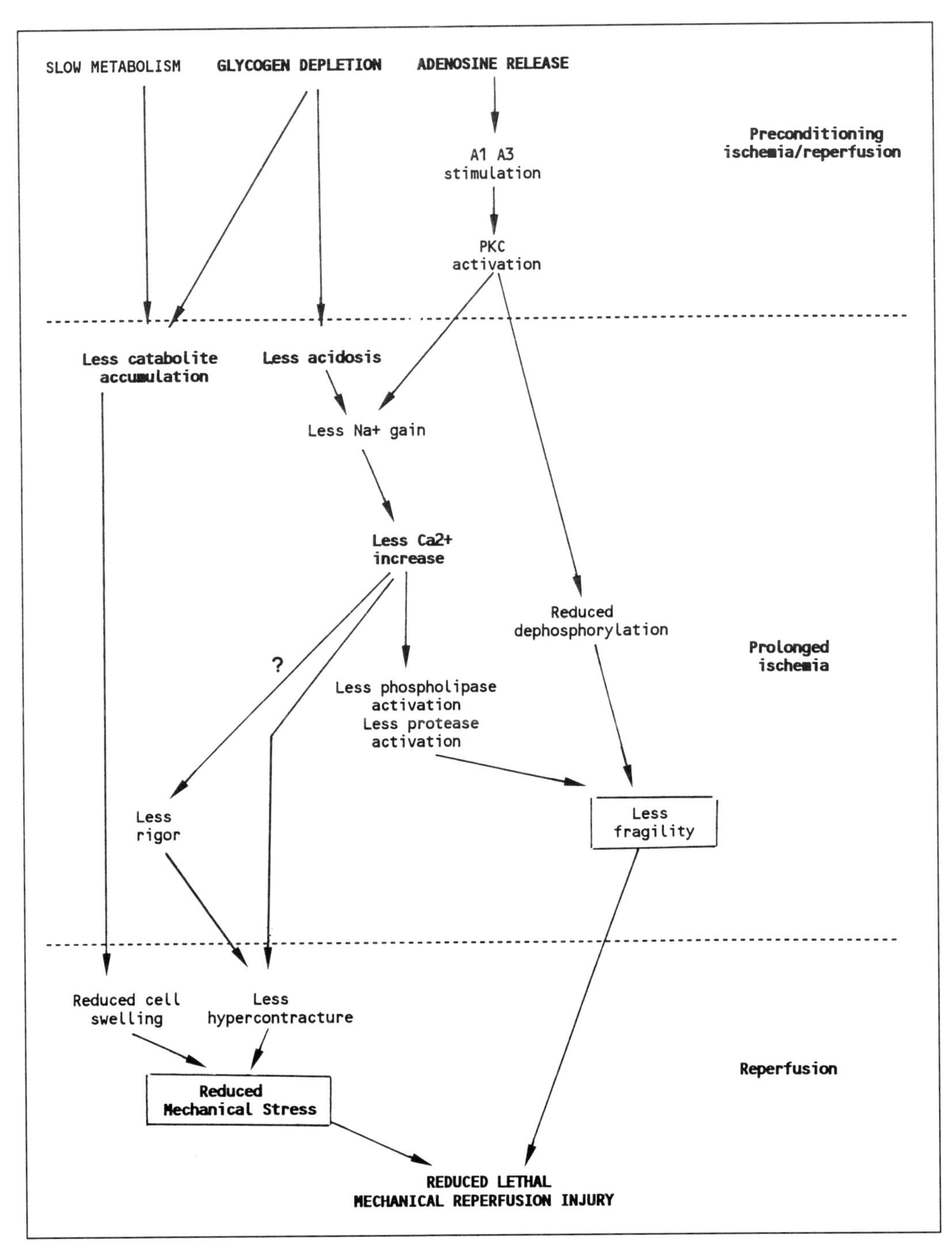

Fig. 2.2. Diagram summarizing the main mechanisms by which ischemic preconditioning may limit lethal mechanical injury during myocardial reperfusion. Note the key roles of reduced lactate accumulation and limited elevation of cytoplasmic calcium.

production.[2,5,32] During preconditioning ischemia, glycogen stores are rapidly depleted, while preconditioning reperfusion does not allow replenishment of glycogen stores.[2,43] The depletion of glycogen results in less lactate accumulation during subsequent prolonged ischemia. It has been shown that when preconditioning reperfusion is progressively prolonged, replenishment of glycogen stores parallels the loss of the beneficial effect of ischemic preconditioning on infarct size.[43,44]

Several studies indicate the importance of Na^+/H^+ exchange in the genesis of the rise in cytoplasmic calcium during ischemia. Arrest with magnesium chloride, which blocks calcium entry via calcium channels and Na^+/Ca^{2+} exchange, have effects on the increase in cytoplasmic calcium and on myocardial necrosis similar to those of preconditioning.[45] Nonselective Na^+/H^+ exchange inhibitors have the same effect.[46] In a recent study, a highly selective inhibitor of this exchanger dramatically reduced infarct size when given before coronary occlusion but had no effect when given during reperfusion, demonstrating the importance of Na^+/H^+ exchange during the ischemic period.[47] Thus, all these data are consistent with the hypothesis that, during ischemia, increased H^+ is exchanged for Na^+, and that, in turn, Na^+ is then exchanged for Ca^{2+}. This exchange is the main factor responsible for the rise in calcium during ischemia. In addition to glycogen depletion, stimulation of A_1 and A_3 adenosine receptors (by G_i coupled mechanisms) contributes to the beneficial effect of preconditioning on calcium during ischemia, since pretreatment with adenosine mimics the effect of preconditioning on the rise in cytoplasmic calcium.[48,49] These G_i proteins, coupled to the adenosine receptors, could directly slow Na^+/H^+ exchange during ischemia.

Several studies show a consistent relationship between the magnitude of the rise in cytoplasmic calcium and the severity of ischemic injury. Maneuvers other than ischemic preconditioning which delay the rise in cytoplasmic calcium also have a protective effect against ischemic injury. However, although there are multiple mechanisms by which the rise in cytoplasmic calcium may contribute to ischemic injury, the relative contribution of each factor has not been established. Increased calcium may contribute both to mechanical fragility during ischemia and to hypercontracture during reperfusion.

Increased cytoplasmic calcium may induce mechanical fragility by several mechanisms.

1. It may cause sarcolemmal fragility by activating phospholipases able to modify the lipid structure of the sarcolemma. It has been shown that small changes in the lipid composition of the sarcolemma may cause important alterations in sarcolemmal fluidity resulting in decreased resistance to mechanical strength.[50] Treatments aimed to prevent lipid peroxidation have been found to limit osmotic fragility in isolated myocytes subjected to transient anoxia and reoxygenation.[51]

2. Increased cytoplasmic calcium may cause cytoskeletal fragility by activating proteases. It has been shown that certain proteases can be activated by minute increases in calcium, and that these proteases are localized around the cytoskeleton-sarcolemma attachments.
3. Increased cytoplasmic calcium could contribute to the development of rigor bonds. Rigor bonds not only consume ATP,[14,20,52] but also cause rigidity and prevent homogeneous force transduction thus resulting in fragility. Rigor bonds activate cross-bridge activity and enhance the affinity of troponin-C for calcium.

Stimulation of adenosine receptors triggers a cascade of events with potential cardioprotective effects. One of them is activation and translocation of protein kinase (PK) C. The results on the role of PKC in the protection afforded by preconditioning are controversial (reviewed in chapter 11). While several authors have failed to abolish the beneficial effects of preconditioning when PKC is inhibited, others have demonstrated activation of PKC by ischemic preconditioning and a cardioprotective effect of maneuvers leading to PKC activation.[53]

Activated PKC may modify the phosphorylation status of multiple key proteins. The effects of PKC activation secondary to A_1 receptor stimulation on ATP sensitive K^+ channels is a well documented example of the phosphorylating actions of PKC.[48] It has been suggested that mechanical fragility induced by ischemia could be mediated by a modification of the phosphorylation status of several cytoskeletal proteins such as vinculin. Substances with phosphatase activity increase mechanical fragility, whilst drugs with anti-phosphatase properties have the opposite effect.[15] It is thus conceivable that the phosphorylating effects of PKC could help to maintain the physical properties of the cytoskeleton and the cytoskeleton-sarcolemma attachments in preconditioned myocardium. In fact, it has been shown that MARCKS (myristoylated, alanine-rich C kinase substrate) and other actin-binding proteins are specific PKC substrates with an important contribution to the regulation of the cytoskeleton-membrane interactions and of the physical structure of the actin network in certain types of cells such as macrophages.[54]

2.3.2. Mechanical Stress

Ischemic preconditioning may reduce the severity of the mechanical overload imposed by reperfusion after a prolonged coronary occlusion by reducing hypercontracture and osmotic swelling.

a. Hypercontracture

Taking into consideration the aforementioned relationship between increased calcium gain and excessive uncoordinated contractile activity during reperfusion, one of the main consequences of the favorable influence of preconditioning on calcium elevation during ischemia should be the reduction of hypercontracture. Since increased intracellular calcium concentrations at the time of restoration of oxidative metabolism are directly responsible for excessive contraction[21-25] the limited rise in intracellular calcium in preconditioned myocytes should protect them against hypercontracture. It can be speculated that a delayed rise in Ca^{2+} in preconditioned myocardium could result in delayed rigor and less chance of hypercontracture occurring during subsequent reoxygenation or reperfusion, since rigor bonds not only activate cross-bridge activity, but enhance the affinity of troponin-C for calcium.

Several findings support a favorable effect of ischemic preconditioning on hypercontracture. Preconditioning effectively reduces the infarct size induced by relatively brief (less than 2 hours) transient coronary occlusions. These infarcts are almost exclusively composed of contraction band necrosis, the pathological expression of reperfusion induced hypercontracture.[30,55] Myocardial segment length analysis with intramyocardial piezoelectric crystals has also confirmed that preconditioning preserves end-diastolic segment length and reduces myocardial segment shrinkage.[56] Reperfusion-induced myocardial shrinkage has recently been shown to be the consequence of hypercontracture.[57]

b. Reperfusion edema

The depletion of glycogen stores during the preconditioning episodes of ischemia results in the reduced accumulation of lactate and protons during a subsequent prolonged episode of ischemia.[55] The accumulation of P_i and other by-products during ischemia are also reduced in preconditioned hearts due to substrate depletion and slowed metabolic activity. In a recent study, Sanz et al[55] investigated the hypothesis that reduced catabolite accumulation during ischemia in preconditioned myocardium results in less osmotic cell swelling during reperfusion. The effects of preconditioning hearts with 5 minutes of ischemia and 5 minutes of reperfusion on postreperfusion myocardial edema, catabolite content and infarct size, were compared with those produced by washout of catabolites by 5 minutes of anoxic intracoronary perfusion after the first 5 minutes of prolonged ischemia. Both protocols reduced lactate accumulation during ischemia and myocardial edema during reperfusion by approximately one-third, as compared to control animals undergoing an episode of coronary occlusion of the same duration (Fig. 2.3). These results indicate that ischemic preconditioning markedly reduces reperfusion edema, and that this effect may be explained by reduced catabolite accumulation. However, only ischemic preconditioning reduced infarct size, demonstrating that reduced reperfusion edema does not suffice to explain the beneficial effect of preconditioning on lethal ischemic injury.

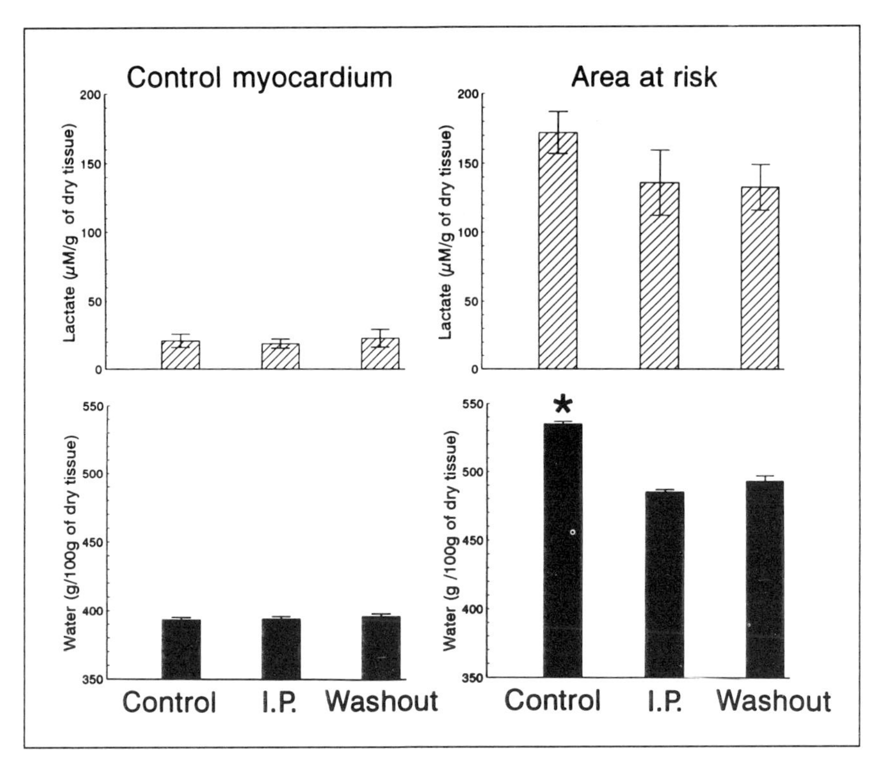

Fig. 2.3. Effect of ischemic preconditioning (I.P), and of anoxic wash-out of catabolites during ischemia on lactate accumulation during 48 minutes of coronary occlusion in the in situ pig heart and on myocardial water content after 30 minutes of reperfusion, as compared to animals with the same occlusion/reperfusion protocol without any additional intervention (Control). No differences in myocardial water or lactate content were observed in control myocardium. In the area at risk, preconditioning and anoxic washout reduced lactate accumulation and myocardial edema in a similar way. Modified from ref. 55.

2.4. CONCLUSION

We consider that there is solid and growing evidence that mechanical overload imposed by hypercontracture, swelling and cell-to-cell mechanical interaction may lethally injure viable, mechanically fragile myocytes during myocardial reperfusion. Hypercontracture and swelling are the consequence of the abrupt normalization of physiological conditions (oxidative metabolism, osmolality and ionic composition) in the presence of severe abnormalities caused by prolonged ischemia (increased H^+_i, Na^+_i and Ca^{2+}_i, catabolite accumulation, mechanical fragility). The information considered so far leads us to formulate the hypothesis that ischemic preconditioning has a marked protective effect against those ischemic changes that are more directly involved in the genesis of lethal mechanical reperfusion injury (Fig. 2.4). This

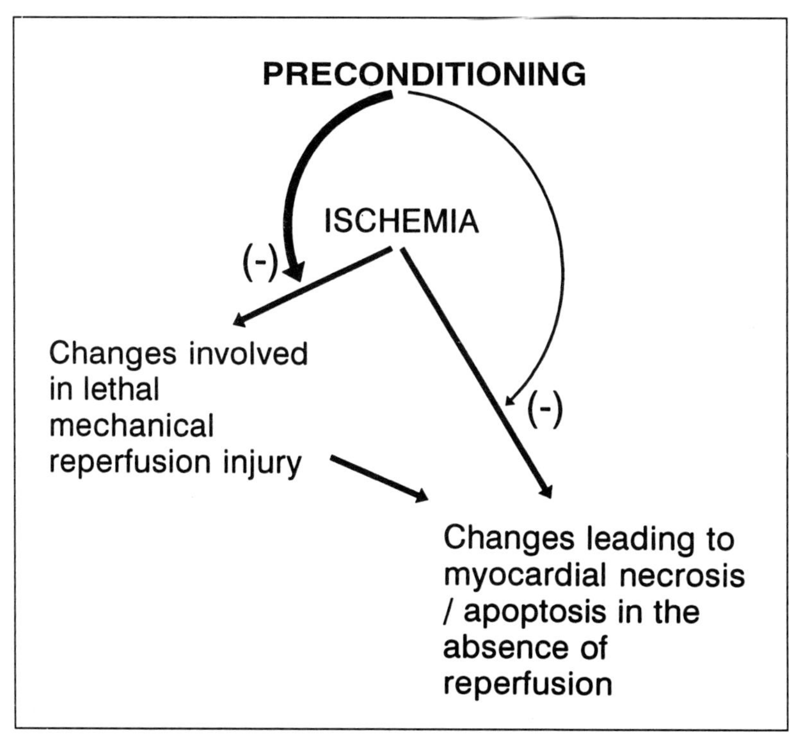

Fig. 2.4. The hypothesis that part of the anti-infarct effect of ischemic preconditioning is due to reduction of lethal mechanical reperfusion injury.

hypothesis should not be interpreted as indicating that the beneficial effect of ischemic preconditioning on infarct size is exclusively, or even mainly, due to limitation of lethal mechanical reperfusion injury. There is much evidence indicating that ischemic preconditioning also has a marked protective effect on derangements not directly implicated in lethal mechanical reperfusion injury. The eventual identification of the mechanisms by which preconditioning protects against mechanical lethal reperfusion injury could represent a major advance towards the development of new therapeutic strategies able to minimize myocardial necrosis secondary to transient ischemia.

Acknowledgment

Partially supported by Grants 93/0668 E from the Fondo de Investigación Sanitaria de la Seguridad Social, and the Concerted Action BMH1CT92/1501, BIOMED I Program from the European Union.

References

1. Murry CE, Jennings RB, Reimer KA. Preconditioning with ischemia: a delay of lethal cell injury in ischemic myocardium. Circulation 1986; 74:1124-1136.

2. Kida M, Fujiwara H, Ishida M et al. Ischemic preconditioning preserves creatine phosphate and intracellular pH. Circulation 1991; 84:2495-2503.
3. Cinca J, Worner F, Carreño A et al. T-Q, S-T segment mapping and hyperemia in reperfused pig heart with ischemic preconditioning. Am J Physiol 1992; 263:H1732-H1738.
4. Janier MF, Vanoverschelde JL, Bergmann SR. Ischemic preconditioning stimulates anaerobic glycolysis in the isolated rabbit heart. Am J Physiol 1994; 267:H1353-H1360.
5. Steenbergen C, Perlman ME, London RE et al. Mechanism of preconditioning: Ionic alterations. Circ Res 1993; 72:112-125.
6. Jennings RB, Murry CE, Reimer KA. Energy metabolism in preconditioned and control myocardium: effect of total ischemia. J Mol Cell Cardiol 1991; 23:1449-1458.
7. Ganote CE, Vander Heide RS. Irreversible injury of isolated adult rat myocytes: Osmotic fragility during metabolic inhibition. Am J Pathol 1988; 132:212-222.
8. Ruiz-Meana M, García-Dorado D, Gonzalez MA et al. Effect of osmotic stress on sarcolemmal integrity of isolated myocytes following transient metabolic inhibition. Cardiovasc Res 1995; 30:64-69.
9. García-Dorado D, Oliveras J. Myocardial oedema: a preventable cause of reperfusion injury? Cardiovasc Res 1993; 27:1555-1563.
10. Steenbergen C, Hill ML, Jennings RB. Volume regulation and plasma membrane injury in aerobic, anaerobic and ischemic myocardium in vitro. Circ Res 1985; 57:864-875.
11. Piper HM, Spahr R, Hütter JF et al. The calcium and the oxygen paradox: Non-existent on the cellular level. Basic Res Cardiol 1985; 80:159-163.
12. Ganote CE. Contraction band necrosis and irreversible myocardial injury. J Mol Cell Cardiol 1983; 15:67-73.
13. Ganote CE. Cell to cell interactions contributing to the "oxygen paradox." Basic Res Cardiol 1985; 80 (Suppl 2):141-146.
14. Steenbergen C, Hill ML, Jennings RB. Cytoskeletal damage during myocardial ischemia: changes in vinculin immunofluorescence staining during total in vitro ischemia in canine heart. Circ Res 1987; 60:478-486.
15. Armstrong SC, Ganote C. Effects of 2,3-Butanedione monoxime (BDM) on contracture and injury of isolated rat myocytes following metabolic inhibition and schemia. J Mol Cell Cardiol 1991; 23:1001-1014.
16. Schlüter KD, Weber M, Schraven E et al. NO donor SIN-1 protects against reoxygenation-induced cardiomyocyte injury by a dual action. Am J Physiol 1994; 267:H1461-H1466.
17. Schlüter KD, Weber M, Ruiz-Meana M et al. Protection of the heart against reoxygenation injury by SIN-1c. Am J Physiol (submitted).
18. Ganote CE, Armstrong S. Ischemia and the myocyte cytoskeleton: review and speculation. Cardiovasc Res 1993; 27:1387-1403.
19. Ladilov Y, Siegmund B, Piper HM. Reoxygenation-induced hypercontracture is the result of an increased susceptibility to calcium. Pflügers

Arch 1995; 429 (Suppl.):R95 (Abstract).
20. Steenbergen C, Murphy E, Watts JA et al. Correlation between cytosolic free calcium, contracture, ATP, and irreversible ischemic injury in perfused rat hear. Circ Res 1990; 66:135-146.
21. Allshire A, Piper HM, Cuthbertson KSR et al. Cytosolic free Ca^{2+} in single rat heart cells during anoxia and reoxygenation. Biochem J 1987; 244:381-385.
22. Bowers KC, Allshire AP, Cobbold PH. Continuous measurements of cytoplasmic ATP in single cardiomyocytes during simulation of the "oxygen paradox". Cardiovasc Res 1993; 27:1836-1839.
23. Steenbergen C, Murphy E, Levy L et al. Elevation in cytosolic free calcium early in myocardial ischemia in perfused rat heart. Circ Res 1987; 60:700-707.
24. Ladilov YV, Siegmund B, Piper HM. Protection of the reoxygenated cardiomyocyte against hypercontracture by inhibition of Na^+/H^+ exchange. Am J Physiol 1995; 268:H1531-9.
25. Saris N-EL, Allshire A. Ca^{2+} transport in mitochondria. In: Fleischer S, Fleischer R, eds. Methods in Enzymology. Vol 174. New York: Academic Press, 1989:68-85.
26. Siegmund B, Ladilov Yu, Piper HM. Importance of sodium for the recovery of Ca^{2+} control in reoxygenated cardiomyocytes. Am J Physiol 1994; 267:H506-H513.
27. Allen DG, Orchard CH. Myocardial contractile function during ischemia and hypoxia. Circ Res 1987; 60:153-168.
28. García-Dorado D, Barrabés JA, Oliveras J et al. Myocardial fibre shrinkage during coronary reperfusion reflects myocyte hypercontracture. Eur Heart J 1994; 15:116.
29. Siegmund B, Klietz T, Schwartz P et al. Temporary contractile blockade prevents hypercontracture in anoxic-reoxygenated cardiomyocytes. Am J Physiol 1991; 260:H426-H435.
30. García-Dorado D, Theroux P, Durán JM et al. Selective inhibition of the contractile apparatus. A new approach to the modification of infarct size, infarct composition and infarct geometry during coronary artery occlusion and reperfusion. Circulation 1992; 85:1160-1174.
31. Schlack W, Uebing A, Schäfer M et al. Regional contractile blockade at the onset of reperfusion reduces infarct size in the dog heart. Pflügers Arch 1994; 428:134-141.
32. Tani M, Neely JR. Intermittent perfusion of ischemic myocardium. Possible mechanisms of protective effects on mechanical function in isolated rat heart. Circulation 1990; 82:536-548.
33. García-Dorado D, Oliveras J, Gili J et al. Analysis of myocardial edema by magnetic resonance imaging early after coronary artery occlusion with or without reperfusion. Cardiovasc Res 1993; 27:1426-1469.
34. García-Dorado D, Theroux P, Muñoz R et al. Favorable effects of hyperosmotic reperfusion on myocardial infarct size. Am J Physiol 1992; 262:H17-H22.

35. Jennings RB, Schaper J, Hill ML et al. Effect of reperfusion late in the phase of reversible ischemic injury. Changes in cell volume, electrolyte, metabolites, and ultrastructure. Circ Res 1985; 56:262-272.
36. Powers ER, Di Bona DR, Powell WJ. Myocardial cell volume and coronary resistance during diminished coronary perfusion. Am J Physiol 1984; 247:H467-H477.
37. Siegmund B, Koop A, Klietz T et al. Sarcolemmal integrity and metabolic competence of cardiomyocytes under anoxia-reoxygenation. Am J Physiol 1990; 258:H285-H291.
38. Frank JS, BradyAJ, Farnsworth S et al. Ultrastructure and function of isolated myocytes after calcium depletion and repletion. Am J Physiol 1986; 250:H265-H275.
39. Tranum-Jansen J, Janse MJ, Fiolet JWT et al. Tissue osmolality, cell swelling, and reperfusion in acute regional myocardial ischemia in the isolated porcine heart. Circ Res 1981; 49:364-381.
40. García-Dorado D, Théroux P, Desco M et al. Cell-to-cell interaction: A mechanism to explain wave-front progression of myocardial necrosis. Am J Physiol 1989; 256:H1266-H1273.
41. Solares J, García-Dorado D, Oliveras J et al. Contraction band necrosis at the lateral borders of the area at risk in reperfused infarcts. Virchows Archiv 1995; 426:393-399.
42. Albuquerque CP, Gerstenblith G, Weiss RG. Importance of metabolic inhibition and cellular pH in mediating preconditioning contractile and metabolic effects in rat hearts. Circ Res 1994; 74:139-150.
43. Wolfe CL, Sievers RE, Visseren FLJ et al. Loss of myocardial protection after preconditioning correlates with the time course of glycogen recovery within the preconditioned segment. Circulation 1993; 87:881-892.
44. Murry CE, Richard VJ, Jennings RB et al. Myocardial protection is lost before contractile function recovers from ischemic preconditioning. Am J Physiol 1991; 260:H796-H804.
45. Murphy E, Perlman M, London RE et al. Amiloride delays the ischemia-induced rise in cytosolic free calcium. Circ Res 1991; 68:1250-1258.
46. Piper HM, Balser C, Ladilov Y et al. The role of Na^+/H^+ exchange in ischemia-reperfusion. Cardiovasc Res (in press).
47. González MA, García-Dorado D, Barrabés JA et al. Inhibition of Na^+/H^+ exchange during coronary occlusion, but not during reperfusion, reduces infarct size in the in situ pig heart model. Circulation (submitted).
48. Grover GJ, Sleph PG, Dzwonczyk S. Role of myocardial ATP-sensitive potassium channels in mediating preconditioning in the dog heart and their possible interaction with adenosine A_1-receptors. Circulation 1992; 86:1310-1316.
49. Ganote CE, Armstrong S, Downey J.M. Adenosine and A_1 selective agonists offer minimal protection against ischaemic injury to isolated rat cardiomyocytes. Cardiovasc Res 1993; 27:1670-1676.
50. Lars Bastiaanse EM, van der Valk-Kokshoorn LJM, Egas-Kenniphaas JM et al. The effect of sarcolemmal cholesterol content on the tolerance to

anoxia in cardiomyocyte cultures. J Mol Cell Cardiol 1994; 26:639-648.
51. Schlüter KD, Jakob G, Ruiz-Meana M et al. Protection of reoxygenated cardiomyocytes against osmotic fragility by NO donors. Am J Physiol (in press).
52. Bowers KC, Allshire AP, Cobbold PH. Bioluminescent measurement in single cardiomyocytes of sudden cytosolic ATP depletion coincident with rigor. J Mol Cell Cardiol 1992; 24:213-218.
53. Mitchell MB, Meng X, Ao L et al. Preconditioning of isolated rat heart is mediated by protein kinase C. Circ Res 1995; 76:73-81.
54. Aderem A. Signal transduction and the actin cytoskeleton: the roles of MARCKS and profilin. TIBS 1992; 17:438-443.
55. Sanz E, García-Dorado D, Oliveras J et al. Dissociation between the antiedema effect and the anti-infarct effect of ischemic preconditioning. Am J Physiol 1995; 268:H233-H241.
56. Ovize M, Rioufol G, Loufoua J et al. Paradoxical persistent regional dilation in reperfused preconditioned pig hearts. Eur Heart J 1994; 15 (Suppl):469 (Abstract).
57. Barrabés JA, García-Dorado D, Ruiz-Meana M et al. Pflügers Arch (in press).

CHAPTER 3

Ischemic Preconditioning Markedly Reduces the Severity of Ischemia and Reperfusion-Induced Arrhythmias: Role of Endogenous Myocardial Protective Substances

Agnes Vegh and James R. Parratt

3.1. INTRODUCTION

It could be argued that an even more important manifestation of ischemic tolerance resulting from ischemic preconditioning than the reduction in infarct size is the profound reduction in the severity of life-threatening ventricular arrhythmias that arise from myocardial ischemia and reperfusion. This is because myocardial ischemic injury can now be limited by a combination of early thrombolysis and appropriate drug therapy in those patients who have survived an acute coronary attack, whereas therapy to prevent the attack itself is less well established. The purpose of this chapter is to outline the evidence for the antiarrhythmic effect of preconditioning and to discuss

Myocardial Preconditioning, edited by Cherry L. Wainwright and James R. Parratt.

the possible mechanisms of this protection with particular reference to studies in a canine model of ischemia and reperfusion.

3.2. The Antiarrhythmic Effect of Ischemic Preconditioning—The Background

Our own interest in the protection against arrhythmias afforded by ischemic preconditioning was stimulated by two unrelated observations. First, the finding that when a coronary artery is occluded, ventricular ectopic activity occurs within a few minutes (phase 1 arrhythmias), and reaches a peak around 5-20 minutes, depending upon the species; the heart then reverts to sinus rhythm after about 30 minutes of ischemia despite the occlusion being maintained.[1] An example of this distribution is illustrated in Figure 3.1. These early phase 1 arrhythmias are perhaps the most likely cause of sudden cardiac death in humans. Later, again depending upon the species, ventricular ectopic activity recommences; these arrhythmias are perhaps similar to those observed in patients who have survived the pre-hospital phase of an acute coronary attack and are admitted to a coronary care unit. Certainly in animal models, the early phase of arrhythmias disappears despite the maintenance of ischemia. One possibility for this spontaneous reversion of ectopic activity is that the initial period of ischemia triggers a defensive response, mechanisms unknown, that then suppresses ventricular activity. This would mean that myocardial cells are able to adapt rapidly to brief ischemic stresses within the timescale of the disappearance of these early arrhythmias, i.e., within 20-30 minutes.

The other finding that stimulated our interest in the antiarrhythmic effect of preconditioning was that of Podzuweit and his co-workers in Bad Nauheim,[2] who demonstrated that the pacemaker activity which occurs when noradrenaline is infused locally into the myocardium is suppressed when a small branch of a coronary artery, well away from the infused area, is subsequently occluded. When the artery is reopened the ventricular tachycardia induced by the intramyocardial administration of noradrenaline resumes within seconds of the release of the occlusion. Podzuweit and his colleagues described this phenomenon as 'the antiarrhythmic effect of ischemia,' and suggested that 'the ischemic myocardium may have previously unrecognized antiarrhythmic properties' and 'that reperfusion arrhythmias might be caused, at least in part, by disappearing ischemic protection.'[2] The evidence that the ischemic myocardium is capable of generating 'endogenous myocardial protective (antiarrhythmic) substances,' that reduce the severity of ongoing and subsequent ischemic insults, has been recently reviewed.[3]

There have been many previous observations (reviewed in refs. 4, 5) of the reduction, by brief periods of ischemia, of the severity of ventricular arrhythmias resulting from a subsequent period of ischemia. These include the findings: (1) that dogs are more likely to survive a coronary artery occlusion if this is performed in two stages; (2) that

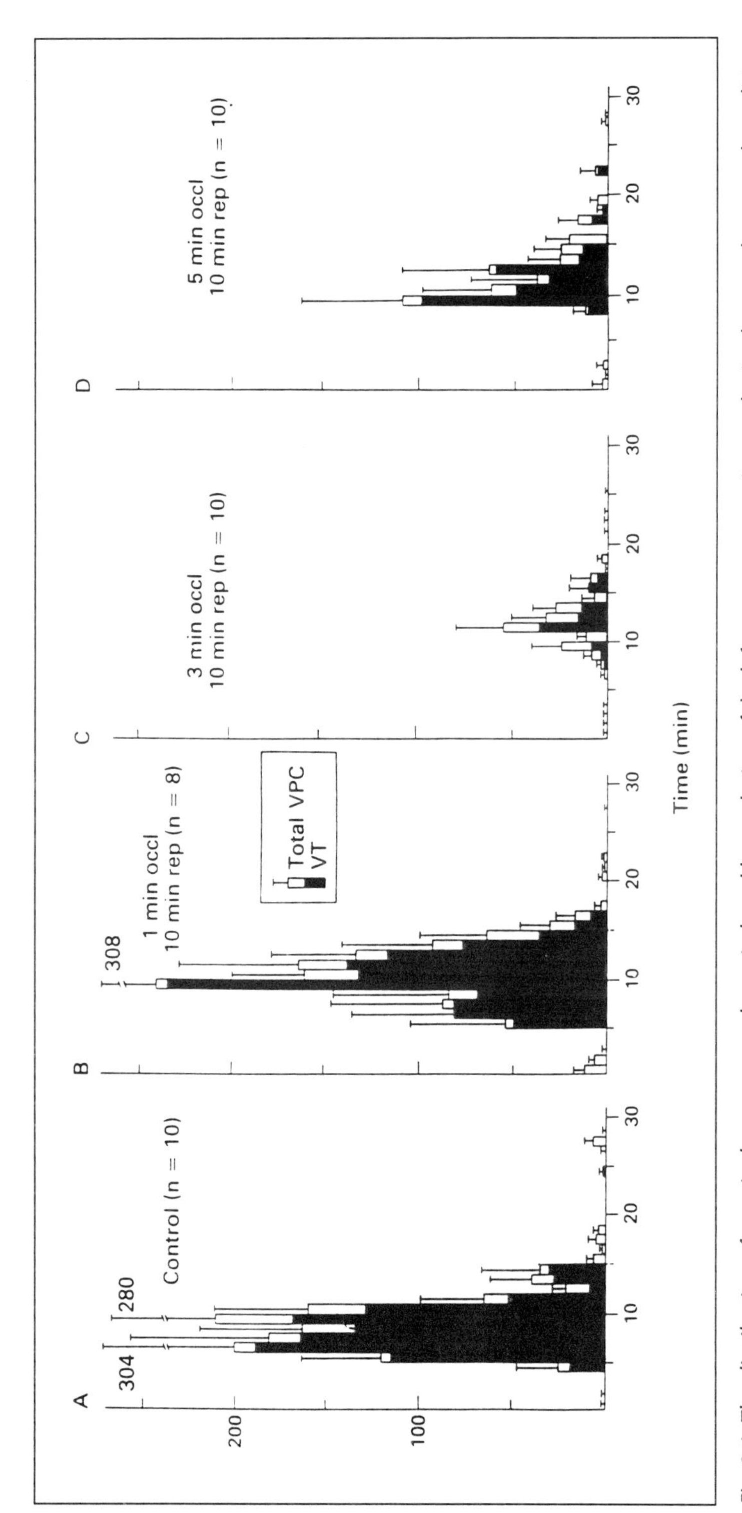

Fig. 3.1. The distribution of ventricular premature beats induced by occlusion of the left coronary artery in anesthetized rats under control conditions (Panel A) and under conditions when the prolonged, 30 minute occlusion is preceded by a short preconditioning occlusion of 1 minute (B), 3 minutes (C) or 5 minutes (D). The reperfusion time in each case was 10 minutes. A preconditioning occlusion of 3-5 minutes markedly reduces the number of premature beats that occurred during a subsequent prolonged occlusion. The optimum preconditioning occlusion time appears to be 3 minutes in this species. VPC, ventricular premature complexes; VT indicates ventricular tachycardia. Reproduced with permission from Vegh A et al, Cardiovasc Res 1992; 26:487-495.

brief repeated periods of ischemia result in less severe ischemic changes during the second and subsequent occlusions (including a reduction in the number of ventricular premature beats); and (3) that epicardial ST-segment changes are more severe during an initial coronary occlusion than they are during subsequent coronary artery occlusions of the same duration.

None of these studies was designed to explore the possibility that *survival from a prolonged ischemic insult* would be modified if it had been preceded by one or more shorter periods of ischemia; in other words to test the hypothesis that ischemic preconditioning results, not only in a reduction in myocardial ischemic damage, but also in a suppression of life-threatening ventricular arrhythmias.

3.3. THE ANTIARRHYTHMIC EFFECTS OF ISCHEMIC PRECONDITIONING—THE PHENOMENON

This was first demonstrated by Sadayoshi Komori in the Strathclyde department using a method[6] for producing ventricular arrhythmias by coronary artery occlusion in anesthetized open-chest rats. Komori and his colleagues showed[7-9] that a 30 minute coronary artery occlusion resulted in marked ventricular ectopic activity which was markedly suppressed if that prolonged occlusion had been preceded, 10 minutes earlier, by a brief 3 or 5 minute (but not a 1 minute) period of ischemia (Fig. 3.1). Because, in this model, reperfusion arrhythmias are severe after the release of a 5 minute occlusion,[9,10] a 3 minute 'preconditioning' period of ischemia appeared to be optimal. This reduction in the severity of arrhythmias by a single preconditioning occlusion was very marked. There was a reduction in the incidence of ventricular fibrillation from 42% (in controls) to only 10% and a reduction in the number of ventricular premature beats during the 30 minute period of ischemia from 1236 ± 262 in the controls from only 200 ± 60 in hearts subjected to preconditioning. This protection was much less pronounced if the interval between the single preconditioning occlusion and the prolonged period of ischemia was increased from 10 minutes to 30 minutes and was almost entirely lost if the reperfusion period was further increased to 1 hour.[9]

This marked antiarrhythmic effect of ischemic preconditioning was subsequently confirmed in larger animal models such as dogs[7,9,11] and pigs.[12] The protection was particularly marked in the canine model.[9] The protocol in these experiments was to occlude the coronary artery for either one or two 5 minute periods and then, at various times afterwards, reocclude that same artery for 25 minutes. This was followed by release of the occlusion and reperfusion of the ischemic area. In brief, these experiments showed that:

1. The protection against arrhythmias was pronounced[9,11] with marked suppression of ventricular ectopic activity (Figs. 3.1 and 3.2) and an absence of ventricular fibrillation during the occlusion period (Fig. 3.3).

2. This was a real protection in that the arrhythmias were not shifted to a different time period. This was shown by prolonging the occlusion from 30 to 60 minutes; even during a 1 hour occlusion period the number of ventricular premature beats was significantly reduced (Fig. 3.2) and, again, there was no ventricular fibrillation over the entire 1 hour occlusion period.[9]

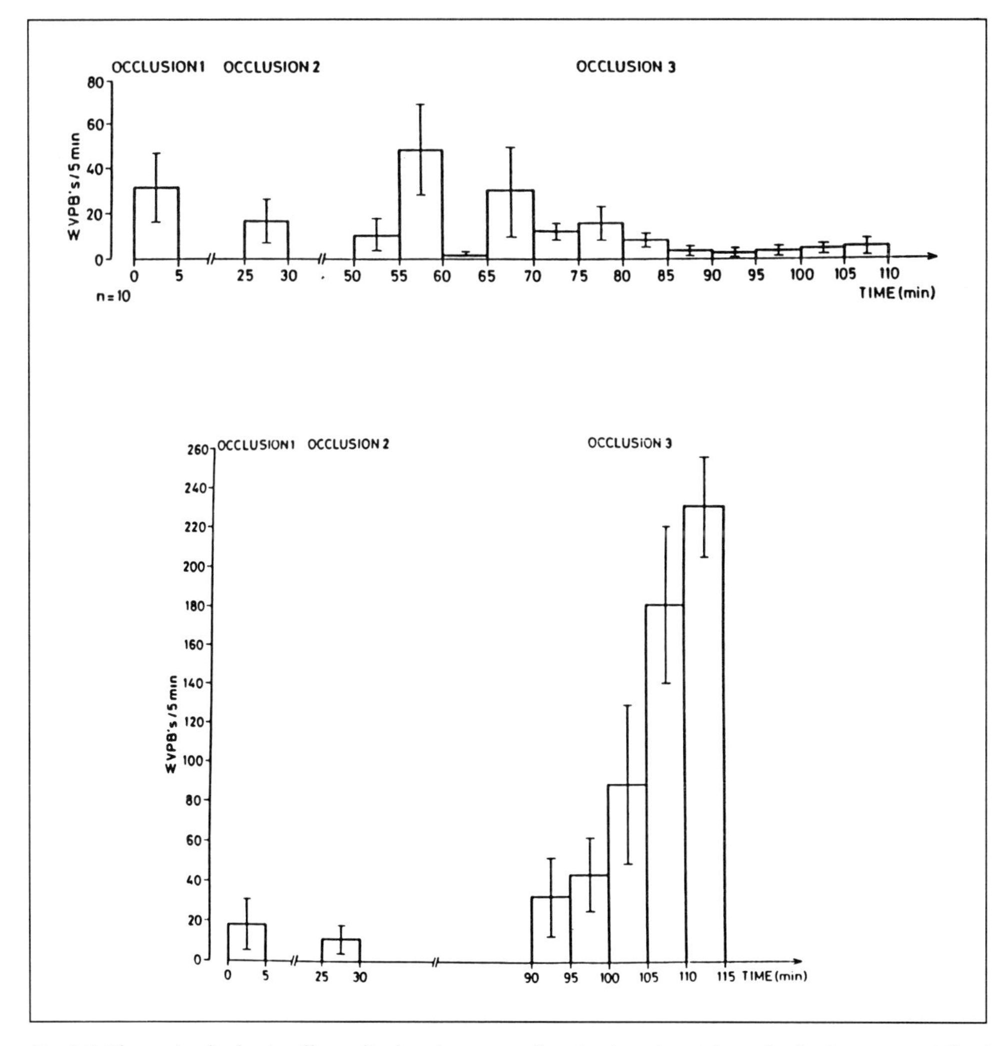

Fig. 3.2. The antiarrhythmic effect of ischemic preconditoning is real and the arrhythmias are not shifted to a later time period. The figure compares ventricular premature beats (in 5 minute intervals) over a 25 minute period in control mongrel dogs not subjected to preconditioning (A) and in dogs first subjected to 2 x 5 minute preconditioning coronary artery occlusions (B). There was a marked reduction in the number of ventricular premature beats (VPBs) during the 25 minute occlusion in those dogs subjected to preconditioning (78 ± 27 over the same 30 minute period and 156 ± 32 over a 60 minute occlusion period compared to 445 ± 140 in the controls). Adapted with permission from Vegh A et al, Cardiovasc Res 1992; 26:487-495.

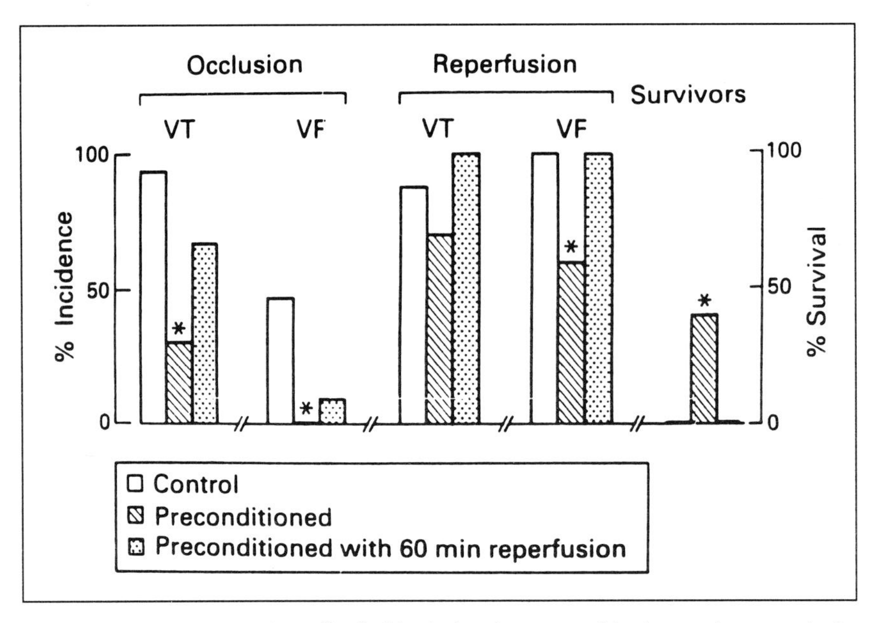

*Fig. 3.3. Transient protection afforded by ischemic preconditioning against ventricular arrhythmias in anesthetized mongrel dogs. The protection is lost if the reperfusion time between the preconditioning occlusions and the prolonged occlusion is increased to 1 hour. The figure shows the incidences of ventricular tachycardia (VT) and ventricular fibrillation (VF) during a 25 minute occlusion of the left anterior descending coronary artery and during subsequent reperfusion. There is a reduction in the incidences of both VT and VF (and an increased survival) in those dogs that were preconditioned by two 5 minute occlusions provided the reperfusion time was 20 minutes but not if the reperfusion time was increased to 1 hour. *P<0.05 v control, nonpreconditioned dogs. Adapted with permission from Vegh A et al, Cardiovasc Res 1992; 26:487-495.*

3. The protection was transient in that if the reperfusion period was increased from 20 minutes to 1 hour most of the protection was lost (Fig. 3.3).
4. There was some evidence for protection against the combined ischemia-reperfusion insult, with survival only in dogs that had been preconditioned (Fig. 3.3). In other words, there was protection both against ischemia and reperfusion induced arrhythmias. Protection by short periods of ischemia against reperfusion arrhythmias occurring following the release of subsequent, also brief, coronary artery occlusions was demonstrated by Shiki and Hearse in 1987.[13]
5. Preconditioning also induced other indices of ischemia severity, such as epicardial ST-segment elevation (Fig. 3.4) and the degree of inhomogeneity of activation within the ischemic area. This was assessed using a 'composite electrode'[14] which gives a summarized recording of R-waves

from 30 epicardial measuring points. In the adequately perfused and oxygenated myocardium all sites are activated virtually simultaneously, resulting in a single large spike. Following occlusion, widening and fractionation of this summarized R-wave occurs, indicating that adjacent fibers are not being simultaneously activated because of inhomogeneity of conduction. This must, in part, reflect local changes in myocardial blood flow immediately beneath the epicardial measuring sites. Figure 3.5 indicates that during occlusion there is about a 60 msec delay in conduction within the prescribed ischemic area; this is markedly reduced (to around 20 msec) in dogs subjected to ischemic preconditioning (Fig. 3.5).

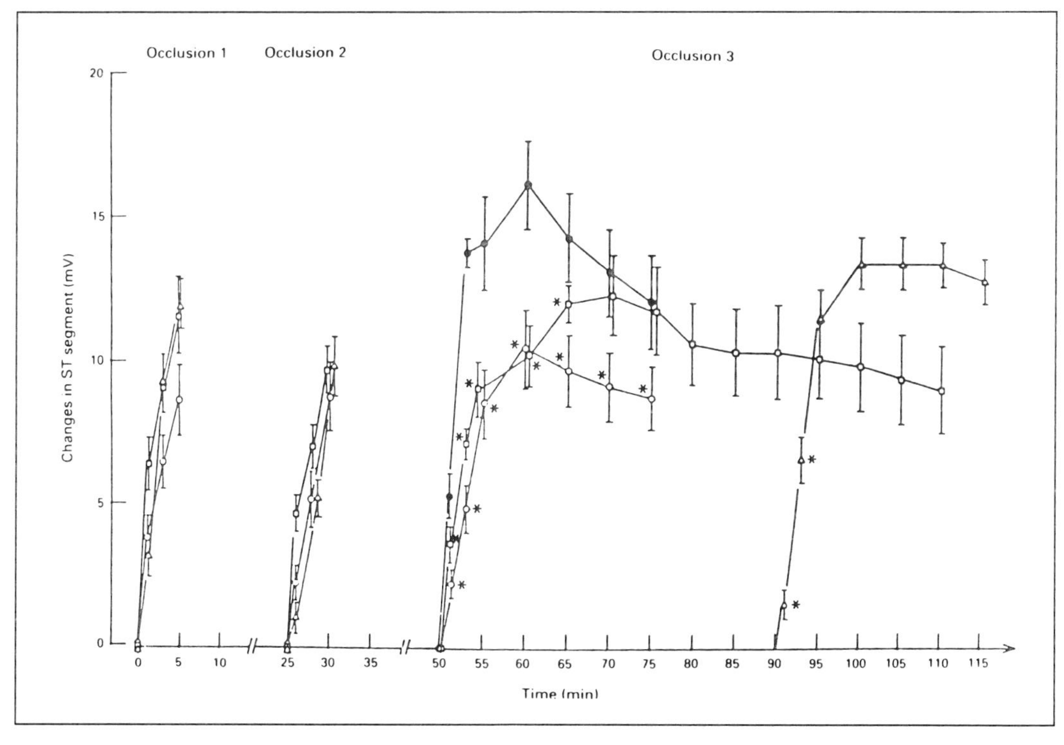

Fig. 3.4. Changes in ST-segment elevation, recorded from epicardial electrodes, in anesthetized dogs subjected to a 25 minute occlusion of the left anterior descending coronary artery (●) and in dogs in which this occlusion (occlusion 3) was preceded by two short 5 minute preconditioning occlusions (occlusions 1 and 2) and in which animals were reperfused at the end of the 25 minute occlusion period (○) or at the end of a 60 minute occlusion period (■). The severity of the ST-segment changes were less pronounced in preconditioned dogs provided the reperfusion time was 20 minutes but not if the reperfusion time between the preconditioning occlusions and the prolonged occlusion was increased to 1 hour (Δ) although, even at this time, there was some delay in the generation of the ST-segment change. Adapted with permission from Vegh A et al, Cardiovasc Res 1992; 26:487-495.

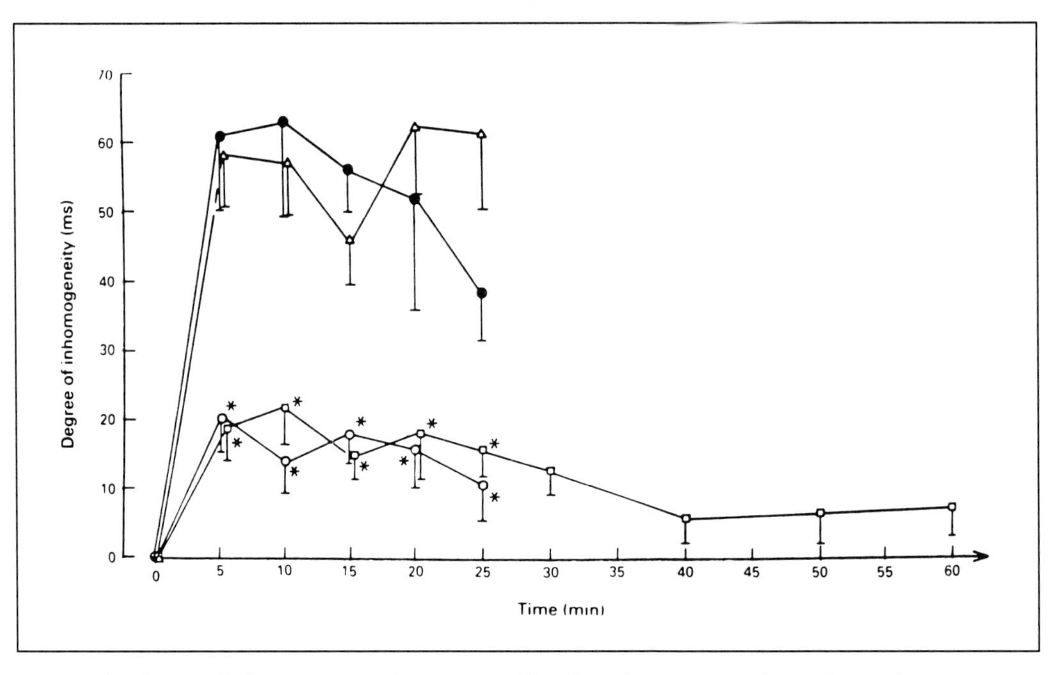

*Fig. 3.5. The degree of inhomogeneity of activation within the ischemic area in dogs subjected to a 25 minute occlusion of the left anterior descending coronary artery (●), and in dogs subjected to two 5 minute preconditioning occlusions (○ and ■). Preconditioning markedly reduces the degree of inhomogeneity, but this is not seen if the artery is reoccluded 1 hour after the last preconditioning occlusion (Δ). *$P<0.05$ versus changes in inhomogeneity during coronary occlusion and at the same time point in dogs that were not preconditioned. Reproduced with permission from Vegh A et al, Cardiovasc Res 1992; 26:487-495.*

There has been some debate as to whether species with variable collateral circulation, such as dogs, can be preconditioned in such a way as to reduce ventricular arrhythmias induced by ischemia. This debate arose because of the failure of the group[15] who first demonstrated the marked reduction in infarct size resulting from preconditioning, to consistently demonstrate a reduction in the incidence of ventricular fibrillation. There are a number of explanations for this failure. First is the use of pentobarbitone anesthesia. This is markedly depressive to the myocardium and suppresses ventricular ectopic activity. The second explanation is the variable weight of the dogs used; in our hands dogs with a mean body weight of less than 17 kg have rather few ventricular ectopic beats when a major coronary artery is occluded.[16] Thirdly, the studies of Jennings and his co-workers were not designed specifically to look at a possible antiarrhythmic effect of ischemic preconditioning and they used longer periods of ischemia to precondition.[15] These longer periods, in our hands, would result in a very high incidence of reperfusion-induced ventricular fibrillation. The study of Lucchesi's group[17] showed, also in dogs, that a single 5 minute preconditioning occlusion greatly reduces the incidence of ventricular fibrillation but that a greater number of preconditioning occlusions (e.g., 12 x 5 minutes) resulted in a higher mortality during a subsequent prolonged occlusion. It seems therefore that the anesthesia, the degree of cardiac sympathetic drive, the body to heart weight ratio, the site of coronary artery occlusion and the number and durations of the preconditioning occlusions all modify the effectiveness of the antiarrhythmic effect of ischemic preconditioning.

In most studies concerned with suppression of ischemia and reperfusion-induced ventricular arrhythmias by brief periods of ischemia, the preconditioning stimulus has been complete occlusion of a major branch of a coronary artery. However, infarct size limitation also results from preconditioning by partial coronary artery occlusions[18,19] by transient ischemia in adjacent myocardial regions[20] or even following ischemia in organs other than the heart.[21,22] Further, an increase in left ventricular wall stretch (produced by acute volume overload) also protects the myocardium against infarction during a subsequent prolonged artery occlusion (reviewed in chapter 5). Rapid cardiac pacing (300 beats/minute for 2 x 2 minute periods or 220 beats/minute for 4 x 5 minute periods) also markedly reduces the severity of arrhythmias that result from a subsequent (in this case the first) coronary artery occlusion.[23,24] Thus, the number of ventricular premature beats was reduced from 528 ± 40 to 136 ± 45, the incidence of ventricular fibrillation from 47% to 0% and the incidence of ventricular tachycardia from 80% to 30%.[23] Changes in ST-segment elevation and in the degree of inhomogeneity of electrical activation within the ischemic area are also less pronounced in dogs subjected to prior cardiac pacing. Pacing not only induces *early* protection against ventricular arrhythmias[25] and infarct size limitation[26] but also protects the heart against the consequences of coronary artery occlusion many hours later.[24,27,28] Cardiac pacing via a stimulating electrode in the right ventricle certainly results in ischemia (as recorded from endocardial electrodes in the left ventricle; Kaszala, Vegh and Parratt, unpublished) but may also result in changes in ventricular wall stretch and the opening of K_{ATP} channels through nonischemic mechanisms.[26] The effect of rapid cardiac pacing on the severity of arrhythmias during the subsequent 5 minute occlusion of the left anterior descending branch of the coronary artery in anesthetized dogs is illustrated in Figure 3.6.

Isolated perfused hearts can also be preconditioned by short periods of coronary artery occlusion and this affords protection against ventricular arrhythmias.[29,30] This is true whether the rats are perfused with a physiological salt solution or with blood from a support animal.[31] The extent of the protection in this model again depends upon the number and duration of the preconditioning occlusions (e.g., it is more effective when the preconditioning occlusion is of 3 minute duration rather than 1 minute) and, again as in vivo, the protection wanes

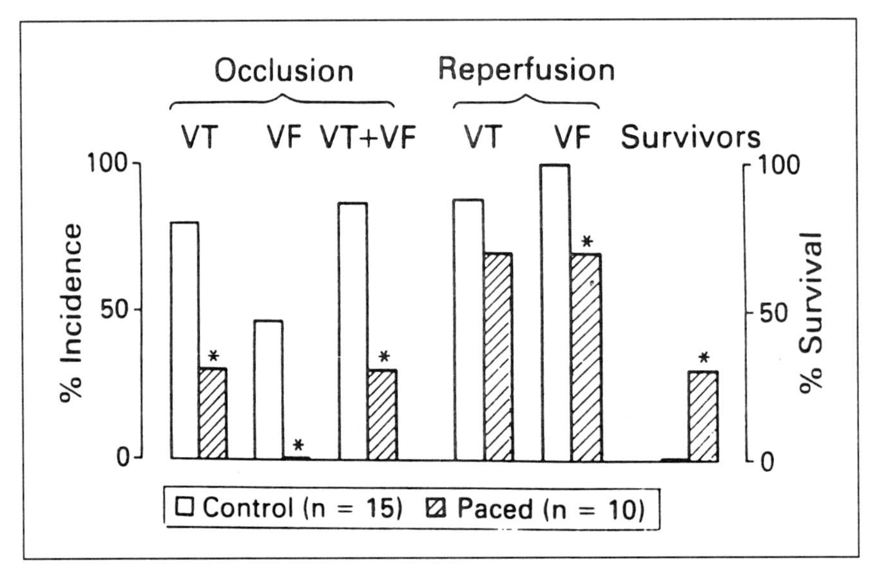

Fig. 3.6. Ventricular arrhythmias during occlusion of the left anterior descending branch of the coronary artery in anesthetized dogs (for 25 minutes) and during subsequent reperfusion and the effect of these arrhythmias of prior rapid ventricular pacing (hatched columns). There is a significant reduction in the severity of these arrhythmias if the occlusion is preceded by a pacing stimulus. Reproduced with permission from Vegh A et al, Cardiovasc Res 1991; 25:1051-1053.

with time, being lost after a 1 hour reperfusion period.[30,32] This is illustrated in Figure 3.7.

3.4. THE ANTIARRHYTHMIC EFFECT OF ISCHEMIC PRECONDITIONING—POSSIBLE MECHANISMS

The hearts of all species so far studied can be preconditioned and a number of possibilities have been suggested to explain this marked cardioprotection.[33] Some of these are outlined in Table 3.1 and are discussed elsewhere in this volume (e.g., chapters 8, 9, 11). The most likely involve alterations in the generation and release of substances from the heart which may either be protective or potentially injurious.[3,34,35] A strong case could be made for the hypothesis that after preconditioning there is an *increased* liberation of substances such as adenosine, nitric oxide and bradykinin, which are protective to the myocardium, and a *reduced* release of potentially injurious substances such as noradrenaline and, perhaps, endothelin.

The role of adenosine in mediating the infarct size limitation associated with ischemic preconditioning seems clear in most species, with the exception of rats, and is also reviewed elsewhere in this volume (chapter 1). The controversies regarding species differences in the protective mechanisms involved in this phenomenon are also reviewed elsewhere (chapter 12). It should also be borne in mind that there

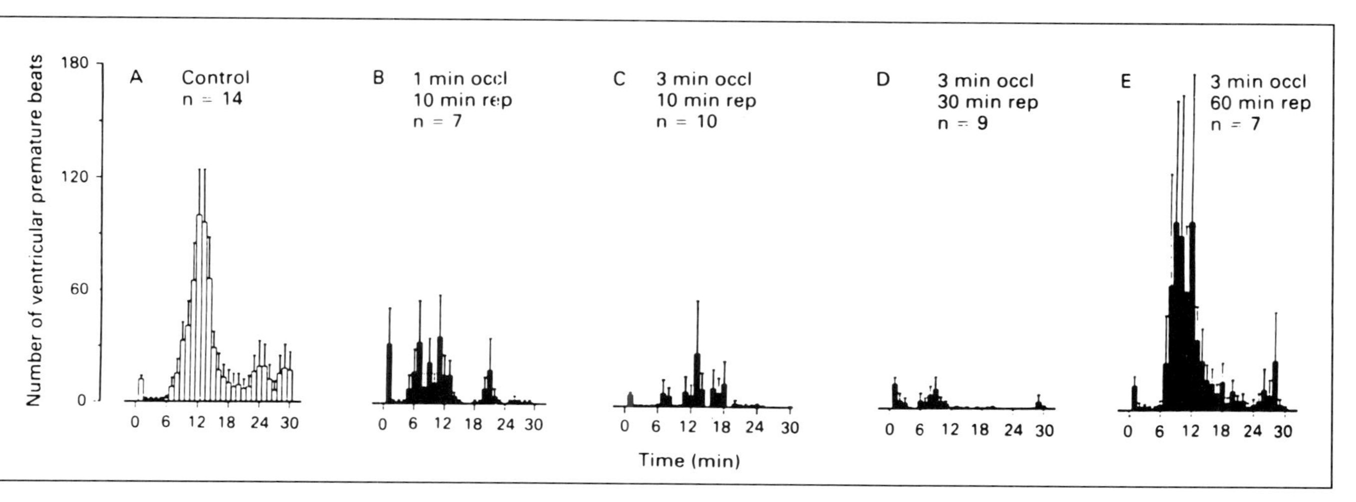

Fig. 3.7. Distribution of ventricular premature beats in rat isolated hearts subjected to a 30 minute coronary artery occlusion in control (nonpreconditioned hearts; A) and in hearts in which this occlusion was preceded by a 1 minute preconditioning occlusion (followed by 10 minute reperfusion; B) or a 3 minute occlusion followed by 10 (C) 30 (D) or 60 (E) minute reperfusion. Reproduced with permission from Piacentini L et al, Cardiovasc Res 1993; 27:674-680.

Table 3.1. Possible mechanisms involved in ischemic preconditioning with potential for pharmacological manipulation

1. Discrete changes in myocardial blood flow.
2. "Energy sparing": depressed myocardial contractility (?) stunning, better maintained ATP levels, improved glycogen synthesis.
3. Limitation of acidosis.
4. Oxygen free radicals.
5. Release of endogenous myocardial protective substances (adenosine, nitric oxide, noradrenaline, etc.) with subsequent involvement of phospholipase enzymes, G proteins, protein kinase C and protein phosphorylation.
6. Reduced release of potentially injurious substances (e.g. noradrenaline).
7. ATP-dependent potassium channels.
8. Induction of protective proteins or enzymes.
9. Any combination of the above.

may be different mechanisms involved in the infarct size limitation effect of ischemic preconditioning and those involved in the antiarrhythmic effect. For example, although the evidence for the role of adenosine in infarct size limitation is strong, there is no evidence that it plays any role in the antiarrhythmic effect of ischemic preconditioning either in rats[29] or in dogs;[3]; this despite the fact that in both these species adenosine, under certain conditions, can certainly act as an 'endogenous antiarrhythmic substance.'[37,38]

3.4.1. The Role of Endothelium-Derived Substances—Evidence from Studies in a Canine Model of Ischemia and Reperfusion

The current hypothesis, for which there is some evidence,[33,39-41] is that during ischemia there is the early release of bradykinin which, acting on B_2 receptors on the surface of endothelial cells, increases intracellular calcium and then stimulates the generation and release of other cardioprotective mediators such as nitric oxide and prostacyclin. Nitric oxide, generated as a result of activation of constitutive nitric oxide synthase within endothelial cells and the subsequent activation of the L-arginine nitric oxide pathway, is thought to diffuse to cardiomyocytes, stimulate soluble guanylyl cyclase, elevate cyclic GMP which then decreases the influx of extracellular calcium,[42] stimulates the cyclic GMP dependent, cyclic AMP phosphodiesterase and reduces myocardial energy demand. This hypothesis is outlined in Figure 3.8 and will now be presented in more detail.

a. Evidence for a role for bradykinin in ischemic preconditioning

This evidence has been recently reviewed[40,41] and will now be briefly summarized.

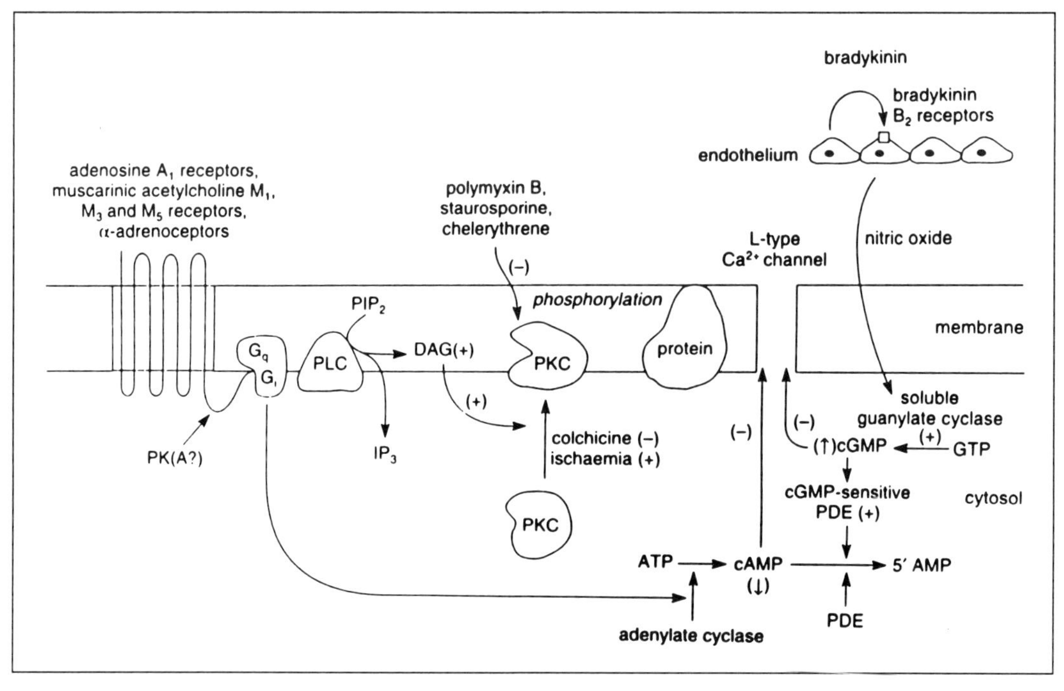

Fig. 3.8. The possible role of endothelium-derived substances in mediating the antiarrhythmic effects of ischemic preconditioning. The hypothesis is that there is early bradykinin release, probably from the endothelium, which then acts on endothelial B_2 receptors to generate nitric oxide. This diffuses to cardiac myocytes, stimulates soluble guanylyl cyclase and elevates cGMP. Another possible mechanism is that bradykinin acts on bradykinin receptors on cardiac myocytes (like adenosine, muscarine and -adrenoceptor agonists) to translocate PKC to the membrane with the subsequent phosphorylation of some protein (?) K_{ATP} channels. These mechanisms are described in more detail in the text.

(i). Kinins are generated under conditions of myocardial ischemia

This was first demonstrated several years ago, occurs both in experimental animals and in man, and has been assessed by changes in plasma kallikrein (or prekallikrein), kininogen or plasma kinins.[40] This generation of kinins during ischemia does not appear to depend upon the presence of blood since it can be demonstrated in isolated hearts perfused with Krebs solution and then subjected to coronary artery occlusion.[43,44] Although the origin of this kinin release under conditions of myocardial ischemia is uncertain, the generation is certainly rapid since in some patients with coronary artery disease subjected to coronary angioplasty, levels in the coronary sinus are elevated up to 50 times[40] during the balloon inflation (Oldroyd, Koide, Zeitlin, Ahmed and Parratt, in preparation). This could perhaps result from a local reduction in pH within endothelial cells following ischemia and the subsequent activation of those acid-optimum kininogenase enzymes which have been demonstrated in canine coronary arteries.[45] The vascular wall, and probably also cardiac myocytes,[46] thus have the equipment (substrate and enzyme) for generating kinins, especially under the kind of acid conditions one might expect during myocardial ischemia.

(ii). Drugs that enhance the release of bradykinin from the ischemic heart reduce myocardial ischemic injury

Potentiation of kinin release is an important mechanism of the cardioprotective effects of angiotensin-converting enzyme (ACE) inhibitors, a subject recently reviewed.[47,48] In brief, the evidence is that levels of kinins released after ischemia are elevated in the presence of ACE inhibitors, that the cardioprotective effects of ACE inhibitors are similar to those of bradykinin itself, and that many of the cardioprotective effects of ACE inhibitors are prevented by the selective inhibitor of the effects of bradykinin at B_2 receptors HOE140 (icatibant).

(iii). Drugs that block bradykinin receptors attenuate the reduction in ischemic injury resulting from the release or administration of bradykinin

The availability of the selective B_2 receptor antagonist icatibant has greatly aided investigations into the possible role of bradykinin in endogenous myocardial protection. The role of bradykinin in the antiarrhythmic effects of ischemic preconditioning has been examined in a canine model in which preconditioning was induced by two short (5 minute) periods of myocardial ischemia, with a 20 minute reperfusion time between each of these preconditioning occlusions and an additional 20 minute reperfusion time between the second of these occlusions and the onset of the prolonged (5 minute) occlusion.[9,11] The pronounced antiarrhythmic effect of ischemic preconditioning is profoundly modified if icatibant is administered either before the preconditioning stimulus, or after the stimulus but before the prolonged coronary artery occlusion.[39]

It proved rather difficult to precondition dogs in the presence of icatibant; the number of ventricular beats during the preconditioning occlusions, and particularly the incidence of ventricular fibrillation, was much more marked in those dogs pretreated with the bradykinin B_2 antagonist.[39] For example, the incidence of ventricular fibrillation either during ischemia or, more commonly, during reperfusion was 50% compared with less than 20% in control animals. Presumably this indicates that bradykinin release, which we have outlined above is probably rapid, occurs even during the preconditioning occlusions and has some protective effect. Those dogs that survived the preconditioning stimulus in the presence of icatibant were much less effectively protected than normal preconditioned dogs. This is illustrated in Figure 3.9 which shows that the number of ventricular premature beats and the incidence and number of episodes of ventricular tachycardia were higher in dogs preconditioned in the presence of icatibant than in those pre-

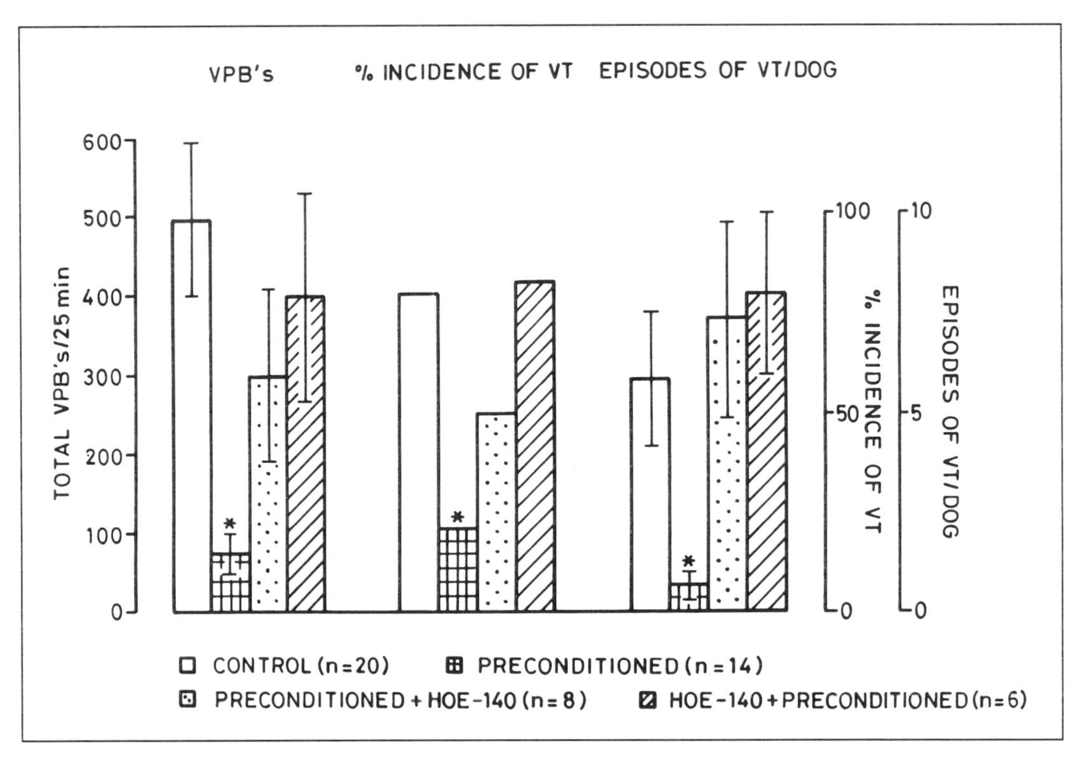

Fig. 3.9. Modification of the antiarrhythmic effects of ischemic preconditioning in dogs by blockade of bradykinin B_2 receptors with icatibant. Icatibant was given either before the preconditioning procedures (hatched columns) or after preconditioning prior to the prolonged 25 minute occlusion (stippled columns). Also shown are the number of ventricular premature beats, the incidence of ventricular tachycardia (VT) and the number of episodes of ventricular tachycardia which occurred during a 25 minute occlusion of the left anterior descending coronary artery in control dogs (open columns) and in preconditioned dogs (hatched columns) in the absence of icatibant. Reproduced with permission from Vegh A et al, Br J Pharmacol 1994; 113:1167-1172.

conditioned in the same way in the absence of the drug.[39] Further, survival from the combined ischemia reperfusion insult (which was 50% in preconditioned dogs) was much less in dogs preconditioned in the presence of icatibant. These modifications in arrhythmia severity were reflected in the changes in ST-segment elevation (Fig. 3.10) and in changes in electrical activation within the ischemic area.[39]

We presume that the proposed protective effects of released bradykinin during preconditioning are similar to those of locally infused bradykinin, which is also profoundly antiarrhythmic.[49] Because the protection afforded by bradykinin itself is markedly attenuated by an inhibitor of the L-arginine nitric oxide pathway[50] the hypothesis is that the primary role of bradykinin under conditions of preconditioning is to stimulate the generation, and subsequent release, of nitric oxide presumably from endothelial cells. However, this does not eliminate the possibility that bradykinin may act as one of the protective mediators involved in this protection as well as a trigger for preconditioning. There is evidence for example that B_2 receptors are present on cardiomyocytes and that when stimulated this results in the translocation of protein kinase C (PKC) from the cytosol to the membrane (see chapter 11). There is also evidence that bradykinin is involved in the infarct size limitation effects of ischemic preconditioning in rabbits.[51]

b. Nitric oxide as a protective mediator of ischemic preconditioning in dogs

The evidence for this, as with bradykinin, is indirect. It depends upon the use of drugs that inhibit the formation of nitric oxide from L-arginine, such as N^G-nitro-L-arginine methyl ester (L-NAME) or of

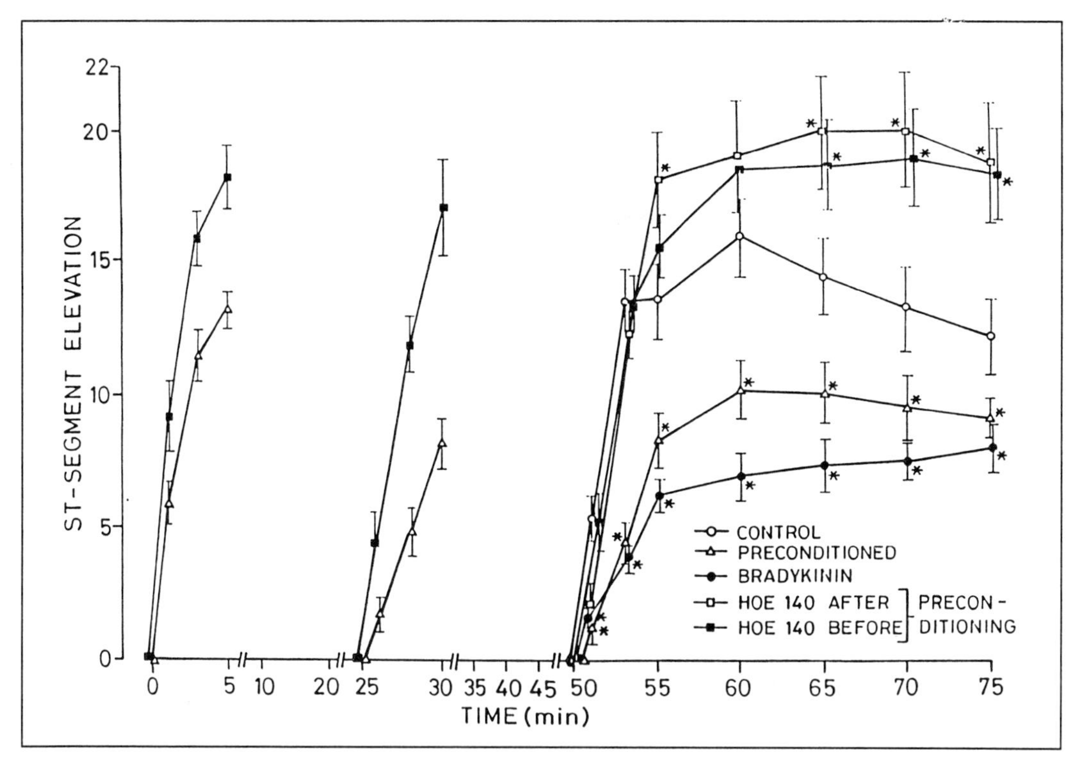

Fig. 3.10. Changes in ST-segment elevation, recorded from epicardial electrodes, during the two 5 minute preconditioning coronary artery occlusions and during the subsequent prolonged (25 minutes) occlusion of that same artery. ST-segment elevation following preconditioning during the prolonged occlusion is markedly reduced (Δ) compared with control, nonpreconditioned dogs (○). This reduction is attenuated by the administration of icatibant (300 μg/kg i.v.) either before the preconditioning procedure (■) or after preconditioning but before the prolonged occlusion (■). Also included are the results obtained following the local intracoronary administration of bradykinin (●) during a 25 minute coronary artery occlusion. Reproduced with permission from Vegh A et al, Br J Pharmacol 1994; 113:1167-1172.

unselective inhibitors of soluble guanylyl cyclase such as methylene blue. By itself, this evidence should be regarded as inconclusive. It can be summarized as follows:

1. Inhibition of the L-arginine nitric oxide pathway, by inhibitors such as L-NAME, when given both before the initial preconditioning occlusion and again before the prolonged occlusion, partially attenuates the protective effects of preconditioning.[16] Thus, there were more ventricular premature beats in those dogs preconditioned in the presence of the inhibitor (220 ± 75) compared to only 78 ± 27 in dogs without the inhibitor, a higher incidence of ventricular tachycardia (60% vs. 20%), more episodes of VT (9.3 ± 4.3 vs. 0.7 ± 0.3) and a higher incidence of ventricular fibrillation on reperfusion. This resulted in a lower survival from the combined ischemia-reperfusion insult (0% vs. 40%). However, the marked protective effects of preconditioning on the incidence of ventricular fibrillation during ischemia were not modified by L-NAME. Changes in epicardial ST-segment elevation and the degree of inhomogeneity of conduction reflected these changes in arrhythmia severity. For example, the increase in inhomogeneity within the ischemic region in dogs preconditioned in the presence of L-NAME was no different from that in control dogs[16] and was about three times higher than that seen in preconditioned dogs in the absence of L-NAME.[9]
2. Among several other actions, methylene blue is able to inhibit the activation of soluble guanylyl cyclase by nitric oxide.[52] The local intracoronary administration of methylene blue, both before and during the preconditioning coronary artery occlusions *and* before and during the prolonged occlusion, abolished the antiarrhythmic effect of preconditioning in the canine model outlined above, and markedly attenuated ischemia-induced changes in ST-segment elevation and in the degree of inhomogeneity of activation within the ischemic area.[53]

c. The possible role of cyclooxygenase products

The evidence for the involvement of cyclooxygenase products as possible protective mediators participating in ischemic preconditioning is less strong than for that for nitric oxide. In brief, the evidence is that:

1. Prostacyclin and thromboxane are released early during ischemia[54] and arrhythmia severity is related to the balance between the release of these substances, i.e., the greater the release of prostacyclin the less severe are early (phase 1A) ischemia-induced arrhythmias.[54]

2. Prostacyclin, infused locally into a branch of a coronary artery in dogs prior to, and during, coronary artery occlusion results in marked suppression of ventricular ectopic activity and in the incidence of ventricular fibrillation.[55] Potentiation of local prostacyclin release with nafazatrom is also markedly antiarrhythmic.[56]
3. When the cyclooxygenase pathway is blocked in dogs there is a marked attenuation of the antiarrhythmic effects of ischemic preconditioning.[11] For example, the number of ventricular premature beats during a 25 minute occlusion period was reduced from 445 ± 140 to 96 ± 22 following preconditioning; in those dogs preconditioned in the presence of sodium meclofenamate, an inhibitor of the cyclooxygenase pathway, the total number of ventricular premature beats (367 ± 95) was no different to the controls and significantly higher than in those dogs preconditioned in the same way but in the absence of the cyclooxygenase inhibitor. There was also a higher incidence of ventricular tachycardia during occlusion and none of the meclofenamate-treated dogs survived reperfusion, in contrast to a 40% survival in dogs preconditioned without the inhibitor. Cyclooxygenase inhibitors do not modify the severity of ischemia or reperfusion-induced ventricular arrhythmias in control, nonreconditioned dogs.[1]

One possible explanation for these results is that a protective prostanoid (prostacyclin?) is involved in the antiarrhythmic effects of ischemic preconditioning. In order to determine this more precisely one would need a selective inhibitor (or antagonist) of prostacyclin release, or activity, since inhibiting the cyclooxygenase pathway also reduces the release of pro-arrhythmogenic substances such as thromboxane.

The relative contribution of cyclooxygenase products and nitric oxide as mediators of the antiarrhythmic effects of ischemic preconditioning in dogs has been examined by simultaneously blocking both cyclooxygenase and L-arginine nitric oxide pathways.[57] As with icatibant, it proved difficult to precondition dogs in the presence of a combination of meclofenamate and L-NAME. Seventy-seven percent of the dogs died during the preconditioning procedure and the remainder fibrillated during the prolonged occlusion (i.e., VF incidence of 100% compared with 0% in normal preconditioned dogs). The results show that nitric oxide and prostacyclin generated during the preconditioning procedure contribute to the antiarrhythmic effects of preconditioning. Further, dual blockade of the cyclooxygenase and L-arginine nitric oxide pathways prevents the protection more effectively than inhibition of either pathway alone.

3.4.2. The Role of Endogenous Myocardial (Antiarrhythmic) Substances

The conclusion from the above studies is that several endogenous protective mediators are released as a result of brief periods of coronary artery occlusion and contribute markedly to the antiarrhythmic effects of preconditioning, at least in the canine model. Which of these mediators is most important is difficult to ascertain but certainly it appears that bradykinin is a key trigger for this protection and may itself also act as a protective mediator. It would be part of the wisdom of the body if more than one protective mediator was released under conditions of ischemia. What is at the moment unclear is what role the endothelium plays in the protection afforded by ischemic preconditioning. If the above hypothesis is correct then endothelial removal should also reduce the effectiveness of ischemic preconditioning. Experiments designed to explore this possibility are difficult (they would involve the removal of endothelial cells, for example from isolated perfused hearts with a detergent or with distilled water) and, unfortunately, there is no evidence that the above mediators are involved in the antiarrhythmic effects of preconditioning in the most suitable animal model for these experiments, the rat. Without such evidence our view remains that the endothelium is intimately involved in this form of myocardial protection, and that damage to this layer would attenuate any protection afforded clinically as a result of preconditioning induced by transient periods of ischemia or by increases in heart rate such as occur during exercise.

The hypothesis outlined above is that bradykinin and nitric oxide release leads to an increase in cGMP levels within cardiac myocytes. There is some preliminary evidence that levels are indeed raised during preconditioning.[33] A number of years ago Opie[58] suggested that elevating cardiac cyclic GMP would be an antiarrhythmic procedure, a view later substantiated by the experiments of Billman.[59] Using dogs with a healed myocardial infarct, Billman showed that the administration of 8-bromo-cyclic GMP reduced the incidence of ventricular fibrillation that occurred following a combination of exercise and a brief coronary artery occlusion.

Elevating cyclic GMP could influence arrhythmogenesis in a number of ways. First, elevation of myocardial cyclic GMP would inhibit calcium influx through L-type channels in the sarcolemma.[42] Secondly, stimulation of a cGMP dependent cAMP phosphodiesterase should lead to a decrease in myocardial cAMP levels. An elevation of cAMP has been implicated in the arrhythmias that result from activation of cardiac sympathetic nerves.[60] There is evidence for such a mechanism from studies involving the administration of either prostacyclin, or the stable derivative 7-oxo-prostacyclin, which has both early and late appearing cardiac protective effects.[61,62] This substance induces various isoforms of phosphodiesterase several hours after its administration and this probably contributes to the late myocardial protection afforded.[63] Thirdly, elevating cGMP depresses myocardial contractility; this is responsible for the depression of cardiac function that results from sepsis and endotoxaemia (reviewed in ref. 64).

A combination of these effects would adequately explain the suppression of ventricular ectopic activity during ischemia by mechanisms that liberate bradykinin and nitric oxide.

3.4.3. The Role of Inhibition of Cardiac Sympathetic Responses

Several of these endogenous myocardial protective substances outlined above (e.g., adenosine, nitric oxide, prostacyclin) have the ability to inhibit noradrenaline release from sympathetic nerves. An attractive, albeit unproven, unifying hypothesis to explain many of the protective effects of ischemic preconditioning (suppression of arrhythmias, infarct size limitation) would be inhibition of cardiac noradrenaline release.[5] Under conditions of ischemia the release of this transmitter plays a key role in the generation of arrhythmias and, by increasing myocardial contractility, oxygen demand and heat production,[65] noradrenaline release contributes to the extent and severity of myocardial ischemic damage. It is not easy to take this hypothesis further. Pharmacological manipulations designed to inhibit noradrenaline release (e.g., with adrenergic neurone blocking drugs) or depletion of cardiac noradrenaline stores (e.g., with guanethidine or reserpine) would themselves reduce the severity of ischemia-induced arrhythmias. It may not be easy to demonstrate an additional antiarrhythmic effect of ischemic preconditioning under these conditions. Further, measurements of noradrenaline release are notoriously difficult, requiring inhibition of neuronal uptake (which itself modifies arrhythmia severity) and local coronary venous sampling from the ischemic area, which is only feasible in large animal models. Recent evidence, in the same canine model as was used to demonstrate the antiarrhythmic effects of ischemic preconditioning, reveals that inhibition of noradrenaline release with the dopamine agonist Z-1046 is profoundly antiarrhythmic.[66]

A possible objection to the above hypothesis is that noradrenaline administration can itself mimic ischemic preconditioning in a variety of animal models (reviewed in chapter 10) and can also reduce the severity of ventricular arrhythmias when infused either intravenously[67] or when given locally into a side branch of a coronary artery.[68] In the canine model we do not know which receptor mediates this protection. How is it possible to reconcile the proposed role of inhibition of noradrenaline release as a major mechanism of ischemic preconditioning with those experiments demonstrating that *administered* noradrenaline can itself mimic preconditioning? One possible explanation is that administered noradrenaline could suppress endogenous noradrenaline release from nerves by a presynaptic effect on α-adrenoceptors; it would act then in the same way as adenosine, prostacyclin or Z-1046. It would be of interest to discover whether or not ischemic preconditioning is still effective in *chronically* denervated hearts.

3.5. CONCLUSION

Ischemic preconditioning, induced either by brief periods of coronary artery occlusion or by rapid ventricular pacing, markedly suppresses the severity of arrhythmias that result immediately after coronary artery occlusion.[69,70] Because such arrhythmias are almost certainly involved in the mechanism of sudden cardiac death in clinical situations it would be important to discover the precise mechanisms of this protection. At the moment, these remain unclear. The most likely possibilities involve the release of endogenous myocardial protective substances or the reduced release of some arrhythmogenic factors such as noradrenaline. If the release of endogenous protective mediators are involved the most likely explanation for the reduction in arrhythmia severity is an alteration in cGMP/cAMP levels.

ACKNOWLEDGMENTS

Most of our own experiments described and reviewed in this chapter were carried out in the laboratories of the Department of Pharmacology at the Albert Szent-Györgyi Medical University of Szeged and we wish to thank the two chairmen involved over this period (1988 to the present), Professors Laszlo Szekeres and Julius Gy Papp, for their encouragement and support. We also wish to acknowledge the financial support of initially, the Royal Society, the Scottish Home and Health Department and the Wellcome Trust and currently the British Council, the European Economic Community (Scientific Network grant No. CIPA CT92 4009) the Hungarian National Committee for Technical Development and the Hungarian State Government (OTKA). In the Glasgow department we are particularly indebted to Mrs. Margaret Laird for producing the manuscript. Among our younger co-workers we particularly appreciate the dedication and enthusiasm of Dr. 'Charlie' (K) Kaszala and Adrienn Kis.

REFERENCES

1. Wainwright CL, Parratt JR. Failure of cyclooxygenase inhibition to protect against arrhythmias induced by ischaemia and reperfusion; implications for the role of prostaglandins as endogenous myocardial protective substances. Cardiovasc Res 1991; 25:93-100.
2. Podzuweit T, Binz K-H, Nennstiel P et al. The antiarrhythmic effects of myocardial ischemia. Relation to reperfusion arrhythmias. Cardiovasc Res 1989; 23:81-90.
3. Parratt JR. Endogenous myocardial protective substances. Cardiovasc Res 1993; 27:693-702.

4. Vegh A, Szekeres L, Parratt JR. Preconditioning: An early protective response to myocardial ischaemia. In: Parratt JR, ed. Myocardial Response to Acute Injury. London: Macmillan Press 1992:110-127.
5. Parratt JR. The discovery of ischaemic preconditioning. Cardiovasc Res 1993; 27:688.
6. Clark C, Foreman MI, Kane KA et al. Coronary artery ligation in anaesthetised rats as a method for the production of experimental dysrhythmias and for the determination of infarct size. J Pharmacol Methods 1980; 3:357-368.
7. Komori S, Parratt JR, Szekeres L et al. Preconditioning reduces the severity of ischaemia and reperfusion-induced arrhythmias in both anaesthetised rats and dogs. J Physiol 1990; 423:16P (Abstract).
8. Komori S, Fujimaki S, Ijili H et al. Inhibitory effect of ischemic preconditioning on ischemic arrhythmias using a rat coronary artery ligation model. Japanese J Electrocardiol 1990; 10:774-782.
9. Vegh A, Komori S, Szekeres L et al. Antiarrhythmic effects of preconditioning in anaesthetised dogs and rats. Cardiovasc Res 1992; 26:487-495.
10. Kane KA, Parratt JR, Williams FM. An investigation into the characteristics of reperfusion-induced arrhythmias in the anaesthetised rat and their susceptibility to antiarrhythmic agents. Br J Pharmacol 1984; 82:349-357.
11. Vegh A, Szekeres L, Parratt JR. Protective effects of preconditioning of the ischaemic myocardium involve cyclooxygenase products. Cardiovasc Res 1990; 24:1020-1023.
12. Wainwright CL, Parratt JR. Electrocardiographic and haemodynamic effects of myocardial preconditioning in pigs. J Mol Cell Cardiol 1990; 22 (Suppl III):PF65 (Abstract).
13. Shiki K, Hearse DJ. Preconditioning of ischemic myocardium; reperfusion-induced arrhythmias. Am J Physiol 1987; 253:H1470-H1476.
14. Vegh A, Szekeres L, Udvary E. Effect of the blood supply to the normal noninfarcted myocardium on the incidence and severity of early post-occlusion arrhythmias in dogs. Basic Res Cardiol 1987; 82:159-171.
15. Murry CE, Jennings RB, Reimer KA. Preconditioning with ischemia: a delay of lethal cell injury in ischemic myocardium. Circulation 1986; 74:1124-1136.
16. Vegh A, Szekeres L, Parratt JR. Preconditioning of the ischaemic myocardium; involvement of the L-arginine nitric oxide pathway. Br J Pharmacol 1992; 107:648-652.
17. Li GC, Vasquez JA, Gallagher KP et al. Myocardial protection with preconditioning. Circulation 1990; 82:609-619.
18. Ovize M, Przyklenk K, Kloner RA. Partial coronary stenosis is sufficient and complete reperfusion is mandatory for preconditioning the canine heart. Circ Res 1992; 71:1165-1173.
19. Koning MMG, Simonis LAJ, De Zeeuw S et al. Ischaemic preconditioning by partial occlusion without intermittent reperfusion. Cardiovasc Res 1994; 28:1146-1151.
20. Przyklenk K, Bauer B, Ovize M et al. Regional ischemic preconditionings

protects remote virgin myocardium from subsequent sustained coronary occlusion. Circulation 1993; 87:893-899.
21. McClanahan TB, Nao BS, Wolke LJ et al. Brief renal occlusion and reperfusion reduces myocardial infarct size in rabbits. FASEB J 1993; 7:A176 (Abstract).
22. Gho BC, Shoemaker RG, van der Lee C et al. Myocardial infarct size limitation in rat by transient renal ischemia. Circulation 1994; 90:I476 (Abstract).
23. Vegh A, Szekeres L, Parratt JR. Transient ischaemia induced by rapid cardiac pacing results in myocardial preconditioning. Cardiovasc Res 1991; 25:1051-1053.
24. Vegh A, Papp J Gy, Kasala K et al. Cardiac pacing in anaesthetized dogs preconditions the heart against arrhythmias when ischaemia is induced 24 h later. J Physiol 1994; 480:89P (Abstract).
25. Kaszala K, Vegh A, Parratt JR et al. Time course of pacing induced preconditioning in dogs. J Molec Cell Cardiol 1995; 27:A145 (Abstract).
26. Koning MMG, Gho BCG, Van Klaarwater E et al. Rapid ventricular pacing produces myocardial protection by non-ischaemic activation of K^{+}_{ATP} channels. Circulation (in press).
27. Vegh A, Papp JGy, Parratt JR. Dexamethasone prevents the marked antiarrhythmic effects of preconditioning induced 20 h after rapid cardiac pacing. Br J Pharmacol 1994; 113:1081-1082.
28. Vegh A, Papp J Gy, Szekeres L et al. Antiarrhythmic effects of ischaemic preconditioning during the 'second window of protection.' J Molec Cell Cardiol 1994; 26:A346 (Abstract).
29. Piacentini L, Wainwright CL, Parratt JR. The antiarrhythmic effect of preconditioning in rat isolated hearts does not involve A_1 receptors. Br J Pharmacol 1992; 107:137P (Abstract).
30. Piacentini L, Wainwright CL, Parratt JR. The antiarrhythmic effect of ischaemic preconditioning in isolated rat hearts involves a pertussis toxin sensitive mechanism. Cardiovasc Res 1993; 27:674-680.
31. Lawson CS, Coltart DJ, Hearse DJ. The antiarrhythmic action of ischaemic preconditioning in isolated blood perfused rat hearts does not involve functional Gi proteins. Cardiovasc Res 1993; 27:681-687.
32. Lawson CS, Coltart DJ, Hearse DJ. 'Dose-dependency' and temporary characteristics of protection by ischaemic preconditioning against ischaemia-induced arrhythmias in rat hearts. J Mol Cell Cardiol 1993; 25:1391-1402.
33. Parratt JR. Possibilities for the pharmacological exploitation of ischaemic preconditioning. J Mol Cell Cardiol 1995; 27:991-1000.
34. Curtis MJ, Pugsley MK, Walker MJA. Endogenous chemical mediators of ventricular arrhythmias in ischaemic heart disease. Cardiovasc Res 1993; 27:703-719.
35. Parratt JR. Protection of the heart by ischaemic preconditioning: mechanisms and possibilities for pharmacological exploitation. Trends Pharmacol Sciences 1994; 15:19-25.
36. Vegh A, Papp JGy, Parratt JR. Pronounced antiarrhythmic effects of pre-

conditioning in anaesthetised dogs: is adenosine involved? J Mol Cell Cardiol 1995; 27:349-356.
37. Wainwright CL, Parratt JR. An antiarrhythmic effect of adenosine during myocardial ischaemia and reperfusion. Eur J Pharmac 1988; 145:183-194.
38. Parratt JR, Boachie-Ansah G, Kane K A et al. Is adenosine an endogenous antiarrhythmic agent under conditions of myocardial ischaemia? In: Paton DM ed. Adenosine and Adenine Nucleotides. London: Taylor and Francis, 1988:157-166.
39. Vegh A, Papp J Gy, Parratt JR. Attenuation of the antiarrhythmic effects of ischaemic preconditioning by blockade of bradykinin B_2 receptors. Br J Pharmacol 1994; 113:1167-1172.
40. Parratt JR, Vegh A, Papp J Gy. Bradykinin as an endogenous myocardial protective substance with particular reference to ischemic preconditioning—a brief review of the evidence. Canad J Physiol Pharmacol 1995; 73:837-842.
41. Parratt JR, Vegh A, Papp J Gy. Nitric oxide generation, following activation of bradykinin (B_2) receptors, mediates the pronounced antiarrhythmic effects of ischaemic preconditioning in anaesthetised dogs. In: Moncada S ed. Biology of Nitric Oxide, Volume 3: Physiology and Clinical Aspects. London, Portland, 1994:69-74.
42. Tohse N, Sperelakis N. cGMP inhibits the activity of single calcium channels in embryonic chick heart cells. Circ Res 1991; 69:325-331.
43. Baumgarten CR, Linz W, Kunkel G et al. Ramiprilat increases bradykinin outflow from isolated hearts of rat. Br J Pharmacol 1993; 108:293-295.
44. Koide A, Zeitlin IJ, Parratt JR. Kinin formation in ischaemic heart and aorta of anaesthetised rats. J Physiol 1993; 467:125P (abstract).
45. Zeitlin IJ, Fagbemi SO, Parratt JR. Enzymes in normally perfused and ischaemic dog hearts which release a substance with kinin like activity. Cardiovasc Res 1989; 23:91-97.
46. Xiong W, Chen L-M, Woodley-Miller C et al. Identification, purification and localization of tissue kallikrein in rat heart. Biochem J 1990; 267:639-646.
47. Parratt JR. Cardioprotection by angiotensin converting enzyme inhibitors. Cardiovasc Res 1994; 28:183-189.
48. Pahor M, Gambassi G, Carbonin P. Antiarrhythmic effects of ACE inhibitors: a matter of faith or reality? Cardiovasc Res 1994; 28:173-182.
49. Vegh A, Szekeres L, Parratt JR. Local intracoronary infusions of bradykinin profoundly reduce the severity of ischaemia-induced arrhythmias in anaesthetised dogs. Br J Pharmacol. 1991; 104:294-295.
50. Vegh A, Papp J Gy, Szekeres L et al. Prevention by an inhibitor of the L-arginine-nitric oxide pathway of the antiarrhythmic effects of bradykinin in anaesthetised dogs. Br J Pharmacol 1993; 110:18-19.
51. Wall TM, Sheehy R, Hartman JC. Role of bradykinin in myocardial preconditioning. J Pharmacol Exp Ther 1994; 270:681-689.
52. Martin W, Villani GM, Jothianandan D et al. Selective blockade of endothelium-dependent and glyceryl trinitrate-induced relaxation by

haemoglobin and by methylene blue in the rabbit aorta. J Pharmacol Exp Ther 1985; 232:708-716.
53. Vegh A, Papp J Gy, Szekeres L et al. The local intracoronary administration of methylene blue prevents the pronounced antiarrhythmic effect of ischaemic preconditioning. Br J Pharmacol 1992; 107:910-911.
54. Coker SJ, Parratt JR, Ledingham IM et al. Thromboxane and prostacyclin release from ischaemic myocardium in relation to arrhythmias. Nature 1981; 291:323-324.
55. Coker SJ, Parratt JR. Prostacyclin-antiarrhythmic or arrhythmogenic? Comparison of the effects of intravenous and intracoronary prostacyclin and ZK 36374 during coronary artery occlusion and reperfusion in anesthetised greyhouds. J Cardiovasc Pharmacol 1983; 5:557-567.
56. Coker SJ, Parratt JR. The effects of nafazatrom on arrhythmias and prostanoid release during coronary artery occlusion and reperfusion in anaesthetised greyhounds. J Mol Cell Cardiol 1984; 16:43-52.
57. Kis A, Vegh A, Papp JGy et al. Dual blockade of the cyclooxygenase and L-arginine-nitric oxide pathway prevents the antiarrhythmic effect of preconditioning. J Molec Cell Cardiol 1995; 27:A159 (Abstract).
58. Opie LH. Role of cyclic nucleotides in heart metabolism. Cardiovasc Res 1982; 16:483-507.
59. Billman GE. Effect of carbachol and cyclic GMP on susceptibility to ventricular fibrillation. FASEB 1990; 4:1668-1673.
60. Krause E-G, Ziegelhoffer A, Fedelsova M et al. Myocardial cyclic nucleotide levels following coronary artery ligation. Adv Cardiol 1978; 25:119-129.
61. Szekeres L, Szilvassy Z, Udvary E et al. 7-OXO PGI_2 induced late appearing and long lasting electrophysiological changes in the heart in situ of the rabbit, guinea-pig, dog and cat. J Mol Cell Cardiol 1989; 21:545-554.
62. Szekeres L, Pataricza J, Szilvassy Z et al. Cardioprotection: endogenous protective mechanisms promoted by prostacyclin. Basic Res Cardiol 1992; 87:215-221.
63. Borchert G, Bartel S, Beyerdorfer I et al. Long lasting anti-adrenergic effect of 7-oxo-prostacyclin in the heart: a cycloheximide sensitive increase of phosphodiesterase isoform I and IV activities. Molec Cell Biochem 1994; 137:57-67.
64. Parratt JR. Nitric oxide and cardiovascular dysfunction in sepsis and endotoxaemia—an introduction and an overview. In: Schlag G, Redl H, eds. Shock, Sepsis and Organ Failure, Fourth Bernard Wiggers Conference. Berlin: Springer 1995:1-21.
65. Parratt JR. The effect of adrenaline, noradrenaline, and propranolol on myocardial blood flow and metabolic heat production in monkeys and baboons. Cardiovasc Res 1969; 3:306-314.
66. Parratt JR, Vegh A, Semeraro C et al. Beneficial effects of Z1046, a selective dopamine receptor agonist, during myocardial ischaemia. Br J Pharmacol 1995; 116:286P.

67. Parratt JR, Campbell C, Fagbemi O. Catecholamines and early post-infarction arrhythmias: the effects of alpha and beta-adrenoceptor blockade. In: Delius W, Gerlach E, Grobecker H et al, eds. Catecholamines and the Heart. Springer, Berlin 1981:269-284.
68. Vegh A, Papp J Gy, Parratt JR. Intracoronary noradrenaline suppresses ischaemia-induced arrhythmias in anaesthetised dogs. J Molec Cell Cardiol 1994; 26:Abstract 343.
69. Parratt JR, Vegh A. Pronounced antiarrhythmic effects of ischaemic preconditioning. Cardioscience 1994; 5:9-18.
70. Parratt JR, Vegh A, Papp J Gy. Pronounced antiarrhythmic effects of ischaemic preconditioning—are there possibilities for pharmacological exploitation? Pharmacol Res 1995; 31:225.

CHAPTER 4

The Protective Effects of Preconditioning on Postischemic Contractile Dysfunction

Alison C. Cave

4.1. INTRODUCTION

Ischemic preconditioning is an important phenomenon by which significant myocardial protection can be achieved, paradoxically, by first subjecting the heart to an additional, but brief, period of ischemia and reperfusion.[1-4] Studies during the past decade have established that ischemic preconditioning can significantly decrease the extent of postischemic contractile dysfunction[1] in addition to reducing the extent of tissue necrosis[2] and ischemia-[3] or reperfusion-induced arrhythmias.[4] The purpose of this chapter however, is to discuss primarily the protection observed against postischemic contractile dysfunction.

Most studies investigating preconditioning against contractile dysfunction have been undertaken in the isolated perfused rat heart,[5-10] although there have been a small number of studies performed in the rabbit,[11,12] dog[13-15] and pig.[16] In the majority of these studies,[5-11] contractile function was assessed via a fluid filled, thin walled balloon placed in the left ventricle which provided the ability to assess both left ventricular systolic and diastolic pressure.

4.2. THE PRECONDITIONING PROTOCOL

The protocol used to induce preconditioning varies considerably between laboratories. In some laboratories, preconditioning may be

Myocardial Preconditioning, edited by Cherry L. Wainwright and James R. Parratt.

induced by a single 5 minute period of ischemia followed by 5 minutes of reperfusion[1] or even a single 2 minute ischemic period[17] while in others multiple bursts of ischemia and reperfusion appear to be necessary.[7] It seems that in order to induce preconditioning, a threshold of stimulation must be achieved by the preconditioning protocol and differences between individual laboratories in terms of this protocol simply reflect variations in the durations of ischemia necessary to reach this threshold in the particular model or species studied. This will be discussed in more detail later in this chapter. Once induced, preconditioning has been shown to provide significant protection against both postischemic systolic and diastolic dysfunction[6] (Fig. 4.1). The protection induced appears to be an all or nothing response, at least in terms of developed pressure, in that additional bursts of ischemia and reperfusion, over and above that necessary to induce preconditioning, do not seem to confer significantly greater protection.[5] Some slight advantage is seen however, in terms of the rate of recovery of diastolic function.[5] This result is in marked contrast to preconditioning induced protection against arrhythmias where a clear "dose-dependent" response is observed.[3] It should also be noted that it is possible to induce preconditioning against contractile dysfunction with other forms of injury such as hypoxia[15,18,19] or low flow ischemia.[18]

Clearly, such preconditioning protocols will themselves cause temporary contractile dysfunction, i.e., myocardial stunning. Following such short episodes of ischemia, no irreversible damage occurs and yet contractile function is significantly reduced[6] (Fig. 4.1). Murry and colleagues[20] proposed that this reduction in contractile function, and the subsequent reduction in energy demands during the initial few minutes of the sustained ischemic period, could significantly delay ATP depletion and thus explain the protection observed upon reperfusion. In order to investigate this, hearts were arrested with high potassium solution at the beginning of the sustained ischemic period to remove the possibility of differences in energy demand at this time between preconditioned and control hearts. Preconditioning-induced protection was, however, unaffected by this procedure, indicating that myocardial stunning and the possibility of reduced energy demands were not prerequisites of preconditioning. In addition, we have investigated whether preconditioning-induced protection provided additional protection over and above that provided by cardioplegia.[21] Control rat hearts were subjected to 2 minutes of hypothermic cardioplegic arrest (20°C) prior to sustained hypothermic ischemia (20°C). Preconditioned hearts were subjected to 5 minutes of normothermic (37°C) ischemia and 5 minutes of reperfusion prior to the hypothermic cardioplegic arrest and ischemia. The results demonstrated that preconditioned hearts recovered better than control hearts and had significantly less creatine kinase release upon reperfusion. These results therefore lend further support to the conclusion that energy demands during contractile arrest are

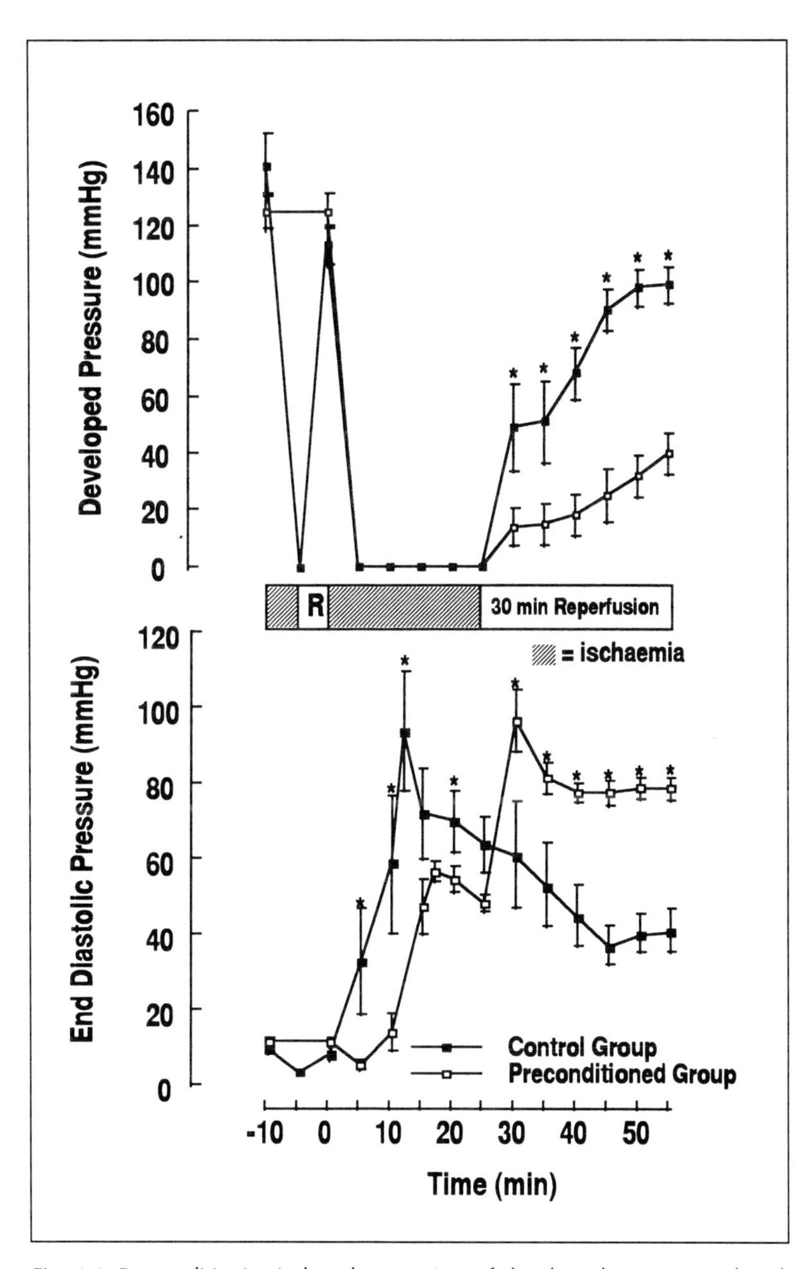

*Fig. 4.1. Preconditioning-induced protection of developed pressure and end-diastolic pressure following a sustained period of no flow global ischemia. Preconditioned hearts were subjected to 5 minutes of ischemia and 5 minutes of reperfusion prior to the sustained ischemic period. Control hearts are represented by the empty squares and preconditioned hearts by the filled squares. Cross-hatched areas represent periods of ischemia; R, reperfusion. Data are presented as means ± standard error of the mean; n = 6/group. *P<0.05 vs. control (2 way repeated measures anova and Fischers test). Reproduced with permission from Cave AC et al, J Mol Cell Cardiol 1994; 26:1471-1486.*

not important in the mechanism of preconditioning and in addition demonstrate that preconditioning may be valuable as an adjunct to cardioplegia during surgery.

A further important characteristic of preconditioning is its "memory." If the intervening reperfusion period between the preconditioning protocol and the prolonged period of ischemia is too long, the protective effect of preconditioning is lost.[4,22,23] In most models, preconditioning-induced protection against tissue necrosis is lost if the intervening reperfusion period is extended for greater than 30-60 minutes, the exact time varying between species. Protection can however be re-instated if a further burst of preconditioning ischemia and reperfusion is imposed just prior to the prolonged ischemic period.[24]

4.3. FUNCTIONAL AND METABOLIC CHANGES DURING THE SUSTAINED ISCHEMIC PERIOD

4.3.1. Effects of Preconditioning on Ischemic Contracture

Historically, the extent of ischemic contracture has been assumed to be directly related to ischemic damage and thus inversely proportional to recovery upon reperfusion, that is, the greater the ischemic contracture and the earlier its onset, the worse is postischemic recovery.[25] Preconditioning however, paradoxically accelerates ischemic contracture and yet improves postischemic recovery.[6] The reduction of myocardial ATP availability below a critical level necessary to maintain compliance is currently one of the most popular theories to explain the onset of contracture during ischemia. Recently, Kolocassides et al[26] have correlated the extent of ischemic contracture with adenosine triphosphate (ATP) content and intracellular pH in preconditioned and control hearts. These authors demonstrated that although ATP appeared to decline quicker in preconditioned hearts this was merely due to the fact that the ATP content was lower in preconditioned hearts at the beginning of the prolonged ischemic period (Fig. 4.2). In fact, the rate of decline in ATP content was the same in preconditioned and control hearts and the onset of contracture occurred when ATP content was between 60-70% in both groups of hearts. Interestingly, 50% of maximum contracture was achieved when ATP reached approximately 50% of its pre-ischemic value in both control and preconditioned hearts. This occurred at approximately 6 minutes in preconditioned hearts and 12 minutes in control hearts. Thus, differences in pre-ischemia total ATP content may explain the acceleration in ischemic contracture in preconditioned hearts. Perhaps as important as the total ATP content may be the ATP content of critical subcellular compartments. ATP from anaerobic glycolysis for example, has been shown to be important in the maintenance of diastolic compliance[27,28] and may play a key role in the process of preconditioning.

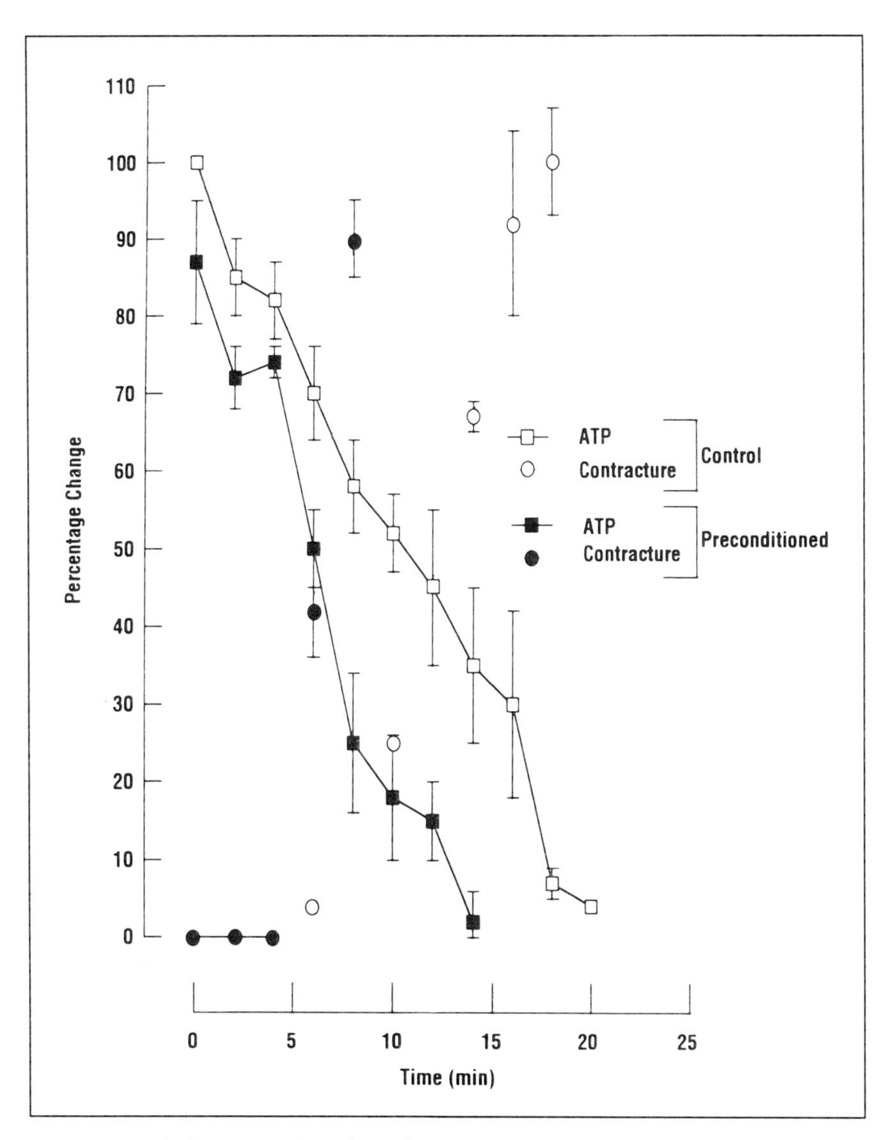

Fig. 4.2. Graph illustrating the effect of preconditioning on ATP content and ischemic contracture during the sustained period of global, no flow ischemia. Preconditioned hearts were subjected to 3 minutes of ischemia and 3 minutes of reperfusion + 5 minutes of ischemia and 5 minutes of reperfusion prior to the 35 minutes of sustained ischemia. Control hearts are represented by the empty symbols and preconditioned hearts by the filled symbols. ATP content is represented by the solid lines and squares and ischemic contracture by the dotted lines and circles. Data are presented as a percentage change ± standard error of the mean. n = 6 hearts/group. Adapted with permission from Kolocassides KG et al, Circulation 1994; 90(4,2):I-476 (Abstract).

4.3.2. Effects of Preconditioning on Anaerobic Glycolysis

Under conditions of oxygen deprivation, myocardial metabolism switches away from aerobic ATP production by the mitochondria to anaerobic production via glycolysis. The end products of this pathway, lactate and protons, thus accumulate during severe ischemia and production of the former may be as used as a marker for glycolytic flux. In preconditioned hearts, lactate release upon reperfusion is decreased by approximately 50% when compared with control hearts[29] (Fig. 4.3). Thus glycolytic flux, the metabolic marker that seems to correlate best with the time to onset of ischemic contracture,[27,28,30] and thus glycolytic ATP production must be markedly reduced in these hearts. There are several studies in the literature in which increased glycolysis has been shown to considerably delay the onset of contracture.[27,28] Thus although the reduced rate of glycolysis appears to

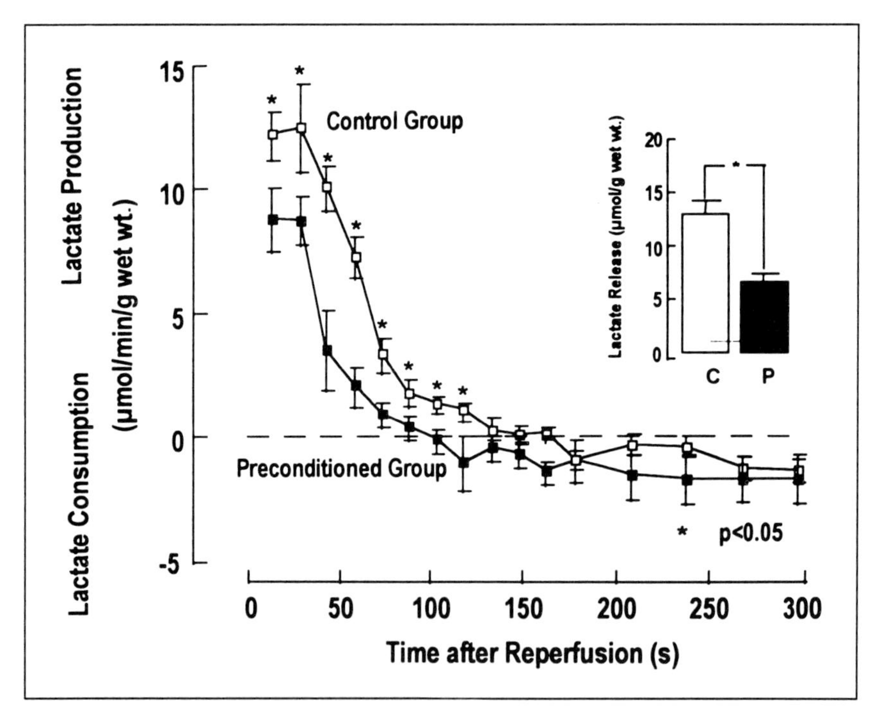

*Fig. 4.3. Graph illustrating the profile of lactate release during the first 5 minutes of reperfusion in control and preconditioned hearts. Preconditioned hearts were subjected to 5 minutes of ischemia and 5 minutes of reperfusion prior to the sustained ischemic period. Inset graph illustrates total lactate release over the 5 minute period of reperfusion in control and preconditioned hearts. Control hearts are represented by the empty squares and preconditioned hearts by the filled squares. C represents control hearts and P represents preconditioned hearts in the inset graph. Data are presented as means ± standard error of the mean; n = 6/group. *P<0.05 vs. control (two repeated measures of anova and Fischers test). Reproduced with permission from J Mol Cell Cardiol 1995; 27:969-976.*

have little effect overall on total ATP content,[1] the reduction in glycolytic ATP production specifically may result in the acceleration of ischemic contracture in preconditioned hearts. Reduced glycogen content,[29] as a result of the preceding ischemic episodes in the preconditioning protocol, are the most likely explanation for the reduced rate of glycolysis. Previous studies have demonstrated that the pre-ischemic glycogen content correlates with the extent of lactate production during a subsequent ischemic period[31] and also the severity of intracellular acidosis.[30,32]

4.3.3. Effects of Preconditioning on Intracellular pH

Several authors have now demonstrated that acidosis is attenuated in preconditioned hearts during the sustained ischemic period.[7,26,33] Kolocassides et al[26] demonstrated that initially, the rate of fall of intracellular pH was similar in control and preconditioned hearts, but that preconditioning attenuated the eventual extent (pH plateau was 5.9 vs. 6.4 in control vs. preconditioned hearts; Fig. 4.4). It seems

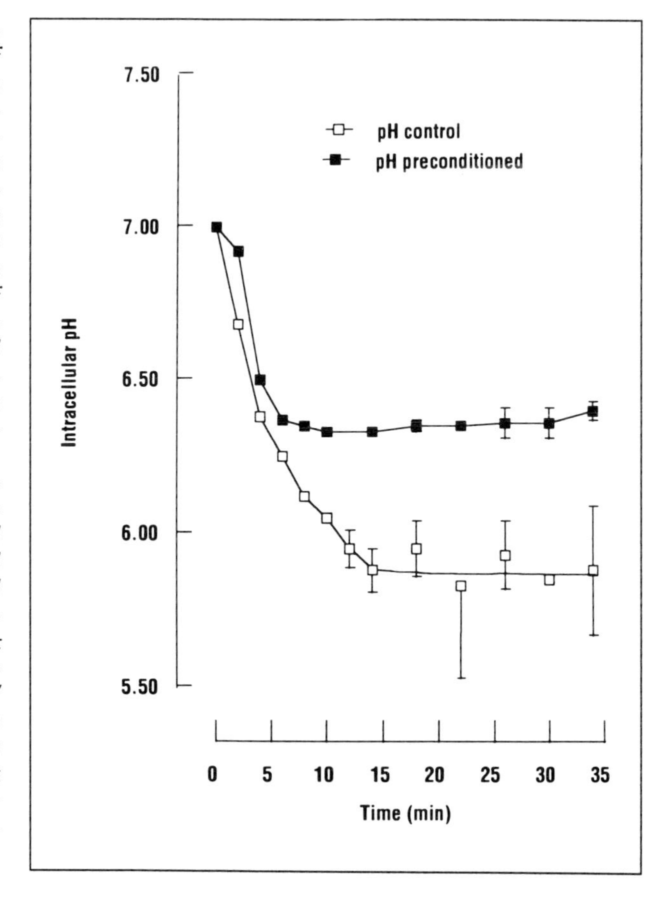

Fig. 4.4. Graph illustrating the effect of preconditioning on intracellular pH during the sustained period of global, no flow ischemia. Preconditioned hearts were subjected to 3 minutes of ischemia and 3 minutes of reperfusion + 5 minutes of ischemia and 5 minutes of reperfusion prior to the 35 minutes of sustained ischemia. Control hearts are represented by the empty squares and preconditioned hearts by the filled squares. Data are presented as means ± standard error of the mean; n = 6 hearts/ group. Adapted with permission from Kolocassides KG et al, Circulation 1994; 90(4,2):I-476 (Abstract).

likely that the attenuation of pH in preconditioned hearts is related to the slowed rate of anaerobic glycolysis and hence a reduced proton accumulation from glycolytic ATP breakdown. It should be noted that glycolysis per se does not result in a net gain of protons but rather the protons form as a result of the breakdown of glycolytically derived ATP. However, it should also be noted that ATP content prior to the start of the sustained ischemia is significantly reduced to approximately 70% of controls as a result of the ischemia associated with the preconditioning protocol. Thus, the attenuation of acidosis during the sustained ischemic period will be partly due to a reduced glycolytic rate and partly due to the reduced ATP content (and subsequent reduction in hydrolysis) prior to the ischemic period.

Thus slowed ischemic metabolism in preconditioned hearts may explain two of the most dramatic characteristics of preconditioned hearts, the acceleration of ischemic contracture and the attenuation of acidosis.

4.4. DOES PRECONDITIONING PROTECT AGAINST REPERFUSION INJURY?

Postischemic contractile dysfunction, following a prolonged period of ischemia, may be either permanent, and represent a form of ischemia or reperfusion-induced injury, or may be only temporary such that, over time, a full return to pre-ischemic function occurs, i.e., myocardial stunning.[34] Evidence from Bolli et al[35] suggests that stunning, after short and fully reversible periods of ischemia, is primarily mediated by the generation of free radicals during the first few minutes of reperfusion and is thus a genuine example of reperfusion injury. Recovery following prolonged ischemic periods, however, probably reflects a combination of both this form of injury (which may be fully reversible) and irreversible damage and cell death induced during the preceding ischemic period.

In preconditioned hearts, both systolic and diastolic function are markedly improved at all stages of reperfusion[6] (Fig. 4.1). However, in this study[6] and most others[7,17,36] reperfusion is not extended sufficiently to allow a full return to pre-ischemic function in control hearts. Although, it seems highly unlikely that control hearts would recover fully from such a severe depression in contractile function. That is, a significant fraction of the protection afforded by preconditioning is mediated against cell death as opposed to the alleviation of stunning. In support of this, creatine kinase release, a marker of cell death, is often reduced in preconditioned hearts[6] although not always significantly so.[1,37] As to whether preconditioning is protective against ischemia-induced injury[38] or injury occurring at reperfusion[36] has not to date been carefully investigated. Since a major component of reperfusion injury appears to be due to free radical-induced damage,[35] the possible involvement of free radicals in the mechanism of preconditioning has been studied by several investigators.[39-42] However in the majority of these studies no positive evidence could be found. Turrens et al[42] investigated a range of antioxidant enzymes and demonstrated that no change in activity was observed when comparing nonischemic and postischemic zones in either control or preconditioned rabbit hearts. Although total glutathione was better preserved in preconditioned hearts, it was concluded that this was simply a result of less washout of glutathione because these hearts were less injured. The authors therefore concluded that enhancement of antioxidant defenses was not the mechanism of preconditioning. Thus, this study reinforces the conclusion that preconditioning improves postischemic contractile recovery primarily by alleviating the extent of ischemic injury rather than by protecting against oxidant stress-induced myocardial stunning at the time of reperfusion.

Yet further support for this conclusion is provided by Przyklenek and colleagues[13] who have demonstrated that ischemic preconditioning cannot provide any protection against the short periods of ischemia which result only in myocardial stunning, but not in cell necrosis. Similar results have also been demonstrated in the pig.[16] The lack of protection against stunning in these studies may simply be further evidence that preconditioning provides protection mainly against ischemia-induced injury, rather than that preconditioning cannot protect against contractile dysfunction in these species. Indeed Shizukuda et al[15] have reported that protection against postischemic contractile dysfunction can be demonstrated in the dog, but only after longer periods of ischemia, for example one hour. This period of ischemia is equivalent to the periods of ischemia after which infarct size is assessed[22] and therefore probably represents protection against ischemia induced injury. It should be noted however that Kitikaze et al[43] have presented some evidence for a protective effect of preconditioning upon reperfusion-induced injury. In these studies, if 5'-nucleotidase activity was inhibited upon reperfusion, resulting in subsequent reduction in adenosine release, the protective effect of preconditioning on infarct size was partially blunted.

4.5. MECHANISM OF PRECONDITIONING-INDUCED PROTECTION AGAINST CONTRACTILE DYSFUNCTION

4.5.1. Extracellular Mediators, Receptor Stimulation and the Initiation of Preconditioning

As mentioned earlier, in order to induce preconditioning a threshold of stimulation must be reached by the preconditioning protocol. It also seems that in different species, different receptor pathways are utilized to trigger preconditioning but that all these receptor pathways converge at the activation of protein kinase C. Thus, in order for preconditioning to be induced, the preconditioning protocol must be sufficient to induce release of the agonists which activate protein kinase C.

It also appears that, in many species and models, if one receptor pathway is abolished, another can substitute.

a. Stimulation of adenosine receptors

There is now considerable evidence that, in the rabbit, preconditioning against cell necrosis is induced via the stimulation of adenosine receptors.[44] Downey and colleagues have demonstrated[44] that blockade of A_1-adenosine receptors (or possibly A_3 receptors) abolished preconditioning-induced protection whereas stimulation of these receptors mimicked preconditioning. Activation of 5'-nucleotidase activity may be responsible for enhanced adenosine release in preconditioned hearts.[43] However, in rat isolated buffer-perfused hearts, adenosine does not appear to be involved in preconditioning; blockade of adenosine receptors with the nonselective adenosine receptor antagonist, 8-(p-sulfophenol)theophylline, did not abolish preconditioning induced protection against contractile dysfunction and, likewise, pre-treatment with adenosine was unable to initiate the protection.[37] This result was also subsequently confirmed in the rat with infarct size as the end-point of assessment.[45] It is not that adenosine cannot induce protection against contractile dysfunction, but rather that in the rat adenosine is not the primary pathway utilized to initiate preconditioning. Adenosine can induce protection in the rat against ischemia-induced contractile dysfunction under certain experimental conditions[46] but not in a protocol mimicking preconditioning, i.e., a short infusion period followed by a washout period.[37] Therefore, although the mechanisms for inducing protection with adenosine are present, preconditioning in rats appears to use an alternative pathway. Likewise in the dog, adenosine receptor activation does not appear to be the primary pathway by which preconditioning is induced. Although stimulation of adenosine A_1-receptors is cardioprotective in the dog, the effects appear to be partly mediated via opening of myocardial K_{ATP} channels.[22] Unfortunately, there have been few mechanistic studies in the dog and rabbit in which contractile dysfunction was the end-point of assessment. However, it does appear that in the rabbit, if the primary pathway for inducing preconditioning is absent, preconditioning can be stimulated by another receptor pathway that stimulates protein kinase C. Thus, in the presence of adenosine receptor blockade, preconditioning can be induced chemically by muscarinic receptor stimulation,[47] opening of ATP-dependent potassium channels,[48,49] α_1-receptor stimulation[49] or stimulation of bradykinin receptors.[50]

b. Stimulation of α_1-receptors

Banerjee et al[17] demonstrated in 1993 that stimulation of α_1-receptors in the isolated perfused rat heart was able to mimic preconditioning and conversely, that blockade of α_1-receptors (with a specific

α_1-antagonist, BE2254) was able to abolish preconditioning induced protection against contractile dysfunction. Similarly, reserpine pretreatment (to deplete endogenous catecholamines) abolished ischemic preconditioning. Consequently, these authors concluded that locally regulated noradrenaline release and subsequent stimulation of myocardial α_1-receptors may initiate preconditioning-induced protection in the rat against contractile dysfunction.

In agreement with the study of Banerjee et al,[17] stimulation of α_1-receptors also preconditions rabbit hearts against infarction.[51] Tsuchida et al[51] reported that a 5 minute infusion of phenylephrine, 10 minutes before the prolonged ischemia, significantly reduced infarct size when compared with control hearts. Furthermore, although this protection could be abolished by the α_1-selective antagonist chloroethylclonidine, it could also be prevented by the protein kinase C inhibitor polymyxin B. The adenosine receptor antagonist, 8-(p-sulfophenyl)theophylline, could not prevent the protection when given simultaneously with phenylephrine, but could if it was administered during the prolonged ischemic period. This blockade could be overcome if phenylephrine was re-administered during the prolonged ischemic period. The authors concluded that, provided at least one of the receptors (α_1-adrenoceptor or A_1-adenosine receptor) was activated during the long ischemia, protection would be realized and that α_1-adrenoceptors and adenosine receptors activate parallel pathways within the myocyte to trigger protection.

c. Stimulation of bradykinin receptors

Banerjee and colleagues have recently demonstrated that in addition to α_1-agonists,[17] stimulation of bradykinin receptors can also initiate preconditioning against contractile dysfunction in rats.[52] These authors also demonstrated that both specific bradykinin (B_2) receptor blockade and protein kinase C inhibition can abolish preconditioning induced by ischemia.[52] Thus in the rat, there appears to be at least two receptor pathways through which ischemia can initiate preconditioning against contractile dysfunction.

The above results were also confirmed in the rabbit in terms of preconditioning against infarction. Goto et al[50] demonstrated that pretreatment with bradykinin significantly reduced infarct size to a similar level as ischemic preconditioning. The protective effect of bradykinin was also completely prevented with the protein kinase C inhibitor polymyxin B but not by nitric oxide inhibition.

d. Stimulation of muscarinic receptors

Stimulation of muscarinic M_2 receptors appears to be yet another receptor pathway through which ischemia can initiate preconditioning-induced protection. Liu and Downey[47] demonstrated that acetylcholine, infused for 5 minutes, 10 minutes prior to a prolonged ischemic period, reduced infarct size in a manner similar to a 5 minute episode of ischemia. In the dog, Yao and Gross[53] have also recently demonstrated that acetylcholine can mimic the effect of ischemic preconditioning on infarct size. However, as for other agonists in the dog, for example adenosine,[22,54] the protective effect of acetylcholine appears to be mediated via ATP-dependent potassium channels.[53]

e. Stimulation of ATP-dependent potassium channels

As mentioned above, preconditioning in the dog appears to be dependent upon the activation of ATP-dependent potassium channels.[22,55] In the dog, glibenclamide administered before or immediately after the preconditioning protocol completely abolished the protective effect on infarct size. Potassium channel openers could mimic the effect of ischemic preconditioning. In addition, although other agonists such as acetylcholine[53] and adenosine[22] are able to mimic preconditioning, they appear to work via the ATP-dependent potassium channel since glibenclamide can abolish their protective effects.

In contrast, in the rat Grover et al[56] demonstrated that glibenclamide could not abolish preconditioning-induced protection against contractile dysfunction. Furthermore, although the potassium channel opener cromakalim was protective against contractile dysfunction, it did not have a similar profile of activity to preconditioning-induced protection; that is, it did not accelerate ischemic contracture. Thus, it appears that in the rat, ATP-dependent potassium channels are not involved in the mechanism of protection against contractile dysfunction. The role of ATP-dependent potassium channels in preconditioning-induced protection against contractile dysfunction in other species and models remains to be determined.

4.5.2. Intracellular Second Messengers and the Role of Protein Kinase C

A common feature of all the receptors discussed above is that they are coupled via a inhibitory G protein, to protein kinase C.[57] Protein kinase C is formed following the stimulation of phosphodiesterase which breaks down phosphatidylinositol to 1,2, diacylglycerol and inositol triphosphate. 1,2, diacylglycerol is thought to stimulate protein kinase C whereas inositol trisphosphate releases calcium from the sarcoplasmic reticulum. Downey and colleagues[58,59] have recently demonstrated that in the rabbit protein kinase C activation is necessary during the period of sustained ischemia in order for preconditioning to be effective. They propose that the preconditioning ischemia translocates protein kinase C into the membrane from the cytosol where it remains until the sustained ischemia is initiated (reviewed in chapter 11). During the sustained ischemia the protein kinase C is once again activated and, due to its prior translocation, is in a position to phosphorylate membrane proteins. The fact that protein kinase C is already translocated to the membrane may be essential if protein phosphorylation is to occur before the phosphorylation potential drops too low during ischemia. This theory also explains the "memory" of the preconditioning phenomenon, since the loss of preconditioning-induced protection may reflect the movement of protein kinase C back to the cytosol. In order to strengthen their theory Liu et al[59] demonstrated that both inhibition of the translocation of protein kinase C, by application of colchicine prior to the preconditioning insult, and inhibition of protein kinase C activity before the sustained ischemia abolished preconditioning-induced protection against infarct size in the rabbit. However, inhibition of protein kinase C activity only, prior to the preconditioning ischemia, did not affect preconditioning-induced protection since only translocation of the enzyme occurs at this time.[59] These results have recently been confirmed in the rat[60] but, in preconditioned canine myocardium, Przyklenk and colleagues[61] could find no evidence of protein kinase C activation. Thus, in contrast to the rat and the rabbit, the dog may use a different second messenger system to transduce preconditioning.

All of these studies were performed with infarct size as the endpoint of assessment and therefore in order to determine whether protein kinase C activation was essential for preconditioning-induced protection against contractile dysfunction in the rat, we investigated the effect of the protein kinase C inhibitor polymyxin B, in rat isolated blood-perfused hearts.[5] Polymyxin B was present throughout and completely abolished preconditioning-induced protection against contractile dysfunction. Although little data is currently available on the role of protein kinase on preconditioning-induced protection against arrhythmias, it appears as though the different initiating pathways may eventually converge at protein kinase C in transducing protection. For example, cardiac α_1-receptors are known to activate protein kinase C,[62] as does the muscarinic M_2 receptor.[63]

4.5.3. The End Effector of Preconditioning

At this time the final mediator of protection is unknown. Protein kinase C phosphorylates many proteins in the heart one of which is the sodium-hydrogen antiporter.[64] In addition, there is some evidence in the literature suggesting an involvement for glucose uptake upon reperfusion in preconditioning induced protection against contractile dysfunction.[65] Fralix et al[36] demonstrated that only those adenosine anagonists which also abolished glucose uptake attenuated preconditioning-induced protection. In contrast, Tani et al[66] reported that preconditioning protected against calcium overload in the sarcoplasmic reticulum. More work is required in order to determine what the final steps in the pathway leading to myocardial protection may be.

4.6. CONCLUSIONS

Ischemic preconditioning clearly provides powerful protection against postischemic systolic and diastolic dysfunction. As the use of revascularization techniques such as thrombolysis, angioplasty and coronary artery bypass become more widespread, the alleviation of postischemic contractile dysfunction becomes more and more important. Since in most clinical situations it is clearly impractical to impose a period of ischemia prior to the ischemic insult in order to induce preconditioning, it is essential that the mechanism of protection be identified such that patients at risk might be treated prophylactically. Recent studies suggest that chronic preconditioning results in receptor downregulation or desensitization.[66] However, an alternative therapeutic approach might be revealed if the mechanism by which preconditioning is mediated could be identified.

ACKNOWLEDGMENTS

The advice and discussions with Dr Michael Shattock are gratefully acknowledged.

REFERENCES

1. Cave AC, Hearse DJ. Ischaemic preconditioning and contractile function: Studies with normothermic and hypothermic global ischaemia. J Mol Cell Cardiol 1992; 24:1113-1123.
2. Thornton J, Striplin S, Liu GS et al. Inhibition of protein synthesis does not block myocardial protection afforded by preconditioning. Am J Physiol 1990; 259:H1822-825.
3. Lawson CS, Coltart DJ, Hearse DJ. "Dose"-dependency and temporal characteristics of protection by ischaemic preconditioning against ischaemia-induced arrhythmias in rat hearts. J Mol Cell Cardiol 1993; 25:1391-1402.
4. Shiki K, Hearse DJ. Preconditioning of ischemic myocardium: reperfusion-induced arrhythmias. Am J Physiol 1987; 253:H1470-H1476.
5. Cave AC, Apstein CS. Inhibition of protein kinase C abolishes preconditioning against contractile dysfunction in the isolated blood perfused rat heart. Circulation 1994; 90(4):I-208 (Abstract).
6. Cave AC, Horowitz GL, Apstein CS. Can ischemic preconditioning protect against hypoxia-induced damage?: Studies on contractile function in isolated perfused rat hearts. J Mol Cell Cardiol 1994; 26:1471-1486.
7. Steenbergen C, Perlman ME, London RE et al. Mechanism of preconditioning. Ionic alterations. Circ Res 1993; 72:112-125.
8. de Albuquerque CP, Gerstenblith G, Weiss RG. Importance of metabolic inhibition and cellular pH in mediating preconditioning contractile and metabolic effects in rat hearts. Circ Res 1994; 74:139-150.
9. Locke-Winter CR, Winter CB, Nelson DW et al. cAMP stimulation facilitates preconditioning against ischemia-reperfusion through norephinephrine and $alpha_1$ mechanisms. Circulation 1991; 84(Suppl II):II-433 (Abstract).

10. Cave AC, Downey JM, Hearse DJ. Adenosine fails to substitute for preconditioning in the globally ischemic isolated rat heart. J Mol Cell Cardiol 1991; 23(Suppl III):S76 (Abstract).
11. Omar B, Hanson A, Bose S et al. Reperfusion with pyruvate eliminates ischemic preconditioning in the isolated rabbit heart: an apparent role for enhanced glycolysis. Coronary Artery Disease 1991; 2:799-804.
12. Cohen MV, Liu GS, Downey JM. Preconditioning causes improved wall motion as well as smaller infarcts after transient coronary occlusion in rabbits. Circulation 1991; 84:341-349.
13. Ovize M, Przyklenk K, Hale SL et al. Preconditioning does not attenuate myocardial stunning. Circulation 1992; 85:2247-2254.
14. Przyklenk K, Bauer B, Ovize M et al. Regional ischemic "preconditioning" protects remote virgin myocardium from subsequent sustained coronary occlusion. Circulation 1993; 87:893-899.
15. Shizukuda Y, Mallet RT, Lee S et al. Hypoxic preconditioning of ischaemic canine myocardium. Cardiovasc Res 1992; 26:526-533.
16. Miyamae M, Fujiwara H, Yokota R et al. Ischemic preconditioning accelerates energy production during reperfusion but does not improve stunning in porcine hearts. Circulation 1992; 86(4):I-339 (Abstract).
17. Banerjee A, Locke-Winter C, Rogers KB et al. Preconditioning against myocardial dysfunction after ischemia and reperfusion by an α_1-adrenergic mechanism. Circ Res 1993; 73:656-670.
18. Zhai X, Lawson CS, Cave AC et al. Preconditioning and post-ischemic contractile dysfunction: The role of impaired oxygen delivery versus extracellular metabolite accumulation. J Mol Cell Cardiol 1993; 25:847-857.
19. Walsh RS, Borges M, Thornton J et al. Hypoxia preceding ischemia reduces infarction despite the lack of an intervening period of reoxygenation. Circulation 1992; 86(4):I-342 (Abstract).
20. Murry CE, Richard VJ, Jennings RB et al. Myocardial protection is lost before contractile function recovers from ischemic preconditioning. Am J Physiol 1991; 260:H796-H804.
21. Cave AC, Hearse DJ. Ischemic preconditioning enhances post-ischemic function and reduces creatine kinase leakage in the rat heart even when used in conjunction with hypothermic cardioplegia. Circulation 1992; 84(4):I-30 (Abstract).
22. Yao Z and Gross GJ. A comparison of adenosine-induced cardioprotection and ischemic preconditioning in dogs. Efficacy, time course, and role of K_{ATP} channels. Circulation 1994; 89:1229-1236.
23. Miura M, Adachi T, Ogawa T et al. Myocardial infarct size limitation by preconditioning; its natural decay and "dose-response" relation. J Mol Cell Cardiol 1991; 23(Suppl II):S.38 (Abstract]).
24. Li Y, Kloner RA. The beneficial effects of preconditioning can be recaptured after they are lost. Circulation 1993; 88(4):I-138 (Abstract]).
25. Hearse DJ, Garlick PB, Humphrey SM. Ischemic contracture of the myocardium; mechanisms and prevention. Am J Cardiol 1977; 39:986-993.

26. Kolocassides KG, Seymour AL, Galiñanes M et al. Detrimental effects of ischemic preconditioning on ischemic contracture? NMR studies of energy metabolism and intracellular pH in the rat heart. Circulation 1994; 90(4,2):I-476 (Abstract).
27. Eberli FR, Weinberg EO, Grice WN et al. Protective effect of increased glycolytic substrate against systolic and diastolic dysfunction and increased coronary resistance from prolonged global underperfusion and reperfusion in isolated rabbit hearts perfused with erthrocyte suspensions. Circ Res 1991; 68:446-481.
28. Owen P, Dennis S, Opie LH. Glucose flux rate regulates onset of ischemic contracture in globally underperfused rat hearts. Circ Res 1990; 66:344-354.
29. Wolfe CL, Sievers RE, Visseren FLJ et al. Loss of myocardial protection after preconditioning correlates with the time course of glycogen recovery within the preconditioned segment. Circulation 1993; 87:881-892.
30. Kingsley PB, Sako EY, Yang MQ et al. Ischemic contracture begins when anaerobic glycolysis stops: a ^{31}P-NMR study of isolated rat hearts. Am J Physiol 1991; 261:H469-H478.
31. Neely JR, Grotyohann LW. Role of glycolytic products in damage to ischemic myocardium. Dissociation of adenosine triphosphate levels and recovery of function of reperfused ischemic hearts. Circ Res 1984; 55:816-824.
32. Dennis SC, Gevers W, Opie LH. Protons in ischemia? Where do they come from; Where do they go to? J Mol Cell Cardiol 1991; 23:1077-1086.
33. Tsuchida A, Miura T, Miki T et al. Role of adenosine receptor activation in myocardial infarct size limitation by ischaemic preconditioning. Cardiovasc Res 1992; 26:456-461.
34. Bolli R. Mechanism of myocardial "stunning." Circulation 1990; 82:723-738.
35. Bolli R, Jeroudi MO, Patel BS et al. Marked reduction of free radical generation and contractile dysfunction by antioxidant therapy begun at the time of reperfusion. Evidence that myocardial "stunning" is a manifestation of reperfusion injury. Circ Res 1989; 65:607-622.
36. Fralix TA, Steenbergen C, London RE et al. Metabolic substrates can alter post-ischemic recovery in preconditioned ischemic heart. Am J Physiol 1992; 263:C17-C23.
37. Cave AC, Collis CS, Downey JM et al. Improved functional recovery by ischaemic preconditioning is not mediated by adenosine in the globally ischaemic, isolated rat heart. Cardiovasc Res 1993; 27:663-668.
38. Murry CE, Jennings RB, Reimer KA. Preconditioning with ischemia: a delay of lethal cell injury in ischemic myocardium. Circulation 1986; 74:1124-1136.
39. Iwamoto T, Miura T, Adachi T et al. Myocardial infarct size-limiting effect of ischemic preconditioning was not attenuated by oxygen free-radical scavengers in the rabbit. Circulation 1991; 83:1015-1022.

40. Liu GS, Stanley AWH, Downey JM. Cyclooxygenase products are not involved in the protection against myocardial infarction afforded by preconditioning in the rabbit. Am J Cardiovasc Path 1992; 4:157-164.
41. Richard V, Tron C, Thuillez C. Ischaemic preconditioning is not mediated by oxygen derived free radicals in rats. Cardiovasc Res 1993; 27:2016-2021.
42. Turrens JF, Thornton J, Barnard ML et al. Protection from reperfusion injury by preconditioning hearts does not involve increased antioxidant defenses. Am J Physiol 1992; 262:H585-H589.
43. Kitakaze M, Hori M, Morioka T et al. Infarct size-limiting effect of ischemic preconditioning is blunted by inhibition of 5'-nucleotidase activity and attenuation of adenosine release. Circulation 1994; 89:1237-1246.
44. Liu GS, Thornton J, Van Winkle DM et al. Protection against infarction afforded by preconditioning is mediated by A_1 receptors in rabbit heart. Circulation 1991; 84:350-356.
45. Liu Y, Downey JM. Blocking adenosine receptors or ATP sensitive K^+ channels does not prevent preconditioning in rat heart. Circulation 1992; 86(4):I-341 (Abstract).
46. Lasley R, Rhee J, VanWylen D et al. Adenosine A_1 receptor mediated protection of the globally ischemic isolated rat heart. J Mol Cell Cardiol 1990; 22:39-47.
47. Liu G, Downey JM. Acetylcholine preconditions rabbit heart: further evidence for G_i protein coupling in preconditioning. Circulation 1992; 86(Suppl I):I-174 (Abstract).
48. Thornton JD, Daly JF, Cohen MV et al. Catecholamines can induce adenosine receptor-mediated protection of the myocardium but do not participate in ischemic preconditioning in the rabbit. Circ Res 1993; 73:649-655.
49. Thornton JD, Thornton CS, Sterling DL et al. Blockade of ATP-sensitive potassium channels increases infarct size but does not prevent preconditioning in rabbit hearts. Circ Res 1993; 72:44-49.
50. Goto M, Liu Y, Ardell JL et al. Bradykinin limits myocardial infarction in rabbits by protein kinase C activation and not by nitric oxide synthesis. Circulation 1994; 90(4,2):I-208 (Abstract).
51. Tsuchida A, Liu Y, Liu GS et al. Alpha$_1$-adrenergic agonists precondition rabbit ischemic myocardium independent of adenosine by direct activation of protein kinase C. Circ Res 1994; 75:576-585.
52. Brew EC, Rehring TF, Banerjee A. Bradykinin receptors mediate preconditioning through protein kinase C. Circulation 1994; 90(4,2):I-207 (Abstract).
53. Yao Z, Gross GJ. Acetylcholine mimics ischemic preconditioning via a glibenclamide-sensitive mechanism in dogs. Am J Physiol 1993; 264:H2221-H2225.
54. Yao Z, Gross GJ. Glibenclamide antagonizes adenosine A_1 receptor-mediated cardioprotection in stunned canine myocardium. Circulation 1993; 88:235-244.

55. Gross GJ, Auchampach JA. Blockade of ATP-sensitive potassium channels prevents myocardial preconditioning in dogs. Circ Res 1992; 70:223-233.
56. Grover GJ, Dzwonczyk S, Sleph PG. ATP-sensitive K^+ channel activation does not mediate preconditioning in isolated rat hearts. Circulation 1992; 86(Suppl I):I-341 (Abstract).
57. Thornton JD, Liu GS, Downey JM. Pretreatment with pertussis toxin blocks the protective effects of preconditioning: evidence for a G-protein mechanism. J Mol Cell Cardiol 1993; 25:311-320.
58. Ytrehus K, Liu Y, Downey JM. Preconditioning protects the ischemic rabbit heart by protein kinase C activation. Am J Physiol 1994; (in press).
59. Liu Y, Ytrehus K, Downey JM. Evidence that translocation of protein kinase C is a key event during ischemic preconditioning of rabbit myocardium. J Mol Cell Cardiol 1994; 26:661-668.
60. Speechly-Dick ME, Mocanu MM, Yellon DM. Protein kinase C. Its role in ischemic preconditioning in the rat. Circ Res 1994; 75:586-590.
61. Przyklenk K, Sussman MA, Kloner RA. Fluoresence microscopy reveals no evidence of protein kinase C activation in preconditioned canine myocardium. Circulation 1994; 90(4):I-647 (Abstract).
62. Talosi L and Kranias EG. Effect of α-adrenergic stimulation on activation of protein kinase C and phosphorylation of proteins in intact rabbit heart. Circ Res 1992; 70:1304-1312.
63. Fleming JW, Wisler PL, Watanabe AM. Signal transduction by G proteins in cardiac tissues. Circulation 1992; 85:419-433.
64. Hathaway DR and March KL. Molecular cardiology:new avenues for the diagnosis and treatment of cardiovascular disease. JACC 1989; 13:265-288.
65. Tani M, Asakura Y, Ebihara Y et al. Effect of preconditioning (PC) with anoxic perfusion (HP) on sarcoplasmic reticular Ca^{2+} uptake (Ca^{SR}) in reperfused rat hearts. Circulation 1991; 84 (Suppl II):II-433 (Abstract).
66. Cohen MV, Yang X, Downey JM. Conscious rabbits become tolerant to multiple episodes of ischemic preconditioning. Circ Res 1994; 74: 998-1004.

CHAPTER 5

Reduction of Infarct Size—"Preconditioning at a Distance"

Karin Przyklenk, Peter Whittaker, Michel Ovize and Robert A. Kloner

5.1. INTRODUCTION

In 1986, Murry and colleagues made the seminal observation that brief episodes of myocardial ischemia—too brief in themselves to cause myocyte death—paradoxically rendered the previously ischemic myocardium more resistant to a subsequent sustained period of coronary artery occlusion.[1] This first report of "preconditioning with ischemia" in the anesthetized dog was soon duplicated by others using similar canine preparations,[2-4] and confirmed by numerous investigators to occur in other species including the rat,[5,6] rabbit[7-9] and pig.[10,11] As a result, there is unprecedented agreement among investigators and laboratories world-wide that ischemic preconditioning profoundly limits myocardial infarct size. However, the question which, to date, remains unanswered is: precisely how does brief ischemia protect the myocytes from a later episode of sustained coronary occlusion?

Our laboratory is among the many that have reported and confirmed reduction in infarct size with preconditioning in the dog, rabbit and rat models. In addition, we have, like many others, sought to elucidate the cellular mechanism(s) responsible for preconditioning. But perhaps our most novel and provocative contribution to this field is the concept of "preconditioning at a distance": that is, while brief occlusions of the circumflex coronary artery protect the circumflex bed from a later more sustained circumflex occlusion, and brief occlusions of the left anterior descending (LAD) coronary artery similarly protect

Myocardial Preconditioning, edited by Cherry L. Wainwright and James R. Parratt.

against a subsequent LAD occlusion, we made the remarkable observation that, in the canine model, *brief ischemia in one coronary bed also protects or preconditions remote virgin myocardium against subsequent infarction.*[12] In this chapter, we summarize our experience with ischemic preconditioning, and the observations that prompted us to propose that preconditioning may be mediated by factor(s) activated, produced or transported throughout the heart. We then review the prospective evidence in support of this concept, and provide our speculations concerning the means by which brief regional ischemia may confer global myocardial protection.

5.2. PRECONDITIONING AND INFARCT SIZE

5.2.1. Documenting the Phenomenon

Using the pentobarbital-anesthetized dog, Murry et al were the first to report that infarct size produced by a sustained 40 minute period of circumflex coronary artery occlusion was dramatically reduced when the sustained occlusion was immediately preceded by four 5 minute episodes of circumflex occlusion, each interrupted by 5 minutes of reperfusion.[1] We confirmed this observation using a similar canine model: infarct size was significantly smaller in dogs that received four 3 minute or four 5 minute episodes of brief LAD occlusion prior to 1 hour of sustained LAD ischemia when compared with control dogs that received an equivalent no intervention period prior to the sustained ischemic insult[4,13] (Fig. 5.1a). Reduction of infarct size was subsequently described in preconditioned rabbits,[8] and our laboratory similarly observed that two 5 minute periods of brief coronary occlusion in the rabbit model reduced infarct size caused by 30 or 40 minutes of sustained coronary occlusion[9,14] (Fig. 5.1b). Our laboratory also reported that preconditioning both limited infarct size and reduced the incidence of ischemia-induced arrhythmias in the anesthetized open-chest rat,[5] observations that have since been confirmed by numerous other investigators[6] (Fig. 5.1c). Finally, we have recently reported an apparent clinical correlate to the preconditioning phenomenon. Retrospective analysis of data derived from the "Thrombolysis in Myocardial Infarction" (TIMI)-4 trial revealed that patients with a history of angina prior to acute myocardial infarction had a lower incidence of in-hospital death, congestive heart failure and/or shock, and smaller CK-determined infarct size when compared with those with no prior angina,[15] supporting the concept that brief antecedent ischemia, even in man, renders the heart more resistant to subsequent more sustained coronary occlusion.

5.2.2. Exploring the Properties of Preconditioning

As was the case in many other laboratories, we next sought to expand our understanding of the fundamental properties of

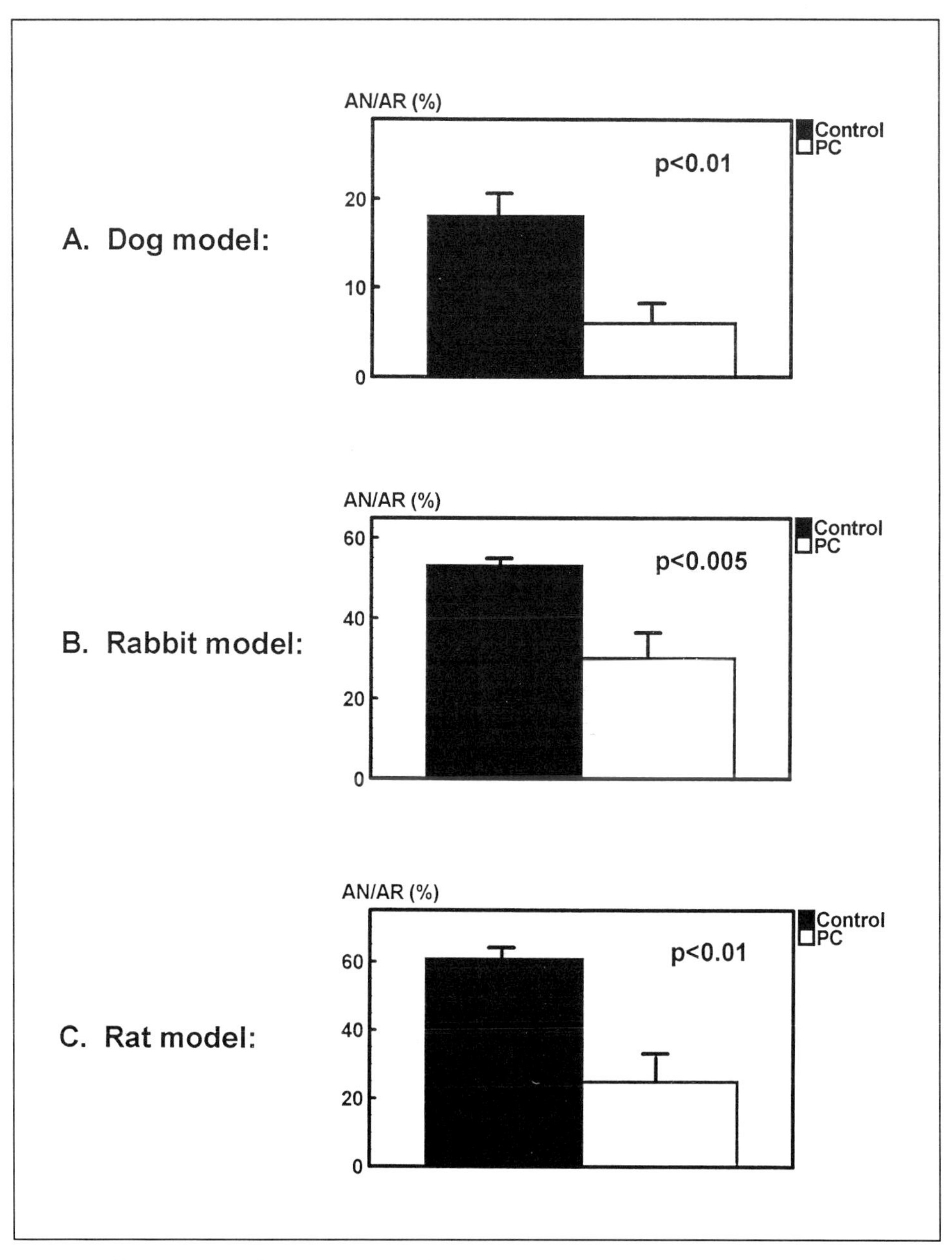

Fig. 5.1. Ischemic preconditioning profoundly limits infarct size. In the canine model (A), infarct size resulting from 1 hour of sustained coronary occlusion is significantly smaller in dogs that received four 5 minute episodes of preconditioning ischemia when compared with controls that received no intervention prior to the sustained ischemic insult. Similarly, two 5 minute episodes of brief ischemia significantly reduced infarct size in rabbits subjected to 40 minutes of sustained occlusion (B), while three 3 minute episodes of preconditioning ischemia protected the rat heart against a subsequent 90 minute coronary occlusion (C). In each example, area of necrosis (AN) in the control and preconditioned (PC) groups is expressed as a percent of the area at risk (AR). Adapted from data presented in references 30, 9, and 5 respectively.

preconditioning. For example, in the rat model, evidence from our laboratory confirmed that, as in the dog,[16] the beneficial effects of preconditioning are transient: a delay of 1 or more hours between the brief preconditioning ischemia and the sustained occlusion resulted in a complete loss of protection against both infarction and arrhythmias.[5] Interestingly, however, protection could be effectively re-established if, after the delay, the preconditioning stimulus was repeated before the sustained ischemic insult.[17] Using the anesthetized open-chest dog, we further found that preconditioning was not dependent upon total mechanical occlusion of the coronary artery: brief and transient episodes of platelet thrombosis,[4] as well as a brief 15 minute episode of subtotal coronary artery stenosis,[18] also reduced infarct size caused by a subsequent 1 hour episode of coronary occlusion.

Importantly, however, our studies in the canine model also revealed that preconditioning is not a panacea. While brief preconditioning ischemia consistently and significantly limited infarct size, preconditioning failed to exert an independent beneficial effect on the acute recovery of contractile function of the salvaged but stunned myocardium.[4,18,19] Second, in our hands preconditioning failed to provide concomitant protection to the coronary vasculature; that is, it did not attenuate the well-described deterioration in endothelium-dependent and endothelium-independent coronary vasodilator reserve known to occur following relief of a sustained coronary occlusion.[13] Finally, in contrast to previous studies using chloralose-anesthetized dogs,[20] and in contrast to the consistent protection against arrhythmias reported in rats,[5,20] we found that preconditioning failed to attenuate the incidence of ectopic beats, tachycardia or lethal ventricular fibrillation in our pentobarbital-anesthetized preparation.[21]

5.2.3. The Search for the Mechanisms(s)

Most recently, emphasis has shifted from describing the characteristics of preconditioning to the challenge of identifying the cellular mechanisms responsible for the cardioprotection. Using a variety of pharmacologic agonists and antagonists, our laboratory has specifically focused on elucidating the potential contributions of various compounds and cellular messengers implicated to play a role, including prostaglandins,[22] adenosine,[23-25] catecholamines,[14,26] muscarinic agonists[13,27] and protein kinase C,[28-30] on the reduction in infarct size achieved with preconditioning.

Interesting and provocative results have been obtained: for example, pharmacologic inhibitors of protein kinase C appear to prevent[28] or attenuate[29] the preconditioning phenomenon in the rat, yet administration of PKC inhibitors in the dog model, even when given in high intracoronary doses, had no effect.[30] These observations are subject to various interpretations. The data may indicate that protein kinase C is an important mediator of infarct size reduction with preconditioning in the rat model, but does not play a role in the dog. Alternatively, it could be argued that: (1) in the dog studies, the dose and route of administration of the agents may not have provided adequate inhibition of protein kinase C within the ischemic/reperfused myocytes; or (2) in the rat protocols, secondary deleterious consequences of the PKC inhibitors (including rupture of cell membranes, altered platelet function, adverse hemodynamic effects, etc.)[30] may have contributed to the larger infarct sizes in the drug-treated preconditioned cohorts. It was our frustration with these inherent uncertainties and limitations of the pharmacologic agonist/antagonist approach that lead us to pursue a new avenue—retrospective data analysis combined with mathematical modeling[31]—in our efforts to elucidate the pivotal mediators of preconditioning.

5.3. PROTECTION VIA NONISCHEMIC TISSUE

5.3.1. Formulating the Hypothesis

Retrospective analysis[31] of data from two previous protocols[5,22] revealed that, in the rat model, 90 minutes of sustained coronary occlusion in control animals resulted in an average infarct size of $61 \pm 3\%$ of the myocardium at risk. Moreover, infarct size in the control cohort was essentially constant, no matter whether the risk region was small (occupying 20% of the total left ventricle) or large (i.e., 70% of the left ventricle; Fig. 5.2).

As expected, mean infarct size in rats preconditioned with three 3 minute episodes of brief ischemia prior to the 90 minute sustained occlusion was significantly smaller than that observed in the controls, averaging only $28 \pm 6\%$ of the myocardium at risk. However, in marked contrast to the results obtained in control animals, infarct size in the preconditioned cohort was highly dependent upon the size of the risk region (Fig. 5.2), with the best-fit relationship between area of necrosis (AN) and area at risk (AR: expressed as a percent of the total left ventricle) being described by the equation:

$$AN/AR \; \alpha \; 1\text{-}[2/AR]^2$$

Most notably, this analysis revealed that protection was in fact lost in those preconditioned rats with a large risk region (Fig. 5.2).[31]

This observation raised a novel possibility: that preconditioning in the rat model might be dependent upon the ability of an as-yet unidentified "mediator" to penetrate into the area at risk from the remaining nonischemic regions of the heart. Thus, if the risk region is small, then the mediator can presumably penetrate the risk region in sufficient concentrations to provide substantial protection during the subsequent sustained ischemia, but if the risk region is large, the available amounts of the mediator are insufficient to provide protection. Based on this observation, we therefore proposed: (1) that brief preconditioning

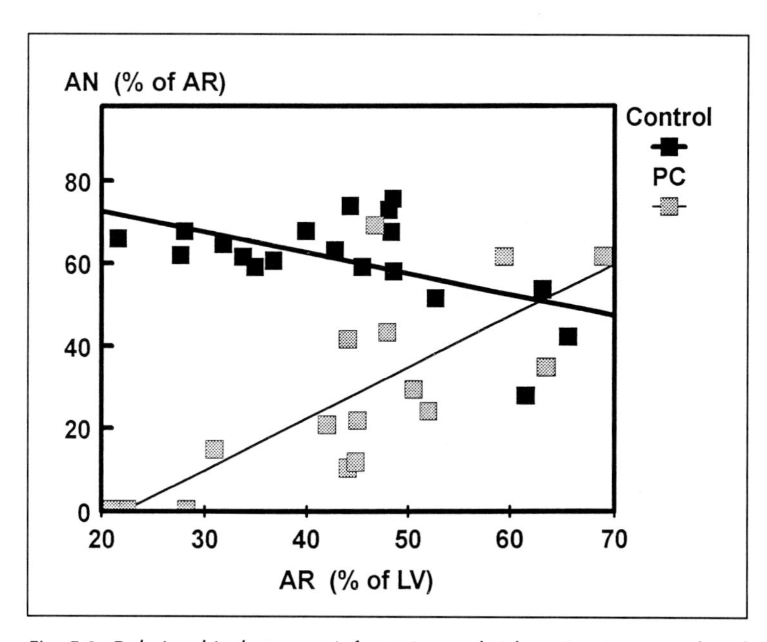

Fig. 5.2. Relationship between infarct size and risk region in control and preconditioned rats. In the control group, area of necrosis (AN: expressed as a percent of the area at risk) was essentially constant over the entire range of risk regions (AR: expressed as a percent of the total left ventricle (LV)). However, in preconditioned animals, the reduction in infarct size achieved with preconditioning waned in those rats with a large risk region. Reproduced with permission from Whittaker P et al, Basic Res Cardiol 1994; 89:6-15.

ischemia activates or produces a protective factor not only in the ischemic bed but throughout the heart; and (2) the protective factor is "consumed" within the risk region during the subsequent sustained occlusion, and may then be transported (perhaps by collateral blood flow or diffusion) into the area at risk from the remaining nonischemic myocardium.[31] As collateral blood flow in the rat model is negligible, we then attempted to fit the raw infarct size and risk region data to a simple diffusion model. The equation derived from the diffusion model was:

$$AN/AR \propto 1-[1/AR]^2$$

remarkably similar to the empirical relationship $AN/AR \propto 1-[2/AR]^2$ derived from the actual experimental results.[31]

5.3.2. Testing the Hypothesis

Mathematical modeling of the preconditioning phenomenon proved to be a challenging and intriguing intellectual exercise but, more importantly, it provided the impetus for the design of new experiments to test the resultant hypotheses. Specifically, it was through this process of mathematical modeling that the concept of the circumflex preconditioning/LAD occlusion protocol was conceived.[12]

We reasoned that if brief ischemia triggered the production of a protective factor throughout the myocardium, then brief ischemia in one region of the heart should effectively protect remote myocardium from subsequent sustained coronary artery occlusion. To test this concept, anesthetized dogs were randomly assigned to undergo four 5 minute occlusions of a circumflex coronary artery branch or an equivalent no intervention period prior to 1 hour of sustained LAD occlusion.[12] Measurement of regional myocardial blood flow confirmed that, as expected, occlusion of the circumflex branch did not render the LAD bed ischemic. Nonetheless, infarct size was significantly smaller in the circumflex-preconditioned group than in the controls ($6 \pm 2\%$ versus $16 \pm 5\%$ of the myocardium at risk respectively; Fig. 5.3a). This reduction in infarct size could not be attributed to differences in the fundamental determinants of infarct size (i.e., collateral blood flow, hemodynamic parameters or area at risk) between groups (Fig. 5.3b), and was, in fact, comparable to that previously reported with circumflex preconditioning/circumflex occlusion[1-3] and LAD preconditioning/LAD occlusion.[4,13,16,18] These results therefore confirmed that, in the canine model, brief ischemia in one region of the heart is able to protect remote myocardium from subsequent infarction, and thus supported the first of the two predictions derived from the mathematical model.

5.4. MECHANISM(S) OF GLOBAL PROTECTION BY BRIEF REGIONAL ISCHEMIA

5.4.1. Myocyte Stretch Plays a Pivotal Role

The first question to arise from these data is: how did the virgin LAD territory "detect" ischemia during the circumflex branch occlusions? Using sonomicrometry, we noted that dogs in the preconditioned group exhibited significant hyperkinesis and dilatation in the normally perfused LAD bed during the brief circumflex occlusions (i.e., systolic shortening and end diastolic length increased to 112% and 110% of their baseline values, respectively)[12] consistent with historical reports of hyperkinesis in remote nonischemic regions of the heart during coronary artery occlusion.[32] Based on these data, we postulated that myocyte stretch during the brief ischemic stimuli might trigger the protection achieved with preconditioning.

We tested this possibility, again in the canine model, by substituting ischemia-induced stretch by rapid volume overload (500 ml of warmed saline infused over 10 minutes into the left atrium) prior to our conventional 1 hour sustained ischemic challenge.[33] Volume overload dilated

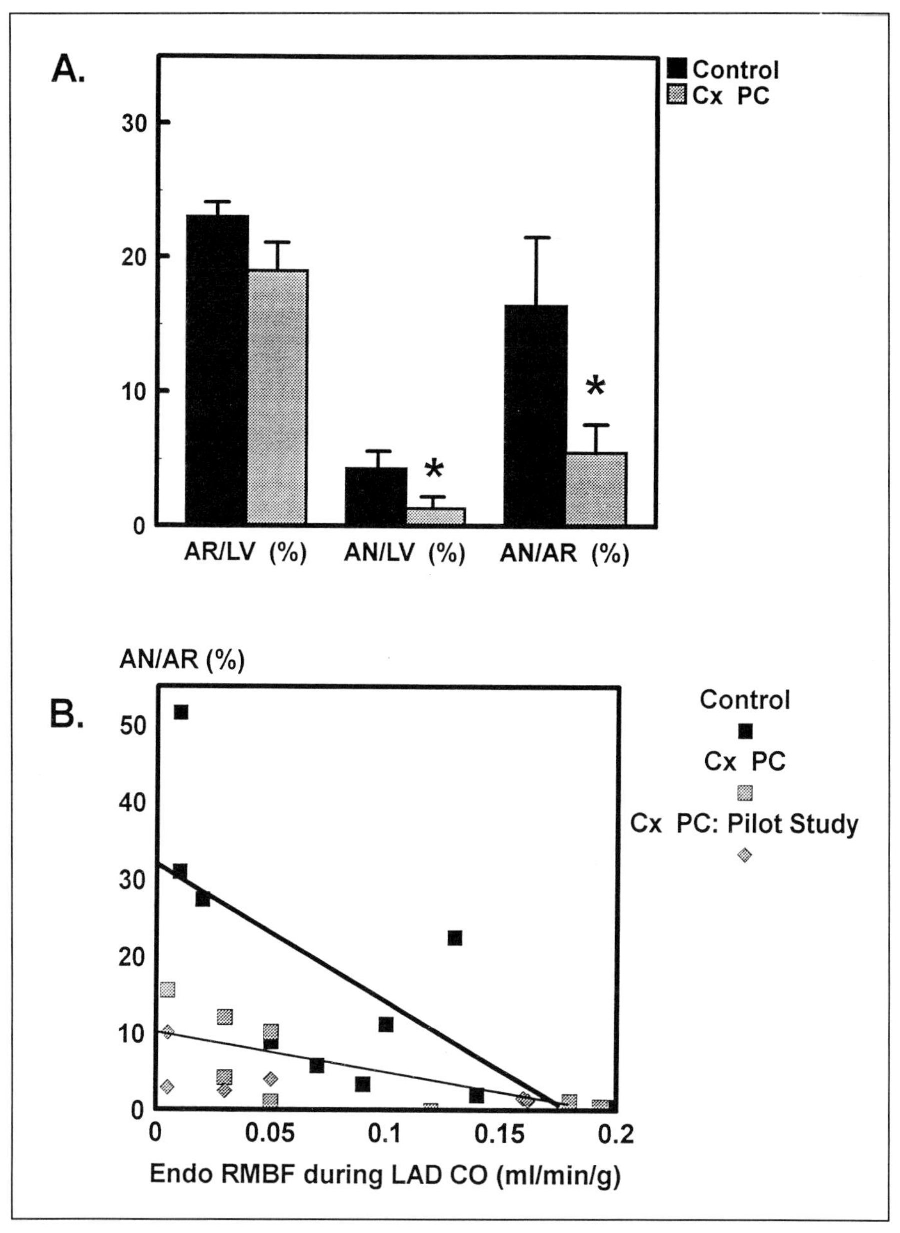

*Fig. 5.3. "Preconditioning at a distance" in the canine model. (A) Mean values of area at risk (AR: expressed as a percent of the total left ventricular (LV) weight) and area of necrosis (AN: expressed both as a percent of the total LV and as a percent of the AR) for dogs randomized to either the control or circumflex preconditioned (Cx PC) groups. *P<0.05 versus control. (B) Infarct size (AN/AR) plotted as a function of subendocardial blood flow (Endo RMBF) during sustained left anterior descending coronary artery occlusion (LAD CO) for control and Cx preconditioned dogs in both the randomized and preliminary protocols. Analysis of covariance revealed a significant downward shift in the regression relation for the Cx preconditioned animals with respect to that obtained for the control group. Reproduced with permission of the American Heart Association from Przyklenk K et al, Circulation 1993; 87:893-899.*

the hearts to approximately 115% of their baseline dimensions, a magnitude of dilatation intermediate between the values of 110% and 119% observed in the LAD bed during brief circumflex occlusions[12] and brief LAD occlusions,[33] respectively. Consistent with our hypothesis, volume overload proved to be protective: mean infarct size was 15 ± 3% of the myocardium at risk, significantly smaller than the value of 32 ± 8% observed in the matched controls that received 1 hour of LAD occlusion alone (Fig. 5.4).

To confirm that this reduction in infarct size was truly due to stretching of the myocytes rather than hemodilution or some other nonspecific effect of volume overload, additional dogs received gadolinium chloride (a potent blocker of stretch-activated ion channels)[34] prior to either rapid infusion of saline, conventional preconditioning,

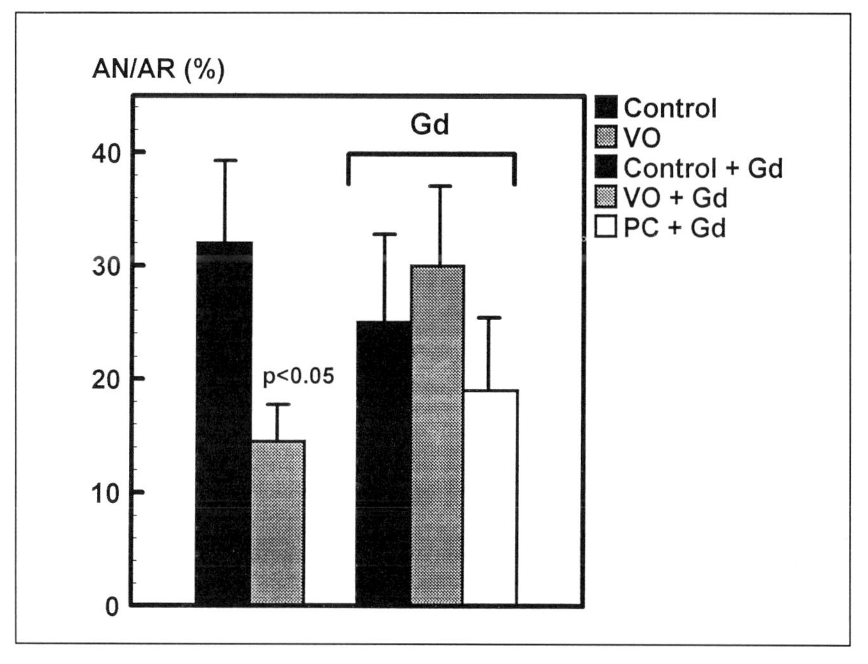

Fig. 5.4. Acute volume overload "preconditions" canine myocardium. Mean area of necrosis (AN: expressed as a percent of the area at risk (AR)) in control dogs subjected to one hour of sustained coronary occlusion, dogs that received volume overload (VO) prior to the sustained occlusion, and dogs treated with gadolinium (Gd) before the control (no intervention) period, volume overload or conventional preconditioning (PC) regimen. Infarct size was significantly reduced in dogs that received volume overload when compared with controls. Treatment with gadolinium, a blocker of stretch-activated ion channels, had no significant effect on infarct size in control dogs, blocked the reduction in infarct size previously obtained with volume overload, and attenuated the expected reduction in infarct size in the preconditioned group. Adapted from data presented in reference 33.

or an equivalent control period. Gadolinium had no effect on infarct size in control animals, but blocked the reduction in infarct size achieved with volume overload and attenuated the expected reduction in infarct size in the preconditioned group (Fig. 5.4).[33]

We have recently obtained further corroborating evidence in both the rat and dog models that myocyte stretch can initiate cardioprotection. Specifically, infarct size in rats subjected to 1 hour of coronary occlusion was significantly reduced when the ischemic insult was immediately preceded by four 0.15-ml injections of saline into the left ventricular free wall ($29 \pm 7\%$ of the myocardium at risk, versus $52 \pm 5\%$ observed in the controls),[25,35] and similar protection has been obtained with multiple intramyocardial saline injections prior to 1 hour of LAD occlusion in our canine preparation.[35] Many factors in addition to myocyte stretch per se, such as focal ischemia at the injection sites and/or global release of catecholamines, may contribute to the reduction in infarct size achieved with the intramyocardial saline injections. However, we found that in the rat model, addition of gadolinium chloride to the injection solution abolished the reduction in infarct size observed with intramyocardial injection of saline alone,[35] again implicating myocyte stretch as playing a pivotal role.

5.4.2. The Next Challenge: Identifying the Cellular Mediator(s)

Our studies to this point have focused on verifying the first consequence arising from the mathematical model: that is, the concept that brief ischemia should trigger the production of a protective substance throughout the heart. This prediction was prospectively confirmed in our circumflex preconditioning/LAD occlusion protocol, and subsequent studies further identified myocyte stretch as an important mediator of the protection achieved with preconditioning.

The next daunting challenge is two-fold: first, to address the second consequence of the mathematical model (i.e., the concept that the unknown mediator is transported, perhaps by diffusion—and/or communication via collateral connections, the lymph system, gap junctions, etc.—into the risk region from the remaining nonischemic myocardium), and, perhaps most importantly, to identify the unknown ion or messenger within the myocyte ultimately responsible for the reduction in infarct size. These two crucial issues remain to be evaluated in future prospective studies.

5.5. CORROBORATING EVIDENCE

5.5.1. Regional Myocardial Ischemia Has Global Consequences

Several lines of evidence, in addition to the results of the "preconditioning at a distance" protocol, support the general concept that virgin myocardium senses and responds to ischemia in a remote region of the heart. For example, brief 5 minute episodes of regional ischemia in the rabbit result in the rapid (i.e., within one hour) expression of heat shock protein (hsp) 70 in both the ischemic and nonischemic myocardium,[36] while 15 minutes of LAD occlusion in the dog model result in small but significant increases in interstitial adenosine concentrations in the remote circumflex territory.[37] In both rabbit and dog myocardium, brief ischemia alters carnitine-linked metabolism in both the ischemic and nonischemic beds,[38,39] while in the porcine model the monophasic action potential is altered in remote nonischemic myocardium during sustained coronary occlusion in both control and preconditioned hearts.[40] In addition, recent preliminary results suggest that in our canine model the enzyme 5'-nucleotidase, responsible for the degradation of adenosine monophosphate to adenosine and implicated as a mediator of ischemic preconditioning,[41] is elevated with preconditioning in both the previously ischemic LAD and remote circumflex territories.[41a] These observations first serve to underscore an important technical consideration: remote nonischemic regions of the heart clearly cannot be assumed to be "normal." However, whether all these changes are simply consequences of remote ischemia, or whether any prove to be actual mediators of cardioprotection, is not known.

5.5.2. Protection Achieved with Preconditioning is Dependent upon Area at Risk

The primary observation which formed the basis of the mathematical model is that, in the rat, the protective effects of preconditioning waned with increasing risk region (Fig. 5.2).[31] Does this observation also hold true for other species?

Two factors confound this type of analysis in other models: variations in collateral blood flow during coronary occlusion, an especially crucial determinant of infarct size in the canine preparation[1-4,12,13,16,30] and an important variable in the rabbit;[9] as well as the uniformly small risk regions typically obtained in the dog, rabbit and pig (i.e., mean values of $18 \pm 2\%$,[30] $31 \pm 2\%$,[9] and $15 \pm 2\%$,[10] respectively, as compared with the risk regions ranging from 23% to 69% that were successfully tolerated in the rat studies). Nonetheless, when these factors are taken into consideration, there is support for the general concept of a loss of preconditioning-induced protection with increasing risk region.

We first addressed this issue[31] in the dog model by relying not on our own data, but rather on the detailed tabulations published by Murry and colleagues.[1,16] When all dogs in both studies were considered, there was no significant correlation between infarct size and risk region in either the control or preconditioned groups. However, when the analysis was restricted to the population of dogs rendered severely ischemic during coronary occlusion (subendocardial blood flow ≤ 0.04 ml/min/g tissue, comparable to the low collateral blood flow characteristic of the rat model), we found that infarct size was indeed dependent upon risk region in the preconditioned group ($p = 0.04$; $n = 6$) but not in the controls ($p = 0.40$; $n = 7$).[31]

We recently performed a similar analysis using data from an ongoing study in the pig (MO, personal communication), a model characterized by low collateral flow and a propensity toward ventricular fibrillation. Despite the fact that risk region only varied from 9% to 26%, the results were once again in general agreement with those obtained in the rat: infarct size was not dependent upon risk region in control animals, but tended to increase with increasing risk regions in preconditioned pigs (Fig. 5.5).

5.5.3. Confirmation of "Preconditioning at a Distance"

While regional ischemia clearly exerts global consequences on the heart, and the protection achieved with preconditioning appears to be dependent upon area at risk, the ultimate question is: can the specific phenomenon of "preconditioning at a distance," predicted from the

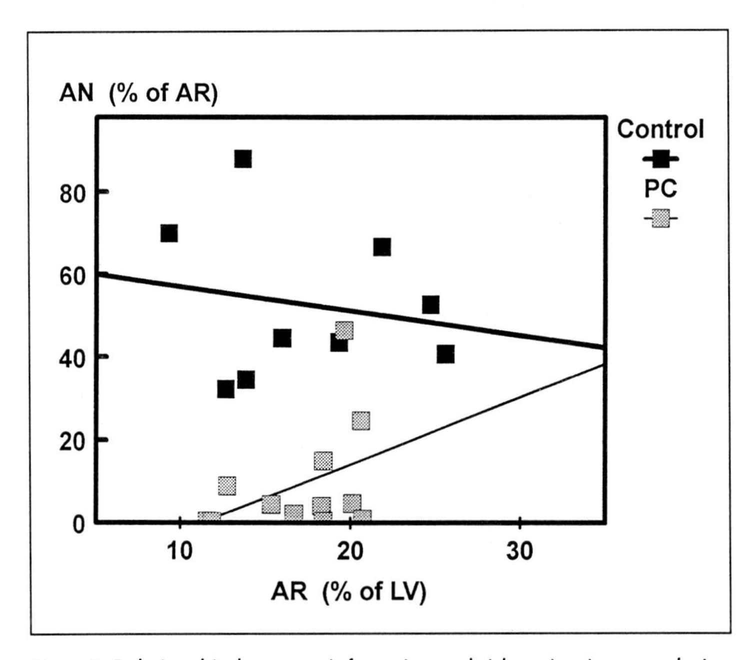

Fig. 5.5. Relationship between infarct size and risk region in control pigs subjected to 40 minutes of sustained coronary occlusion and pigs preconditioned with 1 brief 10 minute occlusion prior to the sustained ischemia. Infarct size (AN: expressed as a percent of the area at risk) tended to increase with increasing risk region (AR: expressed as a percent of the total left ventricle (LV)) in preconditioned pigs, but not in controls.

rat model and confirmed in the dog, be extrapolated to other models and species?

Only three preliminary reports have addressed this issue. Two laboratories have independently made the intriguing observation that, in rabbits[42] and rats,[43] one 10 minute episode of brief renal ischemia protected the rabbit heart from 30 minutes of coronary occlusion,[42] while 15 minutes of renal ischemia protected the rat heart from a subsequent 1 hour coronary occlusion.[43] Whether this "preconditioning at a *great* distance" is a novel phenomenon, or whether it shares a common mechanism with our circumflex preconditioning/LAD occlusion protocol, remains to be determined.

In the third study, Kremastinos and colleagues[44] specifically sought to determine whether, in the rabbit model, a brief 5 minute occlusion of one coronary artery would protect a different vascular bed from a sustained 45 minute ischemic insult. Intermediate results were obtained, with infarct sizes in control, conventionally preconditioned and remote

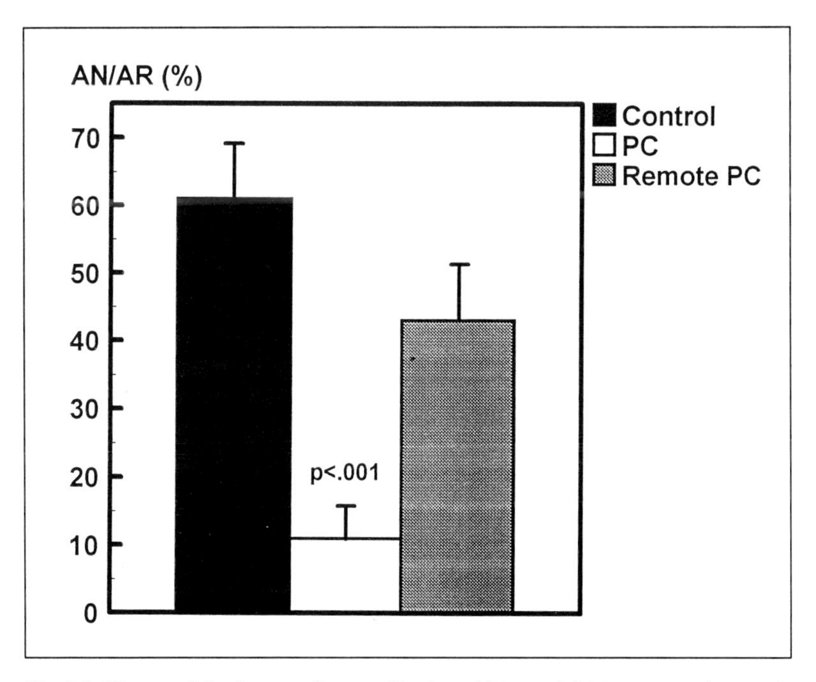

Fig. 5.6. "Preconditioning at a distance" in the rabbit model. Mean area of necrosis (AN: expressed as a percent of the area at risk (AR)) in control rabbits subjected to 45 minutes of sustained coronary occlusion, rabbits preconditioned (PC) with one 5 minute occlusion in the same vascular bed, and rabbits that received one 5 minute occlusion in a remote bed (remote PC). Conventional preconditioning significantly reduced infarct size, while remote preconditioning resulted in a nonsignificant trend toward a reduction in infarct size when compared with controls. Adapted from data presented in ref. 44.

preconditioned group averaging 60 ± 8%, 11 ± 4% and 44 ± 8% of the myocardium at risk, respectively (Fig. 5.6).[44] That is, although there was a trend toward a reduction in infarct size with remote preconditioning, the protection did not approach that achieved with conventional preconditioning.

Numerous technical considerations may contribute to the fact that these preliminary data are only in partial agreement with our observations in the dog model (i.e., one 5 minute ischemic stimulus may be insufficient to achieve protection in a remote vascular territory; results were obtained from small sample sizes of six per group). However, if myocyte stretch is indeed an important mediator of remote preconditioning, then these results obtained in the rabbit model are not surprising. Specifically, rabbit myocardium is highly collagenous (Fig. 5.7), with the percent area occupied by collagen averaging 4.7 ± 0.3% versus the values of 2.5 ± 0.2% and 3.5 ± 0.5% in the left ventricle of rat and dog ($F = 7.9$ and $p<0.01$ by analysis of variance: PW, unpublished observations). This extensive "scaffolding" of stiff and strong collagen in rabbit myocardium would be predicted to substantially limit the capacity for myocyte stretch in the setting of ischemia. In fact, Knowlton et al have reported that end-diastolic length measured within the ischemic territory during four 5 minute episodes of coronary occlusion in the rabbit increased to a maximum of only 104 ± 4% of baseline,[36] considerably less than the values of 120% and 110% observed in the LAD bed of the dog during brief LAD occlusion[33] and remote circumflex occlusions.[12] Thus, the data of Kremastinos and colleagues[44] may be interpreted as providing indirect support for the concept that myocyte stretch is an important mediator by which brief regional ischemia confers global myocardial protection.

5.6. CONCLUSION

Our initial observation that, in the rat, protection achieved with preconditioning is lost with increasing risk region has, as in the classic poem by Frost, lead us down "the road less traveled" in the ongoing efforts to elucidate the mechanisms responsible for ischemic preconditioning. We have, to date, made the observation that brief ischemia in one vascular bed of the canine heart can protect or precondition remote virgin myocardium. These data support the concept that preconditioning may be mediated by factor(s) activated, transported and/or produced throughout the heart, and imply that remote nonischemic tissue may contribute to the protection. In addition, we have taken the first steps in resolving the mechanism(s) of global protection by brief regional ischemia: myocyte stretch appears to play a pivotal role. The most daunting challenge, that of identifying the ion or subcellular messenger ultimately responsible for the reduction in infarct size seen with preconditioning, remains ahead.

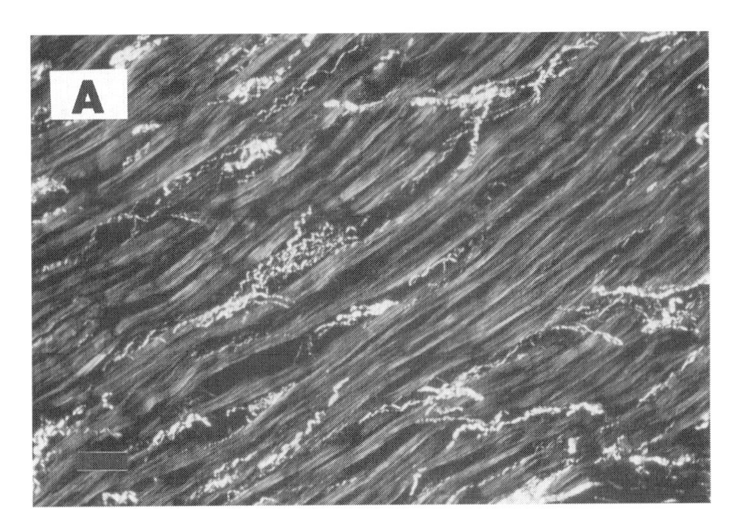

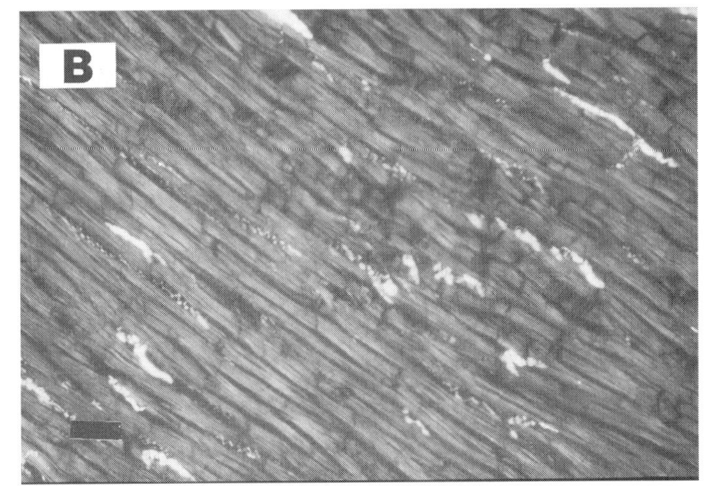

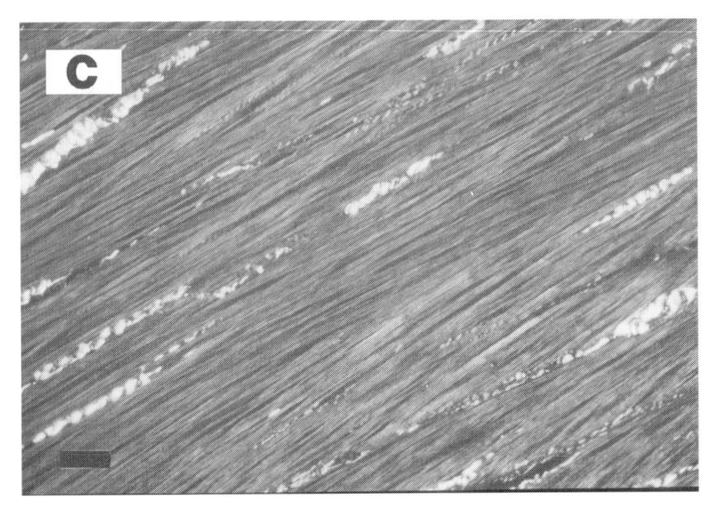

Fig. 5.7. Interspecies differences in myocardial collagen concentration. Histologic sections (5 microns thick) were obtained from the left ventricle of rabbit (A), rat (B) and dog hearts (C), stained with picrosirius red and viewed with polarized light. Cardiac muscle appears green, while the thick collagen fibers appear yellow/orange. Rabbit hearts have a higher collagen concentration than found in either rat or dog myocardium. Bar = 50 microns.

References

1. Murry CE, Jennings RB, Reimer KA. Preconditioning with ischemia: A delay of lethal cell injury in ischemic myocardium. Circulation 1986; 74:1124-1136.
2. Li GC, Vasquez JA, Gallagher KP et al. Myocardial protection with preconditioning. Circulation 1990; 82:609-619.
3. Gross GJ, Auchampach JA. Blockade of ATP-sensitive potassium channels prevents myocardial preconditioning in dogs. Circ Res 1992; 70:223-233.
4. Ovize M, Kloner RA, Hale SL et al. Coronary cyclic flow variations "precondition" ischemic myocardium. Circulation 1992; 85:779-789.
5. Li Y, Whittaker P, Kloner RA. The transient nature of the effect of ischemic preconditioning on myocardial infarct size and ventricular arrhythmia. Am Heart J 1992; 123:346-353.
6. Yellon DM, Alkhulaifi AM, Browne EE et al. Ischaemic preconditioning limits infarct size in the rat heart. Cardiovasc Res 1992; 26:983-987.
7. Iwamoto T, Miura T, Adachi T et al. Myocardial infarct size-limiting effect of ischemic preconditioning was not attenuated by oxygen free-radical scavengers in the rabbit. Circulation 1991; 83:1015-1022.
8. Liu GS, Thornton J, Van Winkle DM et al. Protection against infarction afforded by preconditioning is mediated by A_1 adenosine receptors in rabbit heart. Circulation 1991; 84:350-356.
9. Hale SL, Kloner RA. Effect of ischemic preconditioning on regional myocardial blood flow in the rabbit heart. Coronary Artery Dis 1992; 3:133-140.
10. Schott RJ, Rohman S, Braun ER et al. Ischemic preconditioning reduces infarct size in swine myocardium. Circ Res 1991; 66:1133-1142.
11. Kida M, Fujiwara H, Ishida M et al. Ischemic preconditioning preserves creatine phosphate and intracellular pH. Circulation 1991; 84:2495-2503.
12. Przyklenk K, Bauer B, Ovize M et al. Regional ischemic preconditioning protects remote virgin myocardium from subsequent sustained coronary occlusion. Circulation 1993; 87:893-899.
13. Bauer B, Simkhovich BZ, Kloner RA et al. Does preconditioning protect the coronary vasculature from subsequent ischemia/reperfusion injury? Circulation 1993; 88:659-672.
14. Bankwala Z, Hale SL, Kloner RA. α-Adrenoreceptor stimulation with exogenous norepinephrine or release of endogenous catecholamines mimics ischemic preconditioning. Circulation 1994; 90:1023-1028.
15. Kloner RA, Shook T, Przyklenk K et al. Previous angina alters in-hospital outcome in TIMI-4: A clinical correlate to preconditioning? Circulation 1995; 91:37-45.
16. Murry CE, Richard VJ, Jennings RB et al. Myocardial protection is lost before contractile function recovers from ischemic preconditioning. Am J Physiol 1991; 260: H796-H804.
17. Li Y, Kloner RA. Cardioprotective effects of ischemic preconditioning can be recaptured after they are lost. J Am Coll Cardiol 1994; 23:470-474.

18. Ovize M, Przyklenk K, Kloner RA. Partial stenosis is sufficient, and complete reperfusion is mandatory, to induce preconditioning. Circ Res 1992; 71:1165-1173.
19. Ovize M, Przyklenk K, Hale SL et al. Preconditioning does not attenuate myocardial stunning. Circulation 1992; 85:779-787.
20. Vegh A, Komori S, Szekeres L et al. Antiarrhythmic effects of preconditioning in anaesthetized dogs and rats. Cardiovasc Res 1992; 26:487-495.
21. Przyklenk K, Kloner RA. Preconditioning does not attenuate ventricular ectopy in the canine model. J Mol Cell Cardiol 1994; 26:137 (Abstract).
22. Li Y, Kloner RA. Cardioprotective effects of ischaemic preconditioning are not mediated by prostanoids. Cardiovasc Res 1992; 26:226-231.
23. Hale SL, Bellows SD, Hammerman H et al. An adenosine A_1 receptor agonist, R(-)-N-(2-phenylisopropyl)-adenosine (PIA), but not adenosine itself, acts as a therapeutic preconditioning-mimetic agent in the rabbit. Cardiovasc Res 1993; 27:2140-2145.
24. Li Y, Kloner RA. Adenosine deaminase inhibition is not cardioprotective in the rat. Am Heart J 1993; 126:1293-1298.
25. Whittaker P, Li Y, Kloner RA. Intramyocardial injections of adenosine and saline stimulate a cardioprotective effect in rats. J Am Coll Cardiol 1994; 23:232A (Abstract).
26. Hale SL, Kloner RA. Protection of myocardium by transient, preischemic administration of phenylephrine in the rabbit. Coronary Artery Dis 1994; 5:605-610.
27. Przyklenk K, Kloner RA. Low dose IV acetylcholine acts as a "preconditioning-mimetic" in the canine model. J Cardiac Surgery 1995; 10:389-395.
28. Li Y, Kloner RA. Does protein kinase C play a role in ischemic preconditioning in the rat heart? Am J Physiol 1995; 268:H426-H431.
29. Li Y, Kloner RA. Does inhibition of protein kinase C with H7 block the cardioprotective effects of ischemic preconditioning in the rat? Circulation 1994; 90(Suppl I):I-371 (Abstract).
30. Przyklenk K, Sussman MA, Simkhovich BZ et al. Does ischemic preconditioning trigger translocation of protein kinase C in the canine model? Circulation 1995; 92:1546-1557.
31. Whittaker P, Przyklenk K. Reduction of infarct size in vivo with ischemic preconditioning: Mathematical evidence for protection via non-ischemic tissue. Basic Res Cardiol 1994; 89:6-15.
32. Gallagher KP, Gerren RA, Stirling MC et al. The distribution of functional impairment across the lateral border of acutely ischemic myocardium. Circ Res 1986; 58:570-583.
33. Ovize M, Kloner RA, Przyklenk K. Stretch preconditions canine myocardium. Am J Physiol 1994; 266:H137-H146.
34. Yang XC, Sachs F. Block of stretch-activated ion channels in *Xenopus* oocytes by gadolinium and calcium ions. Science 1989; 243:1068-1071.
35. Whittaker P, Kloner RA, Przyklenk K. Intramyocardial injections and protection against myocardial ischemia: an attempt to explain the

cardioprotective effects of adenosine. Circulation 1996; 93:in press.
36. Knowlton AA, Brecher P, Apstein CS. Rapid expression of heat shock protein in the rabbit after brief cardiac ischemia. J Clin Invest 1991; 87:139-147.
37. Dorheim TA, Hoffman A, Van Wylen DGL et al. Enhanced interstitial fluid adenosine attenuates myocardial stunning. Surgery 1991; 110: 136-145.
38. Simkhovich BZ, Hale SL, Kloner RA. Effect of ischemic preconditioning on carnitine-linked metabolism in the rabbit heart. Coronary Artery Dis 1992; 3:141-147.
39. Simkhovich BZ, Hale SL, Ovize M et al. Ischemic preconditioning and long chain acyl carnitine in the canine heart. Coronary Artery Dis 1993; 4:387-392.
40. Ovize M, Aupetit JF, Riofol G et al. Preconditioning reduces infarct size but accelerates time to ventricular fibrillation in ischemic pig hearts. Am J Physiol 1995; 269:1172-1179.
41. Kitakaze M, Hori M, Morioka T et al. Infarct size-limiting effect of ischemic preconditioning is blunted by inhibition of 5'-nucleotidase activity and attenuation of adenosine release. Circulation 1994; 89: 1237-1246.
41a. Przyklenk K, Zhao L, Kloner RA et al. Monophosphoryl lipid A (MLA) is a "preconditioning-mimetic" in the canine model. Circulation 1995; 92 (Suppl I): I-388 (Abstract).
42. McClanahan TB, Nao BS, Wolke LJ et al. Brief renal occlusion and reperfusion reduces myocardial infarct size in rabbits. FASEB J 1993; 7:A176 (Abstract).
43. Gho BC, Shoemaker RG, van der Lee C et al. Myocardial infarct size limitation in rat by transient renal ischemia. Circulation 1994; 90:I-476 (Abstract).
44. Kremastinos DT, Bouris I, Papadopoulos C et al. Regional preconditioning does not protect remote areas in the collateral deficient species. Eur Heart J 1994; 15:553 (Abstract).

CHAPTER 6

Novel Approaches to Myocardial Preconditioning in Pigs

Ben C.G. Gho, Monique M.G. Koning, René L.J. Opstal, Eric van Klaarwater, Dirk J. Duncker and Pieter D. Verdouw

6.1. INTRODUCTION

The ability of myocardium to adapt to ischemic stress has already been documented in both experimental and clinical studies employing intermittent ischemia some 15 years ago. For instance, it was shown that changes in metabolic markers of myocardial ischemia were much less during the second of two consecutive, but identical coronary blood flow reductions in pigs.[1] Similarly, the second of two consecutive but identical atrial pacing stress tests in patients undergoing a cardiac catheterization for suspected coronary artery disease showed less severe signs of ischemia.[2] At that time, no attempt was made to elucidate the mechanism of the altered response during the second episode of ischemia as the purpose of these studies was to develop models of repeated reversible ischemia with reproducible changes in the various markers of ischemia. In such models it would then be possible to evaluate the effects of pharmacological interventions with the animal or patient serving as its own control. Several subsequent studies on the metabolic and functional consequences of multiple brief periods of ischemia also revealed the absence of cumulative losses in tissue levels of high energy phosphates and in regional contractile function, but these have not always been consistent findings.[3]

It was not until Murry et al showed some 10 years ago that infarct size was reduced from 29% to 7% of the area at risk when in dogs a

Myocardial Preconditioning, edited by Cherry L. Wainwright and James R. Parratt.

45 minute coronary artery occlusion was preceded by four sequences of 5 minute coronary artery occlusion and 5 minutes of reperfusion, that myocardial adaptation to ischemia started to receive wider attention.[4] The ability of brief periods of ischemia to increase the tolerance of the myocardium to the development of irreversible damage ("ischemic preconditioning") has now been shown to occur in a large number of species, including dogs, pigs, rabbits, and rats.[5] With regard to the protection afforded by ischemic preconditioning, several limitations were soon identified. For instance the protective effect of the preconditioning stimuli could not be shown when the duration of the sustained coronary artery occlusion lasted longer than 90 minutes.[4] Also, protection was lost when the duration between the preconditioning stimulus and the sustained occlusion period exceeded two hours.[6]

In the present chapter, we review the results of our studies on ischemic preconditioning in domestic pigs. A number of earlier observations have guided these studies. First, it has been well established that in dogs the relation between infarct size (IS) and the anatomical area at risk (AR) is linear but not proportional, i.e., the regression line describing the relation between IS and AR is linear but has a positive intercept on the AR-axis.[7] Such a relation implies that the IS/AR ratio, which is the index commonly used to express infarct size, is not a constant but depends on AR. As will be shown, this can severely limit the use of IS/AR in assessing infarct size and myocardial protection. Secondly, it is well established that most of the protective effect of a brief period of ischemia is lost when the reperfusion period separating the brief and sustained coronary artery occlusions exceeds two hours.[4] The relation between the extent of protection by ischemic preconditioning and the duration of the intervening reperfusion period is not well established, although some data suggest that the preconditioning stimulus starts to lose some of its efficacy already very early in the intervening reperfusion period.[8] Thirdly, the necessity of a period of reperfusion between the brief and sustained total coronary artery occlusions to elicit ischemic preconditioning is self-evident; otherwise the duration of the sustained coronary artery occlusion will be merely increased. It is less obvious, however, whether a partial coronary artery occlusion can lead to ischemic preconditioning without the need for an intervening reperfusion period. If myocardium can be preconditioned with a partial coronary artery occlusion, the question naturally arises whether this protection is equal for the inner and outer halves of the myocardium as it is well established that a coronary flow reduction affects perfusion of the epicardial layers less dramatically than of the endocardial layers with consequently more severe ischemia in the latter. Thus we assessed the transmural distribution of infarct size after a coronary artery was occluded for 60 minutes immediately following a period in which coronary blood flow was reduced to a fixed percentage of baseline. Finally, Ovize et al[9] have shown that stretch produced by acute volume loading preconditioned the myocardium without the need for ischemia suggesting that ischemia is not obligatory for the induction of preconditioning. We have extended these observations and investigated whether other nonischemic stimuli are capable of preconditioning the myocardium.

6.2. RELATION BETWEEN INFARCT SIZE AND AREA AT RISK IN CONTROL AND PRECONDITIONED PIGS

The effect of ischemic preconditioning is usually evaluated by the ratio of infarct size (IS) and the anatomical area at risk (AR). This index (IS/AR) can be used without restriction only when infarct size development occurs independent of the size of the area at risk. For anesthetized as well as conscious dogs, it has been shown that the linear relation between infarct size and area at risk has a positive intercept on the AR-axis.[7] Because dogs can have an extensive coronary collateral circulation, the positive intercept can be ascribed to collateral blood flow. Consequently, in this species infarct size correlates well with the amount of collateral blood flow to the area at risk. If collateral blood flow is indeed the cause of this positive intercept the relation between infarct size and area at risk should be proportional in pigs, a species with a negligible native coronary collateral circulation. In open-chest anesthetized pigs we have occluded the left anterior descending coronary artery or its branches at different locations, thereby intentionally creating a wide range in the size of the area at risk.[10,11] In 17 animals that underwent a 60 minute coronary artery occlusion we found that the relation between infarct size and area at risk was highly linear ($r = 0.99$, $P<0.001$) and could be described by $IS = 0.97AR - 4.5$, in which IS and AR are both expressed as percent of left ventricular mass (Fig. 6.1, upper panel). Thus, as with findings in dogs, we also observed a positive AR-intercept in a species with minimal coronary collaterals. In support of our findings, Ytrehus et al[12] also recently showed that, in rabbits, the relation between infarct size and area at risk can be described by a linear regression equation with a positive intercept on the AR-axis.

In a subsequent series of experiments we investigated how the relation between infarct size and area at risk is modified by ischemic preconditioning. For this purpose animals were preconditioned with a 10 minute total coronary artery occlusion which was separated from the 60 minute occlusion by 15 minutes of reperfusion.[10,11] In these preconditioned animals the linear relation between infarct size and area at risk was rotated downwards with a lower slope but with a similar positive intercept on the AR-axis as in the control animals ($IS = 0.64AR-4.4$: Fig. 6.1, upper panel). The implication of these findings is that in both control and preconditioned animals IS/AR depends on AR (Fig. 6.1, lower panel). From this figure it can be deducted that, for areas at risk less than 20% of the left ventricular mass in particular, IS/AR is highly sensitive to the size of the area at risk. Figure 6.2 illustrates the infarct size limitation, defined as the difference in infarct sizes (predicted from the regression lines) of the control and the preconditioned animals expressed as a percent of the infarct size of the control animals $[(IS_{control} - IS_{preconditioned})/IS_{control}]$. This figure clearly

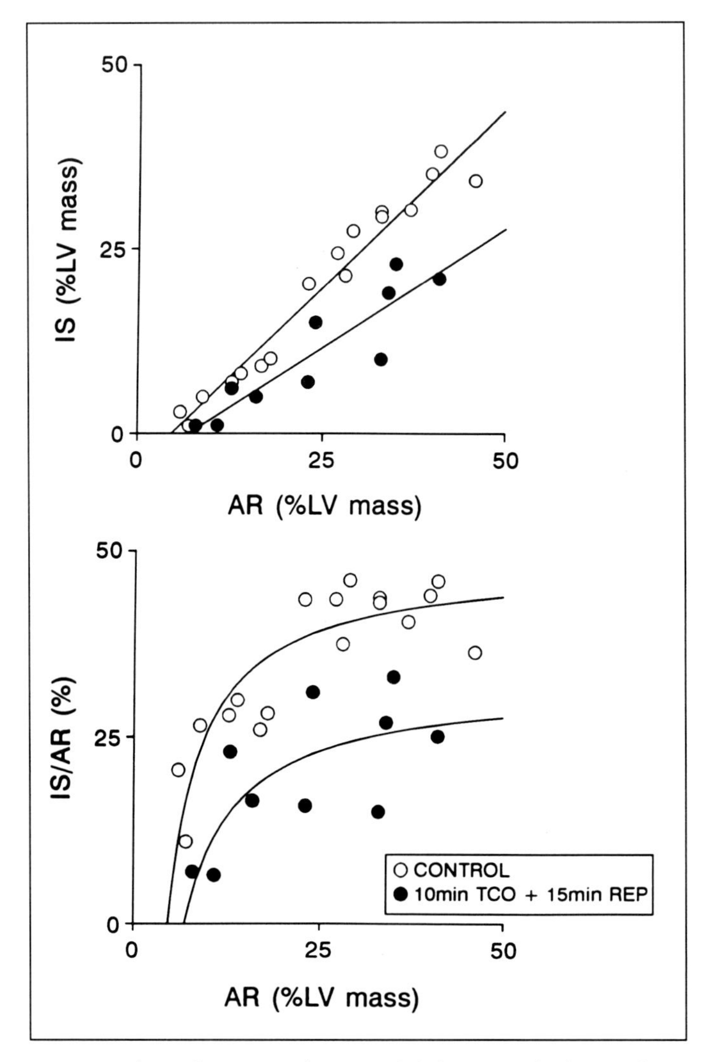

Fig. 6.1. Relation between infarct size (IS) determined 2 hours after a 60 minute total coronary artery occlusion (60 minute TCO), and the area at risk (AR) for a group of control animals and a group of animals preconditioned by a 10 minute total coronary artery occlusion which was separated from the 60 minute TCO by 15 minutes of complete reperfusion (upper panel). Because the relation between IS and AR has a positive intercept on the AR-axis (upper panel), the IS/AR strongly depends on AR for AR less than 20% of the left ventricular mass (LVmass) (lower panel). Adapted from Koning et al.[10,11,16]

demonstrates that an interstudy comparison of the protective effect of ischemic preconditioning can easily lead to erroneous conclusions when different areas at risk have been used. This may be especially true for the pig in which (in order to reduce the incidence of ventricular fibrillation) smaller areas at risk are quite often preferred.

6.3. TRANSMURAL DISTRIBUTION OF INFARCT SIZE IN CONTROL AND PRECONDITIONED PIGS

In order to assess whether the endocardial and the epicardial halves benefit equally from ischemic preconditioning, we divided the myocardium in two layers of equal thickness and related these infarcts to the corresponding areas at risk. The results revealed linear relations with similar positive intercepts on the AR-axis, for the endocardial and epicardial halves of the left ventricle in control pigs subjected to a 60 minute total coronary artery occlusion (Fig. 6.3). In dogs, IS and AR are also linearly related but, in contrast to pigs, the AR-intercept in dogs is considerably higher in the epicardium than in the endocardium.[13] The consequent heterogeneity of transmural infarct size distribution in dogs is likely in part due to the transmural gradient of collateral blood flow in the dog heart. In pigs, in which total coronary artery occlusions result in transmurally homogeneous myocardial blood flow reductions, infarction also progresses from inner to outer layers,[14,15] possibly due to higher energy demands in the inner layers. In agreement with these findings we observed that for a given area at risk the infarct size produced by the 60 minute total coronary artery occlusion was slightly, but significantly, larger in the endocardial than in the epicardial half of the left ventricle.[16]

In the pigs preconditioned by a 10 minute total coronary occlusion and 15 minute reperfusion, the relation between infarct area and area at risk of the endocardial and the epicardial halves was also linear. The AR-intercepts were the same as in the control animal but with lower slopes. The decrease in slope was the same for the endo- and epicardium indicating that the protection was similar in the inner and outer half of the left ventricle.[16]

6.4. THE DURATION OF PROTECTION AFFORDED BY ISCHEMIC PRECONDITIONING

To determine the time course of protection we prolonged the intervening reperfusion period between the brief and sustained coronary artery occlusions from 15 minutes to 60 minutes. We hypothesized that if the preconditioning stimulus already starts to lose its efficacy early in the intervening reperfusion period, the regression line relating infarct size and area at risk would gradually rotate upwards towards the control line when the duration of the intervening reperfusion period is extended. With the 60 minute reperfusion period the relation between infarct size and the area at risk remained virtually unchanged

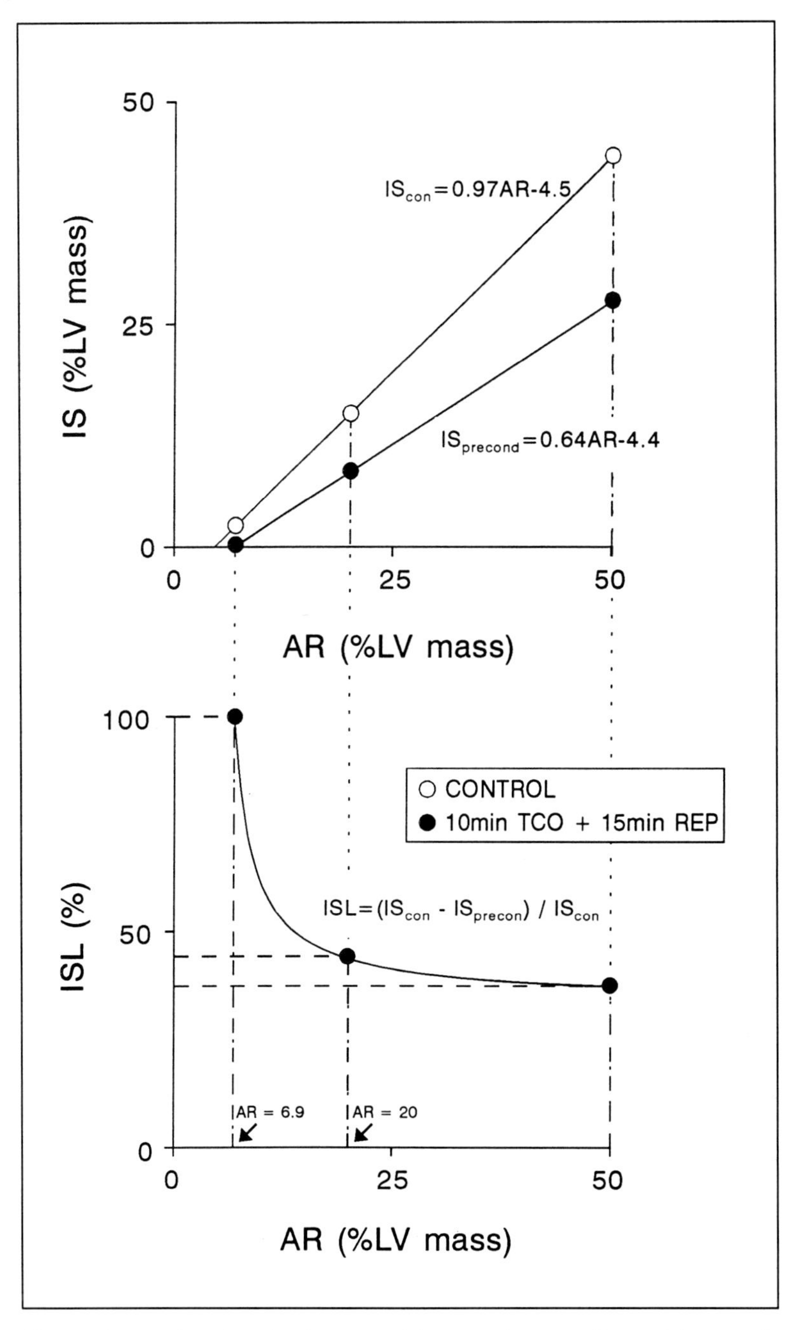

Fig. 6.2. *Regression lines for the control (con) and preconditioned (precond) pigs depicted in Figure 6.1 are shown in the upper panel. The lower panel shows the infarct size limitation (ISL) afforded by the 10 minute TCO + 15 minute REP preconditioning stimulus. ISL (%) was computed as* $(IS_{con} - IS_{precond})/IS_{con}$*, where* IS_{con} *and* $IS_{precond}$ *are the infarct size (IS) of the control group and the preconditioned group, respectively.*

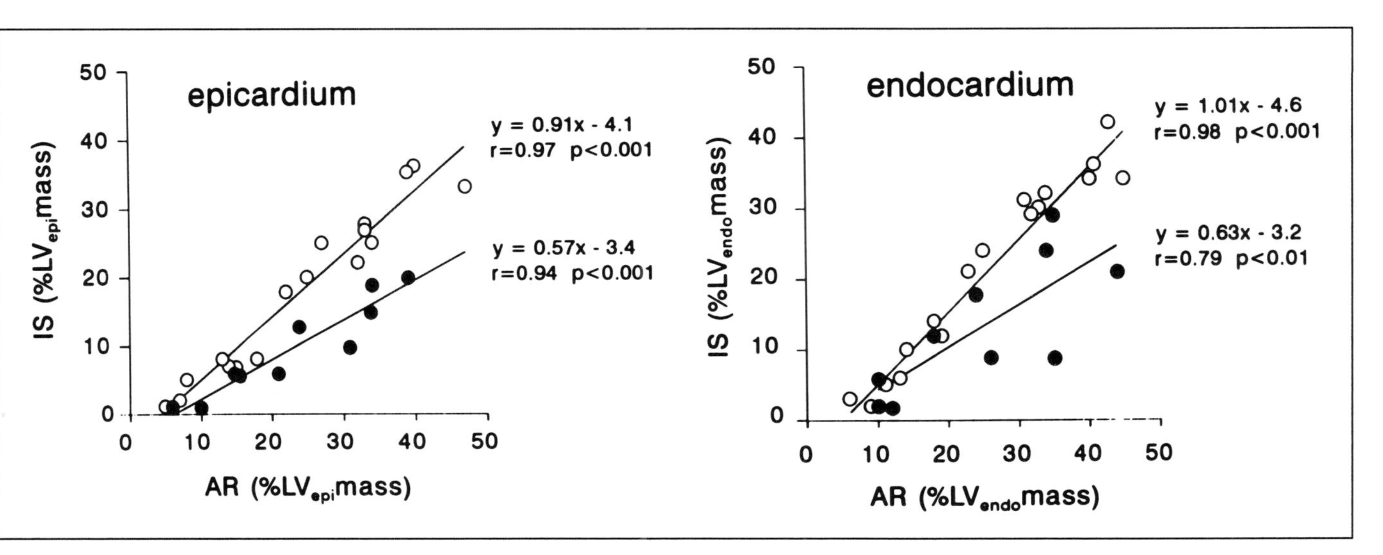

Fig. 6.3. Transmural distribution of infarct size (IS) for control pigs (○, 60 minute TCO only) and for pigs preconditioned by 10 minute TCO + 15 minute REP prior to the 60 minute TCO (●). Note that the infarct size reduction was similar in the epi- and endocardial halves of the left ventricular wall, but that the scatter of the data points around the regression line was considerably less in the epicardial than in the endocardial half of the preconditioned animals. Adapted from Koning et al.[10,11,16]

compared to the 15 minute reperfusion period, indicating that in pigs a 10 minute preconditioning stimulus does not start to lose its efficacy during the first hour of reperfusion.[10] A further prolongation of the intervening reperfusion period up to 2 hours revealed that the infarct size of most animals was no longer different from the control animals. However, the infarct size of a few animals remained well below the regression line of the control animals.[10] Because in the control animals all IS-AR data points are very close to the regression line it appears that the preconditioned animals with the IS-AR data points well below the regression line of the control group were still preconditioned. Infarct size determinations for animals with even longer intervening reperfusion periods showed that a few animals, although a progressively smaller percentage, remained preconditioned up to 5 hours after the brief period of ischemia. It thus appears that in contrast to the earlier observations in rabbits:[8] (1) the efficacy of preconditioning in pigs remains virtually unchanged during the first hour; (2) most pigs have lost protection by ischemic preconditioning at two hours; and (3) a few pigs may remain protected up to five hours.

6.5. ISCHEMIC PRECONDITIONING WITH A PARTIAL CORONARY ARTERY OCCLUSION WITHOUT INTERVENING REPERFUSION

In all "classical" preconditioning experiments unimpeded coronary blood flow was restored after a brief total occlusion before the coronary artery was occluded for a sustained period of time. This period of reperfusion is obviously necessary because otherwise the duration of sustained ischemia will merely be increased. A different situation may arise if myocardium could be preconditioned by a partial coronary artery occlusion. Harris was the first who applied a partial occlusion for 30 minutes (without measuring residual coronary artery blood flow) prior to a complete occlusion which lasted several hours (two-stage coronary artery occlusion) and observed a decrease in the high incidence of ventricular fibrillation usually occurring during the first 30 minutes after the onset of a total coronary artery occlusion.[17] This reduction in the incidence of ventricular fibrillation has been explained by the lesser severity of ischemia during the partial occlusion period. The onset of the subsequent total occlusion (at 30 minutes) was considered beyond the time-point when most lethal arrhythmias occur. It was never considered why these lethal arrhythmias would *not* occur after the artery was occluded completely. Current insight into the effects of ischemic preconditioning[18] offers an alternative hypothesis: the partial occlusion preconditions the myocardium thereby attenuating the incidence of ventricular fibrillation during the subsequent total occlusion. The first attempt to induce myocardial preconditioning by a two-stage coronary artery occlusion was by Ovize et al, who reduced coronary blood flow by 50% for 15 minutes before occluding the coronary artery completely without an intervening reperfusion period between the flow reduction and the complete coronary artery occlusion.[19] Infarct size in these animals was not different from that of a group of control animals which underwent only the sustained total coronary artery occlusion. Increasing the duration of the flow reduction to 25 minutes did not affect the results. Infarct size was markedly reduced, however, when these investigators allowed 10 minutes of complete reperfusion between the 15 minute 50% coronary flow reduction and the sustained artery occlusion. The authors therefore concluded that myocardium could be preconditioned by a partial coronary artery occlusion but that an intervening period of reperfusion was mandatory to elicit the protective action.

We hypothesized that in the study of Ovize et al[19] the flow reduction may not have been severe enough, or that the duration of the flow reduction was too short, for preconditioning to occur. We therefore first reduced blood flow in the left anterior descending coronary by 70% for 30 minutes before occluding the artery for 60 minutes without an intervening reperfusion period. Figure 6.4 shows that all animals which underwent this two-stage coronary artery occlusion had transmural infarct sizes smaller than in the control group. When the duration of the 70% flow reduction period was extended from 30 minutes to 90 minutes before the coronary artery was occluded completely, transmural infarct sizes were still smaller than in the control groups, even though the 90 minute 70% flow reduction per se had already caused some irreversible damage. In contrast, a 30 minute 30% flow reduction preceding the 60 minute total coronary artery occlusion without an intervening period of reperfusion did not limit infarct size.[16] Findings were also negative when the duration of the 30% flow reduction was increased to 90 minutes. It therefore appears that myocardium can be preconditioned by a flow reduction without the need for an intervening reperfusion period, but that the severity of flow reduction is critical.

We also hypothesized that the protection by ischemic preconditioning would most likely not be homogeneously distributed when a partial coronary artery occlusion was used as a stimulus. Because the flow deficit in the endocardial layers would be greater, ischemia would be more severe there than in the epicardial layers. If, as has been proposed, there is a threshold for ischemia before preconditioning can occur, then with moderate flow reductions the endocardium could be preconditioned whilst the epicardium would only be exposed to a subthreshold stimulus. On the other hand, the subendocardium may already start to develop irreversible damage because of the larger flow deficit,[19] as was observed when the coronary artery flow was reduced by 70% for 90 minutes. When we analyzed the endocardial (inner 50%) and the epicardial (outer 50%) infarct size we found that in the animals in which the induction of preconditioning was attempted by

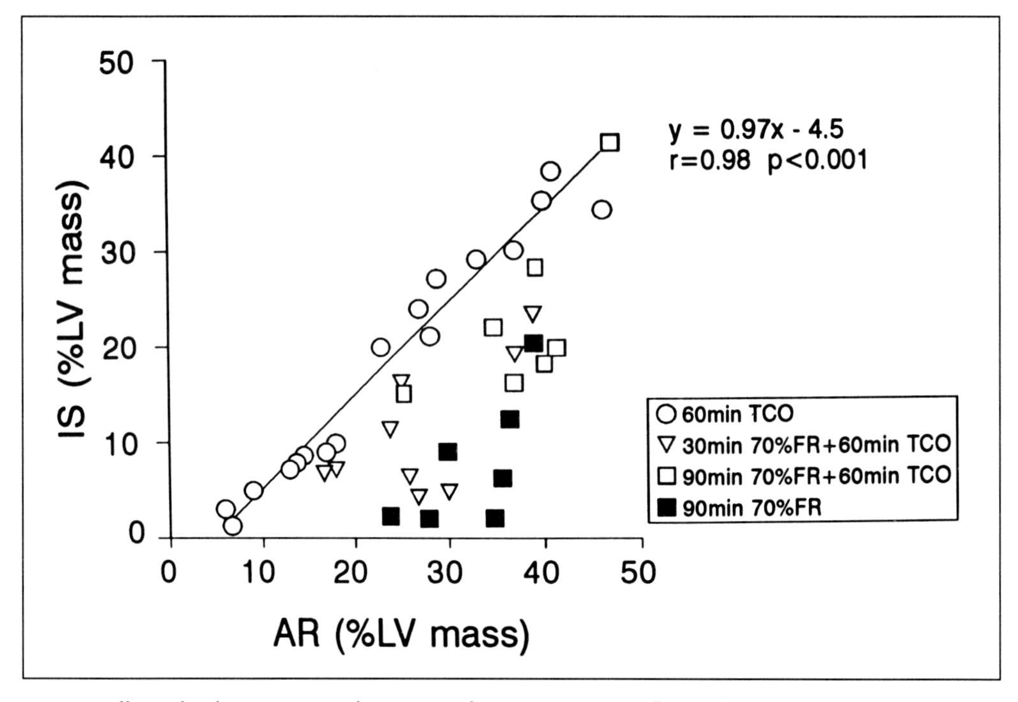

Fig. 6.4. Effect of ischemic preconditioning with a 70% coronary flow reduction (70% FR) without an intervening period of reperfusion on myocardial infarct size produced by a 60 minute total coronary artery occlusion (60 minute TCO). Shown are the regression line and individual data points for the control group (60 minute TCO) and individual data points for animals subjected to a 30 minute 70% FR followed by 60 minute TCO, a 90 minute 70% FR followed by 60 minute TCO and a 90 minute 70% FR without the 60 minute TCO. Note that although the 90 minute 70% FR itself resulted in significant infarction after 90 minutes, it still provided protection against the irreversible damage produced by the subsequent 60 minute TCO. Adapted from Koning et al.[16]

a 30% flow reduction, neither the endocardial nor the epicardial halves were protected.[16] In the animals subjected to the 30 minute 70% flow reduction, infarcts were smaller in the epicardial half than in the endocardial half, but both were less than the infarcts in the corresponding halves of the control animals (Fig. 6.5). The 90 minute 70% flow reduction caused some irreversible damage, but this occurred predominantly in the endocardial region. Nevertheless, the endocardial infarcts after the subsequent 60 minute total coronary artery occlusion were still smaller than in the control group, although the difference was less prominent than in the epicardium. Thus, in contrast to classical preconditioning with a total coronary artery occlusion (which in our hands affords transmurally homogeneous protection), partial occlusions produced preferential epicardial protection. These findings may be of clinical relevance, since a large group of patients suffering from myocardial infarction will have pre-existing coronary artery lesions which may provide variably flow reductions prior to the occlusion by a thrombus.

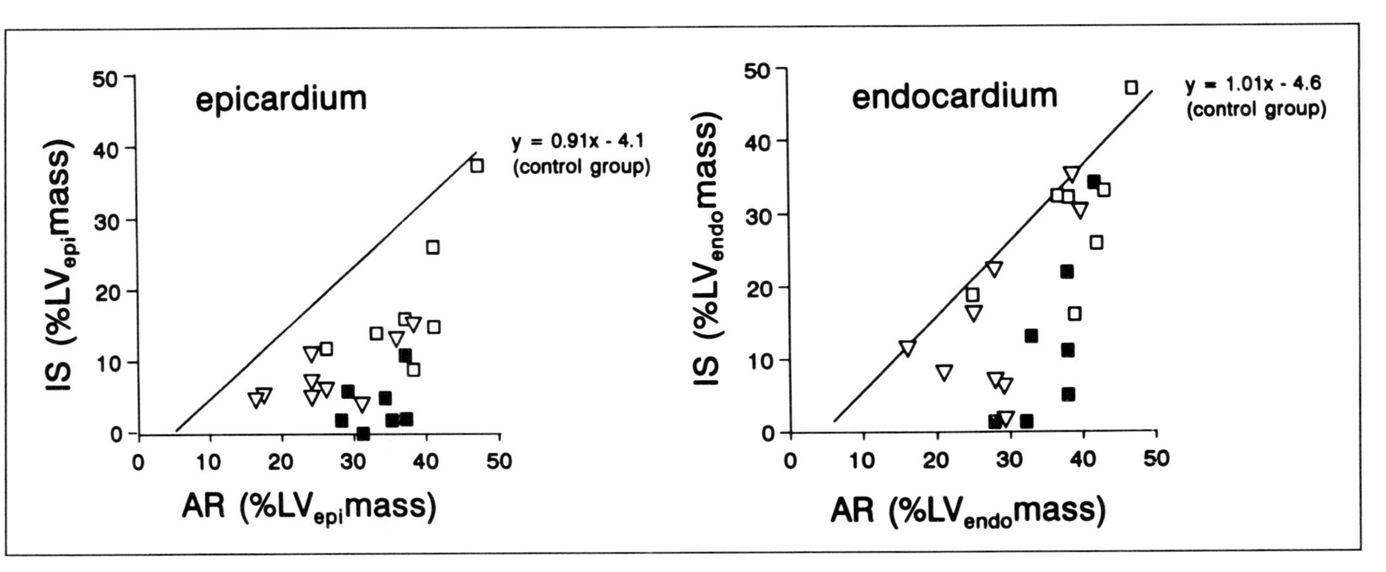

Fig. 6.5. Effect of ischemic preconditioning with a 70% coronary flow reduction (70% FR) without an intervening period of reperfusion on the transmural distribution of myocardial infarct size produced by a 60 minute total coronary artery occlusion (60 minute TCO). Shown are the regression line and individual data points for the control group (60 minute TCO) and individual data points for animals subjected to a 30 minute 70% FR followed by 60 minute TCO, a 90 minute 70% FR followed by 60 minute TCO and a 90 minute 70% FR without the 60 minute TCO. For details of the legends see Figure 6.4. Adapted from Koning et al.[16]

6.6. MYOCARDIAL PROTECTION BY A PERIOD OF RAPID VENTRICULAR PACING

Przyklenk et al have shown that myocardium can be preconditioned by a brief coronary artery occlusion supplying the adjacent myocardium demonstrating that myocardium does not have to become ischemic itself to increase its tolerance to irreversible ischemic damage.[20] This protection could be due to stretch of this virgin myocardium in response to the ischemic dysfunction in the adjacent area. Further support for this hypothesis came from the same group of investigators when they showed that stretching the myocardium by volume loading also attenuated infarct size during a sustained coronary artery occlusion and without the prerequisite of ischemia.[9] The protection by "stretch" could be blocked by pretreatment with gadolinium, an inhibitor of stretch-activated cation channels (chapter 5). In a search for preconditioning the myocardium by other types of stress we have used rapid ventricular pacing. The rationale for this approach is the observation by Vegh et al that rapid ventricular pacing reduced the incidence of ventricular fibrillation during and after subsequent coronary artery occlusion.[21]

In four animals we found that 10 minutes of rapid ventricular pacing at 200 bpm separated from a 60 minute total coronary artery occlusion by 15 minutes of normal sinus rhythm did not limit infarct size. When, in six other animals, the duration of rapid ventricular pacing was increased to 30 minutes there was a small but significant limitation of infarct size in the epicardial half but not in the endocardial half.[22] A much more striking effect was observed when the intervening period of normal sinus rhythm was abolished. Thus, when at the end of a 30 minute period of rapid ventricular pacing, the left anterior descending coronary artery was occluded and the pacemaker immediately switched off, infarct size in both the epicardial and the endocardial halves was significantly reduced compared to that of the control group (Fig. 6.6 and Fig. 6.7). Pretreatment with glibenclamide abolished the protection by ventricular pacing in all but one of the animals, suggesting a role for K^{+}_{ATP} channel activation analogous to the preconditioning by a brief total coronary artery occlusion in pigs.[23-25] It would seem reasonable to assume that ventricular pacing-induced ischemia played a major role in eliciting the protection, since several groups of investigators have shown that ventricular pacing produces changes in the electrocardiogram similar to those observed with ischemia.[21,26] We therefore evaluated the effects of 30 minutes of ventricular pacing on global and regional cardiac performance.[22] In a separate group of pigs we established that ventricular pacing (1) did not affect transmural blood flow and its distribution of the transmural layers; (2) did not widen the arterial coronary venous differences in pH and pCO_2; (3) left high energy phosphate levels and energy charge unchanged; (4) did not cause coronary vasodilation during pacing; (5) did

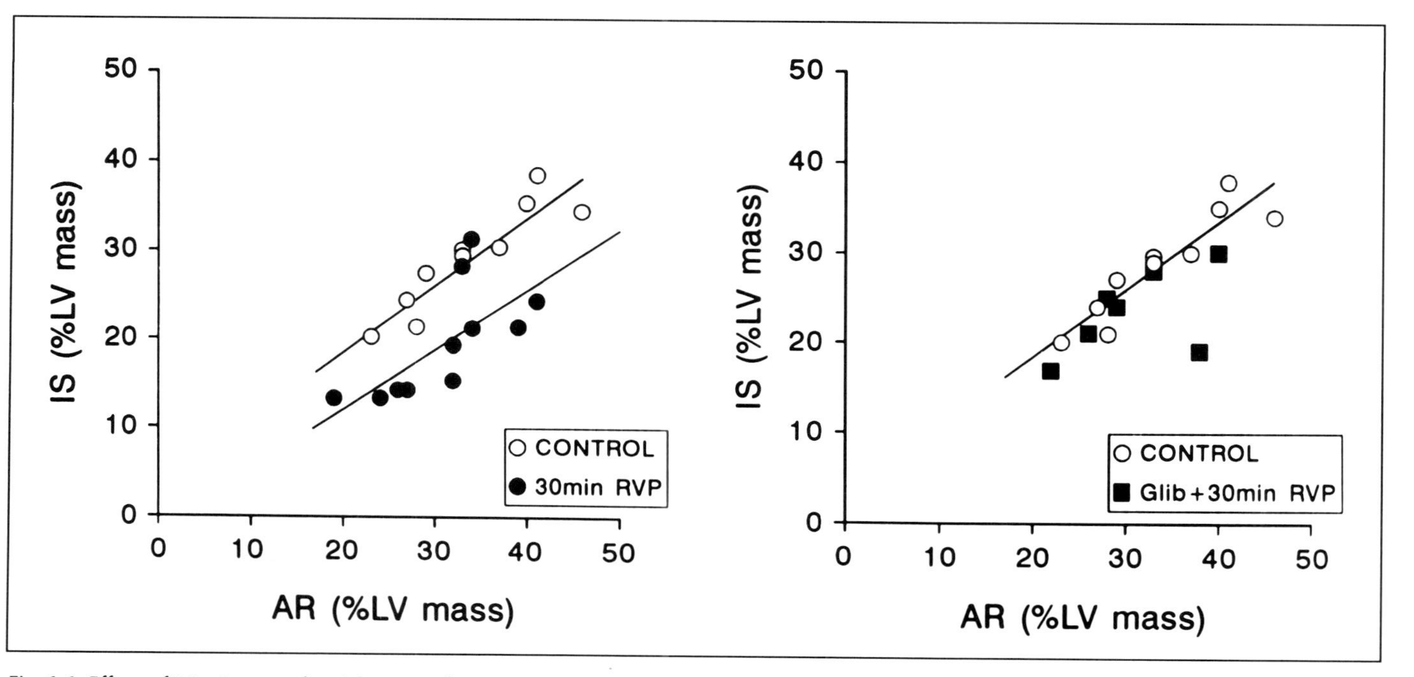

Fig. 6.6. Effect of 30 minutes of rapid ventricular pacing at 200 bpm (30 minute RVP) on myocardial infarct size produced by a 60 minute total coronary artery occlusion. Ventricular pacing was stopped after the onset of the coronary occlusion (left panel). The panel on the right shows that the protective effect of 30 minute RVP was abolished when the 30 minute RVP was performed in the presence of the K^+_{ATP} channel inhibitor glibenclamide (Glib, 1 mg/kg, iv). Adapted from Koning et al.[22]

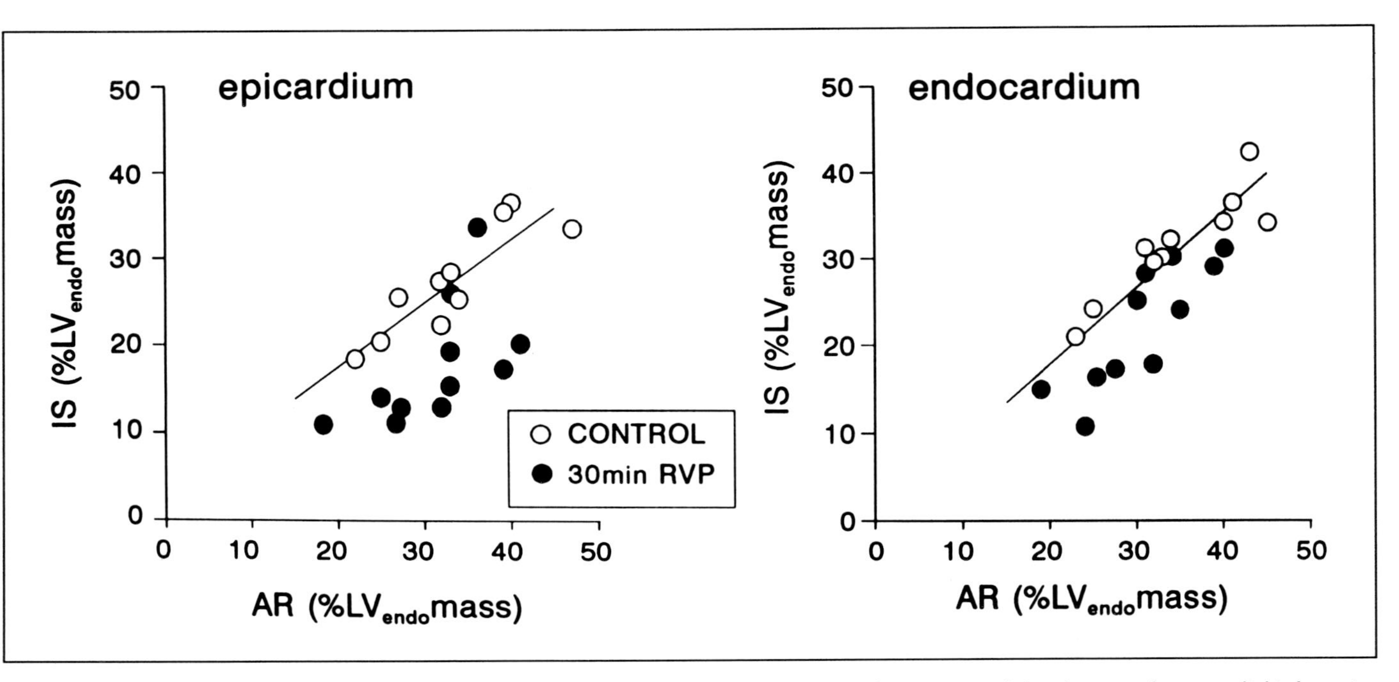

Fig. 6.7. Effect of 30 minutes of rapid ventricular pacing at 200 bpm (30 minute RVP) on the transmural distribution of myocardial infarct size produced by a 60 minute total coronary artery occlusion. Ventricular pacing was stopped after the onset of the coronary occlusion. Note that the protective effect on the epicardial half (left panel) was more pronounced than on the endocardial half (right panel). Adapted from Koning et al.[22]

not result in reactive hyperemia after pacing was stopped; and (6) did not prevent the immediate return of regional cardiac function after pacing was stopped. We feel that these findings fail to indicate the presence of myocardial ischemia. Even if some subendocardial ischemia had gone undetected, it would not have been severe enough to precondition myocardium via a pathway involving ischemia. This conclusion is based on our findings that a 30% reduction in coronary blood flow does not lead to myocardial protection,[16] but produces significant metabolic and functional signs of ischemia.

6.7. CONCLUSIONS

In the present chapter we have presented evidence that care must be taken when the ratio of infarct size and area at risk is used in assessing the amount of protection afforded by ischemic preconditioning, because the value of this ratio depends on the area at risk itself. We have also shown that not only total coronary artery occlusions but also partial coronary artery occlusions can precondition the myocardium. In this model there is no need for an intervening reperfusion period when the flow reduction by the partial coronary artery occlusion is severe enough. The protection produced by this stimulus is considerably greater in the epicardial half than in the endocardial half, which contrasts with the homogeneously distributed protection by a brief total coronary artery occlusion. We have further shown that rapid ventricular pacing is able to precondition myocardium, although the "memory" for protection induced by rapid ventricular pacing is much shorter than that of a brief coronary artery occlusion. Activation of K^+_{ATP} channels by a pathway not involving ischemia appears to play a major role in the mechanism responsible for the pacing-induced protection. Because myocardium can also be preconditioned by other forms of nonischemic stress such as stretch and possibly by transient ischemia occurring in other organs such as the kidney,[27,28] it appears that ischemic precondition is only one mode of stress that protects myocardium against irreversible damage by a sustained coronary artery occlusion. It is premature, however, to conclude that all forms of stress will precondition the myocardium as, for instance, chronic exposure to tobacco smoke will increase infarct size produced by a coronary artery occlusion.[29]

ACKNOWLEDGMENT

This study was supported by grant 92.144 of the Netherlands Heart Foundation and grant CIPA-CT-92-4009 of the European Economic Community.

The authors gratefully acknowledge the excellent secretarial assistance of Mrs. Dineke de Bruyn.

References

1. Verdouw PD, Remme WJ, De Jong JW et al. Myocardial substrate utilization and hemodynamics following repeated coronary flow reduction in pigs. Basic Res Cardiol 1979; 74:477-493.
2. Jackson G, Atkinson L, Oram S. Improvement of myocardial metabolism in coronary artery disease by beta-blockade. Br Heart J 1977; 39:829-833.
3. Reimer KA, Jennings RB. Myocardial ischaemia, hypoxia and infarction. In: Fozzard HA, Haber E, Jennings RBet al, eds. The Heart and Cardiovascular System. Vol II. New York: Raven Press 1992:1875-1973.
4. Murry CE, Richard VJ, Jennings RB et al. Preconditioning with ischaemia: A delay of lethal cell injury in ischaemic myocardium. Circulation 1986; 74:1124-1136.
5. Lawson CS, Downey JM, Preconditioning: state of the art myocardial protection. Cardiovasc Res 1993; 27:542-550.
6. Murry CE, Richard VJ, Jennings RB et al. Myocardial protection is lost before contractile function recovers from ischemic preconditioning. Am J Physiol 1991; 260:H796-H804.
7. Jugdutt BI. Different relations between infarct size and occluded bed size in barbiturate-anesthetized versus conscious dogs. J Am Coll Cardiol 1985; 6:1035-1046.
8. Miura T, Ogawa T, Iwamoto T et al. Infarct size limiting effect of preconditioning: It's duration and "dose-response" relationship. Circulation 1990; 82 (Suppl.III):271 (Abstract).
9. Ovize M, Kloner RA, Przyklenk K. Stretch preconditions canine myocardium. Am J Physiol 1994; 266:H137-H146.
10. Koning MMG, De Zeeuw S, Nieukoop, S et al. Is myocardial infarct size limitation by ischemic preconditioning an "all or nothing" phenomenon? In: Das DK, ed. Cellular, Biochemical and Molecular Aspects of Reperfusion Injury. Annals of the New York Academy of Sciences 1994; 723:336-336.
11. Koning MMG, Simonis LAJ, De Zeeuw S et al. Ischaemic preconditioning by partial occlusion without intermittent reperfusion. Cardiovasc Res 1994; 28:1146-1151.
12. Ytrehus K, Cohen MV, Downey J. Volume of risk zone influences infarct size in rabbits and may account for unexplained variability in infarction studies. Circulation 1993; 88:I137 (Abstract).
13. Koyanagi S, Eastham CL, Harrison DG et al. Transmural variation in the relationship between myocardial infarct size and risk area. Am J Physiol 1982; 242:H867-H874.
14. Fujiwara H, Ashraf M, Sato S et al. Transmural cellular damage in blood flow distribution in early ischaemia in pig hearts. Circ Res 1982; 51:683-693.
15. Klein HH, Schubothe M, Kreuzer H. Temporal and spatial development of infarcts in procine hearts. Basic Res Cardiol 1984; 79:440-447.
16. Koning MMG, Gho BCG, Van Klaarwater E et al. Endocardial and epicardial infarct size after preconditioning by a partial coronary artery

occlusion without intervening reperfusion. Importance of the degree and duration of flow reduction. Cardiovasc Res (in press).

17. Harris AS. Delayed development of ventricular ectopic rhythms following experimental coronary occlusion. Circulation 1950; 1:1318-1328.
18. Shiki K, Hearse DJ. Preconditioning of ischemic myocardium: reperfusion induced arrhythmias. Am J Physiol 1987; 253:H1470-H1476.
19. Ovize M, Przyklenk K, Kloner RA. Partial coronary stenosis is sufficient and complete reperfusion is mandatory for preconditioning the canine heart. Circ Res 1992; 71:1165-1173.
20. Przyklenk K, Bauer B, Ovize M et al. Regional ischemic "preconditioning" protects remote virgin myocardium from subsequent sustained coronary occlusion. Circulation 1993; 87:893-899.
21. Vegh A, Szekeres L, Parratt JR. Transient ischaemia induced by rapid cardiac pacing results in myocardial preconditioning. Cardiovasc Res 1991; 25:1051-1053.
22. Koning MMG, Gho BCG, Van Klaarwater E et al. Rapid ventricular pacing produces myocardial protection by non-ischaemic activation of K^+_{ATP} channels. Circulation (in press).
23. Rohmann S, Weygandt H, Schelling P et al. Effect of Bimakalim (EMD 52692), an opener of ATP-sensitive potassium channels, on infarct size, coronary blood flow, regional wall function, and oxygen consumption in swine. Cardiovasc Res 1994; 28:858-863.
24. Rohmann S, Weygandt H, Schelling P et al. Involvement of ATP-sensitive potassium channels in preconditioning protection. Basic Res Cardiol 1994; 89:563-576.
25. Schulz R, Rose J, Heusch G. Involvement of activation of ATP-dependent potassium channels in ischemic preconditioning in swine. Am J Physiol 1994; 267:H1341-H1352.
26. Szilvassy Z, Ferdinandy P, Bor P et al. Ventricular overdrive pacing-induced anti-ischemic effect: a conscious rabbit model of preconditioning. Am J Physiol 1994; 266:H2033-H2041.
27. McClanahan TB, Nao BS, Wolke LJ et al. Brief renal occlusion and reperfusion reduces myocardial size in rabbits. The FASEB Journal 1993; 7:A118,682.
28. Gho BCG, Schoemaker RG, Van der Lee C et al. Cardioprotection by transient renal ischemia, an in vivo study in rats. Circulation 1994; 90 (Suppl.):I-476 (Abstract).
29. Zhu B, Sun Y, Sievers RE et al. Exposure to environmental tobacco smoke increases myocardial infarct size in rats. Circulation 1994; 89:1282-1290.

CHAPTER 7

Preconditioning in the Human Heart: Fact or Fantasy?

Clive S. Lawson

7.1. INTRODUCTION

The routine management of acute myocardial infarction was revolutionized in the 1980s by the widespread introduction into clinical practice of safe and effective forms of reperfusion therapy. The central importance of reperfusion in any strategy for limiting infarct size has long been recognized in experimental studies. It was not, however, until several large clinical trials had demonstrated that thrombolysis could both achieve reperfusion and, more importantly, reduce mortality[1,2] that such techniques became widely used. More recent studies have shown that direct balloon angioplasty of the acutely occluded coronary vessel has been shown to be even more effective,[3-5] presumably due to reperfusion being achieved more rapidly and more reliably than with thrombolysis. Despite these major advances ischemic heart disease, in its various forms, is still responsible for more deaths in the developed world than any other disease process.

As would be expected from experimental studies, the clinical efficacy of reperfusion therapy declines as the time between the onset of infarction and the initiation of treatment increases. Although clinical studies suggest that reperfusion therapy can be effective even when initiated up to 24 hours after the onset of infarction,[6,7] it is rare for reperfusion to be achieved within the 20-30 minutes of the onset of ischemia that experimental studies would suggest is required to prevent

Myocardial Preconditioning, edited by Cherry L. Wainwright and James R. Parratt.

infarction completely. If, however, the development of ischemic injury could be slowed or delayed, more myocardium could be salvaged by reperfusion therapy initiated at a later time. There is a clear clinical need, therefore, for an effective treatment that slows ischemic injury to use as an adjunct to reperfusion therapy.

Experimental results have suggested that many drugs may have potential to slow the development of acute ischemic damage but only beta-blockers have been shown to reduce mortality in clinical trials.[8] Even with beta-blockade it is far from established that this is due to infarct size reduction (which does not occur in well controlled experimental studies[9]) rather than other mechanisms (such as arrhythmia suppression[10]).

In contrast to the many drug studies published in the 1970s and early 1980s which have frequently produced conflicting results and, in general, rather minor effects on infarct size, ischemic preconditioning reliably and potently delays the development of infarction. Preconditioning protects against infarction in every mammalian species studied to date (i.e., dogs, pigs, rabbits and rats).[11-14] It would seem likely, therefore, that human myocardium will also prove to be amenable to this form of protection. Thus, ischemic preconditioning provides the opportunity for developing a successful adjunctive therapy to improve on the results achieved by reperfusion. However, before a preconditioning-based therapeutic intervention could be introduced into clinical practice a number of conditions would need to be fulfilled:

1. Convincing evidence would need to be provided that preconditioning occurs in man.
2. A form of preconditioning, most likely a pharmacological mimic, would need to be developed.
3. Clinical trials would be needed to demonstrate safety and efficacy.

As will be discussed in detail below even the first of these conditions has yet to be satisfied. Amongst the reasons for this is the ethical difficulty in performing prospective studies of preconditioning and myocardial infarction. However, since its original description the concept of preconditioning has been extended beyond reducing infarct size to a range of other protective effects (the so-called 'nonclassical end-points' for preconditioning). Several studies, especially those employing in vitro models of preconditioning, have reported reduced postischemic contractile dysfunction following preconditioning.[15] However, this effect has only been consistently reported following extended periods of ischemia sufficient to cause irreversible injury. In contrast, improved recovery of contractile function has *not* been consistently reported with shorter ischemic durations that result in exclusively reversible damage[16] or in peri-infarct tissue.[15] Thus, there is currently no convincing evidence that preconditioning can protect against myocardial stunning.

Where preconditioning has been shown to affect reversible injury is in the reduction of arrhythmia severity during early ischemia[17] and following reperfusion[18] (reviewed in chapter 3) and as an attenuation of electrocardiographic[19] and metabolic manifestations of ischemia.[20] It is still not clear precisely to what extent these are related to 'classical' protection against infarction. The mechanism of 'nonclassical' manifestations of ischemic preconditioning may be very different from those involved in limiting necrosis. Nevertheless, to date almost all studies performed in man have used the 'nonclassical' end-points of contractile function, electrocardiography or biochemical evidence of ischemia.

7.2. INSIGHTS FROM STUDIES OF ISOLATED HUMAN MYOCARDIUM

Most in vitro models of preconditioning have employed isolated whole heart preparations. Recently, however, Walker et al[21] have characterized a model of preconditioning using isolated rabbit papillary muscles. Using repeated brief episodes of hypoxia and re-oxygenation to induce preconditioning they were able to protect the heart from a subsequent prolonged episode of hypoxia. Following final re-oxygenation they were able to demonstrate improved recovery of contractile function in the 'preconditioned' tissue. They went on to study strips of human atria taken at cardiac surgery and were able to show a similar pattern of protection.[22] As with 'classical' preconditioning in both the rabbit and human preparations, protection could be mimicked by adenosine A_1 agonists and abolished by adenosine antagonists. This represents quite suggestive evidence that the mechanisms responsible for protection in this model may be similar to those involved in classical preconditioning. The preparation may prove useful for investigating the mechanism of preconditioning in man but it is not an ideal model of 'classical' preconditioning. Several studies have shown that hypoxia can induce preconditioning as potently as ischemia but, as yet, there is no clear evidence that preconditioning can provide protection against hypoxia-induced injury. Furthermore, there is no direct experimental evidence that preconditioning occurs in atrial tissue. Finally, it remains to be established whether the preservation of contractile function reported was due to an effect on reversible or irreversible tissue injury.

Some studies have been performed using isolated human myocytes but, in contrast to the work of Walker et al,[21,22] there is no established model of preconditioning in such a preparation. Recent data have suggested, however, that metabolic manifestations of ischemia can be modulated, with reduced production of protons and lactate and reduced lactate dehydrogenase release.[23] The question of whether such data bears any relevance to the phenomenon of ischemic preconditioning must remain open until more information is available on the processes involved.

7.3. DOES ANGINA PROTECT AGAINST MYOCARDIAL INFARCTION?

Angina is the most common manifestation of reversible myocardial ischemia and is experienced by many patients before infarction.[24] The results of experimental studies would suggest that for episodes of angina to induce preconditioning they would need to be prolonged, lasting at least 2.5 minutes, and occur in the hour or so immediately preceding the onset of infarction. In contrast, in patients with chronic stable angina individual attacks cannot be relied upon to occur in close temporal relationship to the onset of infarction and they are generally short-lived, rarely exceeding 2 minutes. A clinical situation analogous to classical preconditioning does occur, however, when episodes of unstable angina, which can last for many minutes, immediately precede the development of infarction. Although the precise pathophysiological substrate of unstable angina has still not been entirely resolved, the formation (and dispersion) of platelet-rich thrombi on unstable coronary plaques is believed increasingly to be of central importance.[25] Interestingly, experimental evidence suggests that ischemia induced by cyclical variations in coronary blood flow, due to repeated formation and dispersion of platelet thrombi, can induce ischemic preconditioning as effectively as mechanical coronary artery occlusion.[26] Nevertheless, the proportion of infarcts closely preceded by episodes that can be reliably identified as unstable angina is low and this group have never been studied directly in the context of preconditioning.

Interestingly, some reports involving relatively small numbers of patients have suggested that angina may provide some protection against myocardial infarction. Matsuda et al[27] have reported on patients with infarction caused by occlusion of the left anterior descending coronary artery (LAD). They assessed infarct size by the crude measures of global left ventricular ejection fraction and the proportion of the left ventricular wall with normal contraction and found that those with angina for at least 6 weeks prior to infarction had better preserved left ventricular function. However, it is unlikely that this represents a bona fide clinical manifestation of preconditioning for several reasons. The infarct related artery in all of the patients included in this study was chronically occluded and without reperfusion the effect of preconditioning is overcome within a short period.[11] It is much more likely that the infarcts were smaller in patients with angina because of collateral flow recruitment. This inference is borne out by results of a similar study by Hirai et al[28] who found that preservation of contractile function was indeed associated with improved collateral circulation provided angina had occurred for over a week before the onset of infarction. Collaterals were less well developed and left ventricular function was the same in patients with no prior history of angina.

Retrospective analysis of data from the TIMI II study, involving a much larger patient population, suggested that those with angina for more than one week prior to infarction had a less complicated course whilst in hospital, and also a trend towards a lower mortality whilst in hospital.[29] Again, this result may be explained by recruitment of collateral flow. To confound interpretation further, for unexplained reasons the patients with angina prior to infarction were found to be less likely to have re-occluded their infarct-related artery. It is possible that this could have contributed to the trend towards a reduced mortality.

Two recent studies,[31,31a] however, have suggested that collateral flow may *not* explain these beneficial effects. Indeed, of all the retrospective studies of angina preceding infarction that by Ottani et al[30] provides the most convincing evidence for bona fide classical ischemic preconditioning in man. They examined a group of 25 patients presenting within 2 hours of onset of their first anterior infarct who then reperfused (on ECG criteria) within 90 minutes following thrombolytic therapy and had a patent infarct related artery at subsequent angiography. Patients were divided into those without prodromal angina and those with angina in the 24 hours prior to infarction. Any with angina before 24 hours prior to infarction were excluded from the study. There was no difference in time to treatment, time to reperfusion or myocardium at risk between the groups. Nevertheless, compared with those with no angina at all, the patients with angina limited to the 24 hours before infarction appeared to have smaller infarcts as indicated by reduced release of creatine kinase, better ventricular contractile function and less marked residual ECG changes.

Ottani et al[30] made an attempt to address the issue of collateral flow by also excluding patients with angiographically detectable collaterals, although the angiograms were not performed until on average 25 days after infarction. Kloner et al[32] have recently presented retrospective data from the TIMI 4 study which involved a much more heterogeneous group of patients but provided similar results. Patients with prior angina appeared to have smaller and less complicated infarcts whilst in the acute phase with a lower in-hospital mortality, less frequent hemodynamic problems and reduced creatine kinase release. The unique feature of this study is that collateralization was assessed angiographically only 90 minutes after thrombolysis. Interestingly the collateral score was not found to be different between those with and those without prior angina. Despite these very suggestive data caution is still required. It remains debatable to what extent the degree of collateralization during an acute event (when the infarct-related artery is still occluded) is reflected by that present angiographically after reperfusion, especially in view of the insensitivity of angiography for detecting clinically significant amounts of collateral flow. Furthermore, the prodromal phase of both studies is still very long when compared with the time over which protection by "classical" preconditioning is lost, which appears to be at most 2 hours.[11]

There are further difficulties in interpreting the results of these studies. Clinically it can be very difficult to separate unstable angina (i.e., prolonged pain occurring at rest) from the onset of infarction itself when the two occur close together. In addition, leaving aside differences in collateral blood flow the sub-groups in each of these studies are not really comparable. Patients with angina are generally prescribed anti-ischemic medication, which may affect the clinical outcome following infarction,[8] and they are more likely to have multivessel coronary disease. Indeed, studies which have involved several thousands of patients have all indicated that those with angina prior to infarction actually do worse, both in the short and long terms.[31b,32]

7.4. DOES PRECONDITIONING OCCUR DURING BALLOON ANGIOPLASTY?

Superficially, coronary balloon angioplasty appears to provide a convenient clinical model of preconditioning and indeed has been employed as such by several authors. Even though angioplasty (usually!) produces reversible rather than irreversible ischemic injury, and thus results may not be directly related to classical preconditioning, the attractions of this model are obvious. It involves a common clinical procedure with repeated episodes of regional ischemia occurring under (usually!) controlled conditions. Anecdotally, the first suggestion of adaptation during angioplasty was included in the case-report of a patient who developed a coronary occlusion during a procedure.[33] It took 4 minutes to re-open the vessel during which the patient had severe pain and marked ECG changes. Quite remarkably, subsequent balloon inflations of up to 30 minutes were tolerated without pain or ECG changes. It is important to note, however, that the pressure distal to the balloon was higher during the later inflations suggesting increased collateralization.

The landmark study of preconditioning and angioplasty was reported in 1990 by Deutsch et al.[34] They subjected patients to 90 second balloon inflations in the left anterior descending coronary artery (LAD). Several lines of evidence suggested that ischemia was less severe during the second inflation when compared to the first. Chest pain was less intense and, more objectively, there was less ST segment shift on the surface ECG and the pulmonary artery pressure was less elevated. In some of the patients myocardial lactate production was measured and this was also found to be lower during the second inflation. Interestingly, 60 second balloon inflations do not appear to be sufficient to induce protection. Using this duration Oldroyd et al[35] found no difference in either ST-segment changes or lactate production. Taken together the results of these two studies suggest that, as with classical preconditioning, there is a threshold duration of ischemia for the initiation of protection. It is somewhat surprising, however, that this threshold duration is substantially less than the minimum 2.5 minutes required to reduce infarct size.[36]

Again the most likely alternative to preconditioning as the protective mechanism is the opening of preformed collateral vessels as a consequence of the initial balloon inflation. Rather weak evidence against this was provided by Deutsch et al[34] in that coronary sinus blood flow was rather lower during the second inflation than the first. However, the issue has been addressed more directly by Cribrier et al.[37] Their study again involved patients undergoing angioplasty to the LAD but multiple (up to five) balloon inflations were used. In addition these were maintained for 120-370 seconds, much longer than in previous studies. Similar protection to that reported by Deutsch et al[34] was found with less chest pain, less ST-segment elevation, less marked elevation of left ventricular filling pressure and also less impairment of left ventricular ejection fraction. Collateral circulation was assessed both by angiographic grading, a rather insensitive method of assessment, and by measurement of pressure distal to the inflated balloon. The results suggested that much of the adaptation during angioplasty may be explained by increased collateral flow. Collateralization on angiography was higher in over half of the patients and the distal coronary pressure was also increased. Thus, at least for a proportion of patients, adaptation occurring during PTCA may be explained by recruitment of collateral blood flow. It remains possible, however, that preconditioning-mediated protection against reversible injury may also occur. Separation of this from effects of collateral flow will require very carefully designed studies.

7.5. THE 'WARM-UP' PHENOMENON AND PRECONDITIONING

Diurnal variations occur in many of the clinical manifestation of myocardial ischemia.[38,39] Approximately one fifth of patients with angina report a 'warm-up' phenomenon with severe symptoms in the morning, especially shortly after rising from bed, and an improvement later in the day.[40] Several explanations for this pattern have been suggested, including circadian variations in autonomic tone.[41] However, the description of ischemic preconditioning has shed new light on the 'warm-up' phenomenon because of the possibility that myocardial ischemia produced by activity shortly after rising might precondition the heart and increase its tolerance to activity later in the day. In a number of studies patients have undergone repeated exercise stress testing, and several have reported that they tolerate a second period better than the first.[42-44] The adaptive mechanisms involved, however, have not been clearly established.

In one such study Okazaki et al[43] examined patients with chronic stable angina and single vessel coronary disease involving the LAD. With sequential treadmill exercise tests adaptation was demonstrated with less severe angina on the second exercise period and a higher workload achieved before the development of ECG changes. The myocardial oxygen consumption was also lower during the second exercise period but not at rest. This suggests that the adaptive process involved an increase in metabolic efficiency, a feature typical of preconditioning. Interestingly, release of adenosine was higher during the second exercise period. This is consistent with the increase in 5'-nucleotidase activity reported to occur in preconditioned myocardium.[45]

Williams et al[42] have reported a study using repeated episodes of pacing-induced ischemia. They found that during the second period of pacing both myocardial oxygen consumption and lactate production were reduced, again typical of ischemic preconditioning.[20] As with other possible clinical manifestations of preconditioning, effects on collateral blood flow need to be considered. Although providing only rather indirect evidence, in neither of the above studies[42,43] was coronary sinus blood flow increased, suggesting no increase in overall myocardial perfusion. It is possible that the 'warm-up' phenomenon may prove to be a clinical manifestation of preconditioning. Final judgment will need to be reserved until direct data on collateral blood flow and the further data on the temporal aspects of 'warm-up' become available.

7.6. PRECONDITIONING AND CARDIAC SURGERY

Yellon et al[46] have studied preconditioning during cardiac surgery with repeated episodes of ischemia induced by cross-clamping the aorta. This approach has a major advantage over other clinical studies of preconditioning in that ischemia is global, thereby overcoming influences of collateral blood flow. As with other prospective clinical studies, ethical considerations prevent deliberate induction of irreversible myocardial injury. They examined, therefore, the effect on high energy phosphate metabolism during a 10 minute ischemic episode of two preceding 3 minute ischemic cycles separated by 2 minutes of reperfusion. Myocardial biopsies showed, as expected, that the 3 minute episodes resulted in some degree of ATP depletion. Compared with controls, however, these "preconditioned" hearts had a higher ATP content following the subsequent 10 minute ischemic insult.

A number of experimental studies using both NMR[47] and biochemical techniques[20] have shown a similar pattern of ATP preservation. Preconditioning itself results in some depletion of high energy phosphates but the rate of depletion during a subsequent prolonged ischemic period is slowed. This results in a short time period, after approximately 15 minutes of ischemia, during which the ATP content of preconditioned myocardium is transiently higher than in nonpreconditioned tissue. Thus the results of Yellon et al,[46] although limited in scope, suggest that preconditioning might occur during cardiac surgery. High energy phosphate preservation alone, however, is not sufficient to provide conclusive evidence of preconditioning. To be of wider scientific relevance, and of real clinical benefit, evidence is required of reduced intra-operative myocardial necrosis, better recovery of contractile function, shorter patient recovery times following operation and, ultimately, reductions in procedural and long-term mortality. Unfortunately the experimental studies performed to date do not provide a particularly good precedent for this. In one study, preconditioned porcine hearts reperfused during reversible ischemia again had better preserved ATP content which was sustained through a prolonged reperfusion period. ATP preservation was not, however, found to be associated with improved recovery of contractile functional.[48]

7.7. WHERE MIGHT PRECONDITIONING FIND A ROLE IN CLINICAL MEDICINE?

A major question that requires answering before any therapeutic exploitation is possible is whether any of the clinical phenomena described above are bona fide manifestations of preconditioning. All of the above studies are limited because of either confounding effects of collateral blood flow, the study end-points involving manifestations of reversible rather than irreversible ischemic injury or the study design being retrospective (and therefore subject to bias). Nevertheless, an intervention based on preconditioning might prove beneficial in a large number of clinical settings where anti-ischemic protection is required.

7.7.1. Myocardial Infarction

A reduction of the mortality from acute myocardial infarction could be achieved by two distinct mechanisms. Firstly, delaying the development of necrosis (a cardinal feature of preconditioning) would allow more myocardium to be salvaged by reperfusion therapy. It is well established that the principal determinant of prognosis following infarction is the extent of myocardial damage.[49] Thus the potential exists for an effective form of adjuvant therapy to improve the efficacy of thrombolysis. Secondly, although there have, to date, been no studies to establish whether preconditioning has antiarrhythmic activity in man, realization of only a fraction of the antiarrhythmic potency of preconditioning apparent from animal studies[17,18] would be of enormous therapeutic benefit. The majority of patients who die of infarction do so of fatal ventricular arrhythmias occurring within the first few minutes. Thus, the antiarrhythmic action of preconditioning may ultimately prove to be of greater clinical significance than protection against necrosis.

7.7.2. Chronic Stable Angina

For many patients current anti-anginal medications provide incomplete relief of symptoms. A pharmacological form of preconditioning ('preconditioning in a bottle') might overcome some of the limitations of current agents by acting directly on the ischemic process. Interestingly nicorandil, which acts by opening potassium channels, has recently been licensed for use in the treatment of angina. It is not clear how much of its anti-anginal effect can be attributed to direct effects on myocardial ischemic processes as opposed to its vascular actions.

7.7.3. Cardiac Surgery and Coronary Angioplasty

Preconditioning-based interventions could increase the safety of cardiac surgery and coronary angioplasty which, by their nature, necessarily involve inducing episodes of myocardial ischemia. In routine practice, however, relatively few major complications occur during these procedures. Preconditioning, in whatever form, would increase the complexity of the procedures and the potential risk of a preconditioning phase of treatment would need to be weighed against any possible benefit. It is possible that preconditioning might be useful in high risk cases where complications might be anticipated.

A few notes of caution are warranted. Firstly, all experimental studies of preconditioning to date have been performed with normal myocardium. It is far from clear that diseased human myocardium would react to brief episodes of ischemia in the same manner. Furthermore, if the 'warm-up' phenomenon does indeed prove to be a manifestation of preconditioning it is possible that patients might precondition themselves simply by their normal day-to-day activities. If this was the case they would be unlikely to gain further benefit from an additional preconditioning-based intervention. In addition, if they were ever to be widely adopted, preconditioning-based interventions would need to be simple to apply and this would be likely to involve a pharmacological rather than physical method of inducing the preconditioned state. Although many alternative methods of inducing preconditioning have been described (e.g., infusions of A_1 or M_2 agonists, hypoxia, rapid pacing, etc., reviewed in ref. 50) none is really suitable for routine clinical practice. Perhaps of greater importance is that to be of general clinical use the preconditioned state would need to be maintained. This would appear to be a difficult goal to achieve in view of the short time-course of classical preconditioning. Although it appears possible to re-initiate preconditioning once it has worn off there is currently no experimental evidence to indicate that a chronically preconditioned state can be maintained. However, recent studies have shown that protection against infarction returns 24 hours after its initial loss,[51,52] (reviewed in chapter 13) albeit somewhat less potently. Although the temporal characteristics of this 'second window of protection' have not been fully elucidated it is likely to be much longer lived than classical preconditioning.

7.8. CONCLUSIONS

Although protective adaptation occurs in the human heart in a number of clinical settings it is not clear if any of these are due to ischemic preconditioning. Prospective clinical studies have, necessarily, involved manifestations of reversible ischemia and there is certainly no conclusive evidence for 'classical' ischemic preconditioning in man. It would be very surprising, however, if the human heart proved to be the first of the many species studied to date to be incapable of being preconditioned. Currently there is increasing confidence that, as more data becomes available on the scope of the phenomenon and the underlying molecular mechanisms, preconditioning might provide novel therapeutic possibilities for clinical cardiology.

References

1. GISSI. Long term effects of intravenous thrombolysis in acute myocardial infarction: final report of the GISSI study. Lancet 1987; 2:871-874.
2. ISIS-2 collaborative study group. Randomised trial of intravenous streptokinase, oral aspirin, both, or neither amongst 17,187 cases of suspected acute myocardial infarction. Lancet 1988; 2:349-360.
3. Gibbons RJ, Holmes DR, Reeder GS et al. Immediate angioplasty compared with the administration of a thrombolytic agent followed by conservative treatment for myocardial infarction. N Engl J Med 1993; 328:685-691.
4. Grines Cl, Browne KF, Marco J et al. A comparison of immediate angioplasty with thrombolytic therapy for acute myocardial infarction. N Engl J Med 1993; 328:673-679.
5. Zijlstra F, de Boer MJ, Hoorntje JCA et al. A comparison of immediate coronary angioplasty with intravenous streptokinase in acute myocardial infarction. N Engl J Med 1993; 328:680-684.
6. EMERAS Collaborative Group. Randomised trial of late thrombolysis in patients with suspected acute myocardial infarction. Lancet 1993; 342:767-772.
7. Late Study Group. Late assessment of thrombolytic efficacy (LATE) study with alteplase 6-24 hours after onset of acute myocardial infarction. Lancet 1993; 342:759-766.
8. Yusuf S, Peto R, Lewis J et al. Beta blockade during and after myocardial infarction: an overview of the randomized trials. Prog Cardiovasc Dis 1985; 27:335-371.
9. Jennings RA, Reimer KA. Effect of beta-adrenergic blockade on acute myocardial ischemic injury. In: Gross F ed. Modulation of sympathetic tone in the treatment of cardiovascular diseases. Berne: Hans Huber, 1979:103-114.
10. Rossi P, Yusuf S, Ramsdale D et al. Reduction of ventricular arrhythmias by early intravenous atenolol in suspected acute myocardial infarction. Br Med J 1983; 293:506-510.
11. Murry CE, Jennings RB, Reimer KA. Preconditioning with ischemia: a delay of lethal cell injury in ischemic myocardium. Circulation 1986; 74:1124-1136.
12. Schott RJ, Rohmann S, Braun ER et al. Ischemic preconditioning reduces infarct size in swine myocardium. Circ Res 1990; 66:1133-1142.
13. Cohen MV, Liu GS, Downey JM. Preconditioning causes improved wall

motion as well as smaller infarcts after transient coronary occlusion in rabbits. Circulation 1991; 84:341-349.
14. Li YW, Whittaker P, Kloner RA. The transient nature of the effect of ischemic preconditioning on myocardial infarct size and ventricular arrhythmia. Am Heart J 1992; 123:346-353.
15. Ovize M, Kloner R, Przyklenk K. Preconditioning and myocardial contractile function. In: Przyklenk K, Kloner R, Yellon D eds. Ischemic preconditioning: the concept of endogenous cardioprotection. Massachusetts: Kulwer Academic Publishers 1994:41-60.
16. Ovize M, Pryzklenk K, Hale SL et al. Preconditioning does not attenuate myocardial stunning. Circulation 1992; 85:2247-2254.
17. Lawson CS, Coltart DJ, Hearse DJ. 'Dose'-dependency and temporal characteristics of protection by ischaemic preconditioning against ischaemia-induced arrhythmias in rat hearts. J Mol Cell Cardiol 1993; 25:1391-1402.
18. Shiki K, Hearse DJ. Preconditioning of ischemic myocardium: reperfusion-induced arrhythmias. Am J Physiol 1987; 253:H1470-H1476.
19. Vegh A, Komoro S, Szekeres L et al. Antiarrhythmic effects of preconditioning in anaesthetised dogs and rats. Cardiovasc Res 1992; 26:487-495.
20. Murry CE, Jennings RB, Reimer KA. Ischemic preconditioning slows energy metabolism and delays ultrastructural damage during a sustained ischemic episode. Circ Res 1990; 66:913-931.
21. Walker DM, Marber MS, Walker JM et al. Preconditioning in isolated superfused rabbit papillary muscles. Am J Physiol 1994; 266: H1534-H1540.
22. Walker DM, Walker JM, Pattison C et al. Preconditioning protects isolated human muscle . Circulation 1993; 88(Suppl I):138 (Abstract).
23. Ikonomidis JS, Tumiati LC, Mickle DAG et al. Preconditioning protects human cardiac myocytes from ischemic inury. Circulation 1993; 88(Suppl I):I-570 (Abstract).
24. Alonzo AM, Simon AB, Feinleib M. Prodromata of myocardial infarction and sudden death. Circulation 1975; 52:1056-1061.
25. Davies MJ, Thomas AC, Knapman PA et al. Intramyocardial platelet aggregation in patients with unstable angina suffering sudden ischaemic cardiac death. Circulation 1986; 73:418-423.
26. Ovize M, Kloner RA, Hale SL et al. Coronary cyclic flow variations "precondition" ischemic myocardium. Circulation 1992; 85:779-789.
27. Matsuda Y, Ogawa H, Moritini K et al. Effect of the presence or absence of preceding angina pectoris on left ventricular function after acute myocardial infarction. Am Heart J 1984; 108:955-958.
28. Hirai T, Fujita M, Yamanishi K et al. Significance of preinfarction angina for the preservation of left ventricular function in acute myocardial infarction. Am Heart J 1992; 124:19-24.
29. Muller DW, Topol EJ, Califf RM et al. Relationship between antecedent angina pectoris and short-term prognosis after thrombolytic therapy for acute myocardial infarction. Thrombolysis and Angioplasty in Myocardial Infarction (TAMI) Study Group. Am Heart J 1990; 119:224-231.

30. Ottani F, Galvani M, Ferrini D et al. Ischemic preconditioning: prodromal angina limits myocardial infarct size. J Am Coll Cardiol 1993; 21:149A (Abstract).
31a. Ottani F, Galvani M, Ferrini D et al. Prodromal angina limits infarct size. A role of ischemic precoditioning. Circulation 1995; 91:291-297.
31b. Barbash GI, White HD, Modan M et al. Antecedent angina pectoris predicts a worse outcome after myocardial infarction in patients receiving thrombolytic therapy: experience gleaned from the international tissue plasminogen activator/streptokinase mortality trial. J Am Coll Cardiol 1992; 20:36-41.
32. Kloner RA, Shook T, Przyklenk K et al. Previous angina alters in-hospital outcome in TIMI-4. A clinical correlate to preconditioning? Circulation 1995; 91:37-47.
33. Heibig J, Bolli R, Harris S. Initial coronary occlusion improves tolerance to subsequent prolonged balloon inflation. Cathet Cardiovasc Diagn 1989; 16:99-102.
34. Deutsch E, Berger M, Kussmaul WG et al. Adaptation to ischemia during percutaneous transluminal coronary angioplasty. Clinical, hemodynamic and metabolic features. Circulation 1990; 82:2044-2051.
35. Oldroyd KG, Paterson JG, Rumley AG et al. Coronary venous lipid peroxide concentrations after coronary angioplasty: correlation with biochemical and electrocardiographic evidence of myocardial ischaemia. Br Heart J 1992; 68:43-47.
36. Ovize M, Pryzklenk K, Kloner RA. Preconditioning with one very brief episode of ischemia does not enhance recovery of function in peri-infarct tissue. Eu Heart J 1992; 13(Suppl):438 (Abstract).
37. Cribrier A, Korsatz L, Koning R et al. Improved myocardial ischemic response and enhanced collateral circulation with long repetitive coronary occlusion during angioplasty: a prospective study. JACC 1992; 20:578-586.
38. Hausmann D, Lichtlen PR, Nikutta P et al. Circadian variation of myocardial ischemia in patients with stable coronary artery disease. Chronobiol Int 1991; 8:385-398.
39. Rocco MB, Barry J, Campbell S et al. Circadian variation of transient myocardial ischemia in patients with coronary artery disease. Circulation 1987; 75:395-400.
40. MacAlpin RN, Kattus AA. Adaptation to exercise in angina pectoris. Circulation 1966; 33:183-201.
41. Selwyn AP, Raby K, Vita JA et al. Diurnal rhythms and clinical events in coronary artery disease. Postgrad Med J 1991; 67(Suppl V):S44-47.
42. Williams DO, Bass TA, Gewirtz H et al. Adaptation to the stress of tachycardia in patients with coronary artery disease: insight into the mechanism of the warm-up phenomenon. Circulation 1985; 71:687-692.
43. Okazaki Y, Kazuhisa K, Sato H et al. Attenuation of increased regional myocardial oxygen consumption during exercise as a major cause of warm-up phenomenon. JACC 1993; 21:1597-1604.

44. Jaffe MD, Quinn NK. Warm-up phenomenon in angina pectoris. Lancet 1980; 2:934-936.
45. Kitakaze M, Hori M, Takashima S et al. Increased 5'-nucleotidase activity caused by protein kinase C enhances adenosine production in hypoxic cardiomyocytes of rats. Circulation 1991; 84(Suppl II):620 (Abstract).
46. Yellon DM, Alkhulaifi AM, Pugsley WB. Preconditioning the human myocardium. Lancet 1993; 342:276-277.
47. Kida M, Fujiwara H, Ishida M et al. Ischemic preconditioning preserves creatine phosphate and intracellular pH. Circulation 1991; 84:2495-2503.
48. Miyamae M, Fujiwara H, Kida M et al. Preconditioning improves energy metabolism during reperfusion but does not attenuate myocardial stunning in porcine hearts. Circulation 1993; 88:223-234.
49. Pasternak R, Braunwald E, BE S. Acute myocardial infarction. In: Braunwald E ed. Heart Disease. Philadelphia: WB Saunders 1988: 1222-1313.
50. Lawson CS, Downey JM. Ischaemic preconditioning: state of the art myocardial protection. Cardiovasc Res 1993; 27:542-550.
51. Kuzuya S, Hoshida S, Yamashita N et al. Delayed effects of sublethal ischemia on the aquisition of tolerance to ischemia. Circ Res 1993; 72:1293-1299.
52. Marber MS, Latchman DS, Walker JM et al. Cardiac stress protein elevation 24 hours after brief ischaemia or heat stress is associated with resistance to myocardial infarction. Circulation 1993; 88:1264-1272.

CHAPTER 8

Role of ATP-Sensitive Potassium Channels in Myocardial Preconditioning

Gary J. Grover

8.1. INTRODUCTION

The past few years have seen an explosive growth of research on the mechanism of myocardial preconditioning. There has also recently been an increased interest in the role of ATP-sensitive potassium channels (K_{ATP}) in the pathogenesis of myocardial ischemia. These two originally distinct lines of investigation intersected with the publication by Gross and colleagues showing that preconditioning in dogs can be abolished by the K_{ATP} blocker glyburide (glibenclamide).[1] In this chapter I will describe the sequence of studies which led to the K_{ATP} hypothesis for preconditioning and I will present these data in a manner which will inform the reader of the thought processes leading to our current understanding (whether right or wrong) of the role of K_{ATP} in mediating myocardial preconditioning. First, I will describe what is known about cardiac K_{ATP}. I will then review the data describing the pharmacology of the cardioprotective effects of K_{ATP} openers. This will be done to show that the pharmacological profile of cardioprotection by K_{ATP} openers is consistent in many respects to that observed for preconditioning (preconditioning mimetic). I will then describe the studies showing the effects of K_{ATP} blockers on preconditioning. It turns out that the order in which the data are presented in this chapter are also generally in chronological order, therefore giving the reader an historical perspective of the field.

Myocardial Preconditioning, edited by Cherry L. Wainwright and James R. Parratt.

8.2. K_{ATP}: GENERAL CONSIDERATIONS

K_{ATP} were first described in cardiac tissue by Noma[2] as a potassium channel that was inhibited by intracellular ATP. K_{ATP}, under conditions of normal intracellular ATP, have a low open probability, but open probability increases as ATP is reduced. K_{ATP} is an intermediate conductance channel (40-80 pS) which has since been described in insulin-secreting cells, various regions of the brain, skeletal muscle, smooth muscle, and kidneys.[3-5] It is possible that K_{ATP} in different tissues are not identical, although many similarities are evident.

While much needs to be learned about the regulation of K_{ATP}, a clearer picture appears to be emerging (see review in ref. 6). Data supports the presence of an ATP regulatory (inhibitory) site on K_{ATP}. It is hypothesized that ATP binding at this site inhibits channel opening and this inhibition is not dependent on phosphorylation. ADP reduces the sensitivity of K_{ATP} to ATP and it is thought that the ratio of ADP/ATP is important in modulating channel function. It should be pointed out that in the absence of ATP, ADP inhibits K_{ATP} opening. In addition to the inhibitory site, a phosphorylation site is hypothesized as it is known that MgATP primes or sensitizes K_{ATP} for opening. It is thought that phosphorylation of this site is necessary for K_{ATP} opening and that channel run-down is caused by dephosphorylation of this site. The importance and regulation of phosphorylation in the activity of K_{ATP} is currently unclear, but cAMP-dependent protein kinase is thought to be important in modulating channel function and may be part of the signal transduction pathway for K_{ATP} openers.[7,8] Protein kinase C may also have a modulatory role, but the nature of this activity is presently unclear.[9] Some of the difficulties in interpreting the data may stem from the multitude of cell types used in these studies.

K_{ATP} also appears to be regulated through receptor-G protein interactions, best characterized by the interaction of A_1-adenosine receptors with K_{ATP} via a G_i protein.[10] Further discussion of the interaction of adenosine receptor agonists and K_{ATP} will follow later in this chapter. GDP, which is released upon G_i-protein activation, may also couple with a nucleotide diphosphate site which will enhance channel opening (see review in ref. 6).

8.3. PHARMACOLOGY OF THE CARDIOPROTECTIVE EFFECTS OF K_{ATP} OPENERS

8.3.1. OVERVIEW

Structurally diverse agents have been identified which selectively open K_{ATP}. They were originally described for their ability to relax smooth muscle, but subsequently were found to open K_{ATP} in other tissue types, including the myocardium. K_{ATP} openers are thought to relax smooth muscle by hyperpolarizing plasma membranes thereby reducing calcium entry through voltage sensitive calcium channels.[11-13]

This has led to their being dubbed "indirect calcium antagonists," although this probably does not completely explain smooth muscle relaxation.[11-13] It is thought that K_{ATP} openers can inhibit calcium release from intracellular stores, but more work needs to be done to further elucidate mechanisms of action. Examples of the major chemical classes of K_{ATP} openers are shown in Figure 8.1.

The K_{ATP} openers shown in Figure 8.1 are all known to act on cardiac K_{ATP} (see review in ref. 6). Activation of myocardial K_{ATP} would be expected to shorten action potential duration due to enhanced repolarizing or outward potassium currents. They are not particularly effective at hyperpolarizing cardiac myocytes, at least under normoxic conditions. Action potential shortening is also observed during myocardial ischemia[14,15] and several studies have indicated that part or all of this current is through K_{ATP}.[16,17] This outward potassium current is thought to contribute to ST-segment shifts observed during ischemia. Since ST-segment shifts (and outward potassium currents) are correlated with ischemic severity, it was presumed that K_{ATP} openers might be pro-ischemic. It was also possible that inhibition of calcium entry secondary to action potential shortening (or inhibition of ischemic

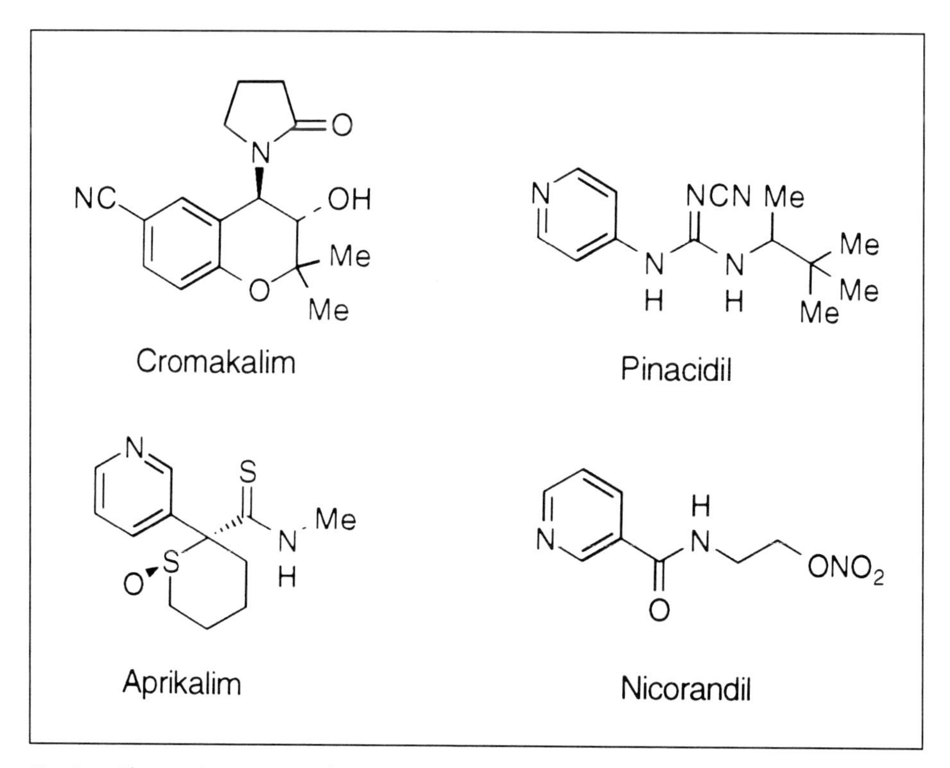

Fig. 8.1. Chemical structures of commonly used K_{ATP} openers. These compounds represent different structural subtypes with the common activity of K_{ATP} activation.

depolarization) by K_{ATP} openers would be beneficial under ischemic conditions. The selective cardiac K_{ATP} openers available at that time were useful tools for studying the role of K_{ATP} in the pathogenesis of myocardial ischemia. This part of the chapter therefore, will be used to show that the cardioprotective profile of K_{ATP} openers is consistent with the cardioprotective profile of preconditioning.

Early studies by Gross and colleagues showed a beneficial effect for the nicotinamide nitrate analog, nicorandil, in canine models of ischemia and reperfusion.[18,19] It was not known at the time that this agent (in addition to its known nitrate-like activity) was a K_{ATP} opener. Even when this information became available, it was still not clear whether this agent was protecting via K_{ATP} activation or nitrate activity. We determined the effect of the more selective K_{ATP} openers pinacidil and cromakalim in a rat isolated heart model of ischemia and reperfusion[20] and found that both agents significantly reduced necrosis and enhanced postischemic recovery of contractile function in this severe model of ischemic injury. The protective effects were concentration-dependent and were best observed when given before the onset of ischemia. The protective effects of pinacidil and cromakalim were found to occur within the concentration range of 1-10 μM. At >10 μM, the protective effects peaked and no further protection was observed. Before ischemia, neither pinacidil nor cromakalim had significant cardiodepressant effects and both significantly increased coronary flow before ischemia and during reperfusion.

This profile of cardioprotective activity is remarkably similar for nearly all other K_{ATP} openers tested, with the one exception being P-1075.[21-25] This compound is cardioprotective in the nanomolar range, and this is in excellent agreement with its high vasorelaxant potency.[22] All of the reference agents are potent coronary dilators, although this probably is not responsible for cardioprotection since isolated hearts perfused under constant flow conditions were also protected.[23] When given only during reperfusion, K_{ATP} openers significantly increased reperfusion coronary flow, but did not protect the hearts, indicating that the enhanced coronary reflow was not necessary for cardioprotection.[25] None of the K_{ATP} openers reduced cardiac function before ischemia within the cardioprotective concentration range. Similar results in in vitro models of ischemia have been found by other investigators.[26-28]

Structurally distinct blockers of K_{ATP} are useful tools for determining the mechanism of action of putative K_{ATP} openers. The cardioprotective effects of K_{ATP} openers have been uniformly found to be abolished by K_{ATP} blockers.[20,26,28] This has been shown for the sulphonylurea glyburide, sodium 5-hydroxydecanoate, and the anthranilic acid meclofenamate.[20,29,30] The ability of meclofenamate to block K_{ATP} was independent of its cyclooxygenase inhibitory activity.[30] The blockers alone have no effect on the severity of ischemia, suggesting that K_{ATP} may not be open under the ischemic conditions studied. These K_{ATP} blockers will not only abolish the cardioprotective effects of the openers, but the combination is often paradoxically pro-ischemic. These data agree with the data suggesting that K_{ATP} blockers do not directly displace the openers from their receptor, but bind to another, allosterically linked receptor.[31-32] This will be discussed in more detail later in this chapter. Glyburide actually loses some of its K_{ATP} blocker activity under ischemic conditions, but still completely abolishes the cardioprotective actions of K_{ATP} openers.[33] Glyburide also abolishes the coronary dilator activity of K_{ATP} openers.[20] Sodium 5-hydroxydecanoate is thought to be most active during ischemia and therefore has little effect on the coronary dilator action (pre-ischemia) of K_{ATP} openers, but completely abolishes their cardioprotective effects.[29] Meclofenamate appears to have a pharmacologic profile which is similar to sodium 5-hydroxydecanoate.[30]

Cardioprotection has also been observed in whole animal models of myocardial ischemia, although the results are somewhat more variable. Nicorandil was shown years ago to reduce necrosis and stunning in canine models of ischemia and reperfusion.[18,19] In Gross' hands, the cardioprotective effects of nicorandil are abolished by glyburide,[34] although in our laboratory, it has direct cardioprotective effects at high concentrations (almost millimolar) which would not be seen with systemic treatment.[35] Nicorandil in our laboratory does not have an identical profile of action compared to other K_{ATP} openers, but K_{ATP} opening probably does account for some of its activity.[35] While there may be some debate with nicorandil, most investigators find cardioprotective effects for more selective K_{ATP} openers. We have shown cardioprotection for cromakalim and pinacidil in dogs, although we had to give the drugs via the intracoronary route to prevent hemodynamic alterations.[21,25] Gross and colleagues have successfully administered aprikalim intravenously and affected cardioprotection in canine models of stunning and infarction.[36,37] Cardioprotection has recently been shown for K_{ATP} openers in pigs.[38] There have been several negative studies in dogs and rabbits in which no protective effects for K_{ATP} openers were found.[39-41] Possible reasons for these negative studies have been listed as coronary steal or hemodynamic disturbances, as well as confounding effects of anesthetics.

Much of the protective effects of the K_{ATP} openers were lost when given only during reperfusion, although this does not completely exclude the possibility that they may directly attenuate reperfusion injury. We have shown K_{ATP} openers to directly protect globally ischemic rat hearts during ischemia per se.[42] Time to the onset of contracture during ischemia is increased in a concentration dependent manner by a variety of K_{ATP} openers in this species.[22,30,42] While this demonstrates that some of the protective effects of K_{ATP} openers occur during ischemia, it also suggests that the protective effect is accompanied by ATP conservation as it is thought that contracture represents rigor bond formation secondary to ATP depletion. This suggests a novel profile of action since K_{ATP} openers caused no changes in cardiac function at cardioprotective concentrations. We then tested the hypothesis that K_{ATP} openers could conserve ATP during ischemia in concentrations not reducing cardiac function and found significant conservation of ATP.[42] These findings were confirmed by Cole's laboratory.[43] These data suggest that energy conservation and cardiac function can be separated. One suggested mechanism is an increased efficiency of energy utilization and data indicating this possibility have been published.[21]

The mechanism of the protective action of K_{ATP} openers is poorly understood. Early hypotheses suggested that action potential shortening or inhibition of ischemic depolarization caused by K_{ATP} activation reduced calcium influx into myocytes. While K_{ATP} openers have been found to reduce intracellular calcium during ischemia,[44] we do not know if this is their primary effect. Significant action potential shortening should be accompanied by cardiodepression, which is neither "necessary nor sufficient" for protection by K_{ATP} openers. Recent work from Yao and Gross[45] showed that, in dogs, a dose of bimakalim could be found which reduced infarct size with minimal effects on epicardial monophasic action potential duration. These results are intriguing, although further work was necessary because of the relative lack of sensitivity of epicardial monophasic action potential determinations. We have shown similar results using monophasic action potential determinations in dogs,[46] but also using intracellular recording techniques in guinea pig papillary muscles.[47] We have also synthesized cromakalim analogs which retain glyburide-reversible cardioprotection while being relatively devoid of vasodilator and action potential shortening activity, again suggesting a lack of correlation between electrophysiologic effects and cardioprotection.[47,48] The mechanism of the cardioprotective effects of K_{ATP} openers is therefore still undecided. Current thinking suggests that an intracellular K_{ATP} may be mediating cardioprotection. A K_{ATP} has been found to be expressed in mitochondrial membranes which may be important in maintaining its electrochemical gradient and mitochondrial volume.[49,50] It has also been hypothesized to control mitochondrial energetics. K_{ATP} openers have been found to open this channel within their cardioprotective concentration range, therefore further work in this area is warranted.[51]

8.3.2. Summary of the Profile of K_{ATP} Openers in Ischemia: Comparison to Preconditioning

K_{ATP} openers appear to exert a direct protective effect on the ischemic myocardium. This protective effect is most consistent in in vitro models of ischemia, probably because of a lack of interference of hemodynamic and anesthetic effects. A reduction in cardiac function is not necessary for their cardioprotection, yet ATP is conserved. This is very similar to the profile to that observed for preconditioning in which reduced myocardial function is not necessary for preconditioning and yet a conservation of ATP is observed (see review in ref. 52). Preconditioning is also not dependent on changes in coronary collateral flow and the same is true for K_{ATP} openers. While these data do not prove that K_{ATP} mediates preconditioning, they are at least consistent with the profile of preconditioning. More definitive proof required the use of K_{ATP} blockers and their effect on precondtioning and these data will be covered shortly. There are, however, several other features of K_{ATP} openers suggesting that their profile of activity is consistent with preconditioning.

One interesting component of preconditioning is the finite window of memory observed following the preconditioning stimulus. Data accumulated by several laboratories showed that the potassium current stimulated by K_{ATP} openers was enhanced with repeated exposure of the preparation to the drug. For example, Escande et al[53] showed that pinacidil enhanced steady state potassium currents in guinea pig cardiac myocytes more effectively with repeated dosing. Between dosing, the pinacidil was allowed to wash out. These results suggest the possibility of memory in the action of K_{ATP} openers. It would be interesting to determine if similar activity can be observed for cardioprotection since sarcolemmal potassium currents may not be predictive of cardioprotection.

In a very important study, Yao and Gross[54] showed that the K_{ATP} opener bimakalim could reduce the threshold for preconditioning in dogs. Ten minutes of preconditioning was found to significantly reduce infarct size following a subsequent prolonged ischemia, while 3 minutes of preconditioning was not sufficient to cause protection. This shows that an insult of a minimal severity is required for preconditioning to be expressed. The investigators then combined the 3 minute preconditioning with a subthreshold dose of bimakalim and were then able to show cardioprotection. This is interesting as one might suspect that a cardioprotective agent would increase the threshold of preconditioning due to a reduction of ischemic stress during the brief ischemia unless, of course, this agent was working via a similar mechanism.

8.4. EFFECT OF K_{ATP} BLOCKERS ON PRECONDITIONING

While the cardioprotective profile for K_{ATP} openers is consistent with preconditioning, this is not definitive proof that K_{ATP} mediates preconditioning. It was necessary to determine the effect of blockers of K_{ATP} on preconditioning. We are fortunate in having several structurally distinct K_{ATP} blockers to use as tools. Glyburide is probably the best known blocker which has use for treating type II diabetes because of its ability to block K_{ATP} in insulin secreting cells. Glyburide is an important tool because the pharmacology of this agent is fairly well understood. What is known is that glyburide will abolish all of the classical activities of K_{ATP} openers, such as smooth muscle relaxation, action potential shortening and cardioprotection. Since glyburide loses some of its blocking activity under ischemic conditions, negative results have to be interpreted with some degree of caution.[14] Sodium 5-hydroxydecanoate is another blocker which appears to be most active under ischemic conditions, without causing hypoglycemia, thus giving us another valuable tool.

It was Gross' group who first determined the effect of K_{ATP} blockers on preconditioning.[1] They preconditioned dog hearts with a 5 minute occlusion of a coronary artery followed by 60 minutes of ischemia and 5 hours of reperfusion. Preconditioning significantly reduced infarct size in this model and the effect was completely abolished by glyburide. Glyburide alone had no effect on infarct size. None of the effects observed were accompanied by changes in collateral blood flow. These data strongly suggested the possibility that K_{ATP} was involved in the mechanism of preconditioning. To further assess the role of K_{ATP} in preconditioning, another study was performed by this group using sodium 5-hydroxydecanoate, in addition to intracoronary glyburide to prevent plasma glucose lowering, in a model similar to that described above.[55] Both glyburide and sodium 5-hydroxydecanoate abolished preconditioning, despite having no effect on infarct size in nonpreconditioned animals. The protective effect of preconditioning was abolished by glyburide whether given before preconditioning or immediately before the prolonged, subsequent ischemia.

Since the publication of these studies, numerous laboratories have shown K_{ATP} blockers to abolish preconditioning. Studies from our laboratory confirmed Gross' studies in dogs showing an inhibitory effect of glyburide on preconditioning.[56] Rohmann et al[38] showed that glyburide abolished the protective effect of 10 minutes of coronary occlusion on a subsequent 60 minute ischemic episode in pigs. In a similar study, Schulz et al[57] showed that glyburide abolished the protective effect of preconditioning (10 minute preconditioning) in pigs but, interestingly, did not reduce adenosine release (this will be discussed in more detail later). They also showed that preconditioning slightly enhanced action potential shortening during the first minutes of the prolonged occlusion, although this effect was minor (6-7% change from sham).

While there seems to be agreement that preconditioning in pigs and dogs is abolished by glyburide, studies in rabbits has been somewhat more controversial. Studies by Toombs et al[58] showed that preconditioning could be abolished by glyburide in an in situ rabbit model. In a similar model, Thornton et al[59] could not show protection. The only difference between the models used by the respective laboratories was the anesthesia, with Toombs et al using ketamine-xylazine and Thornton et al using pentobarbital. When the latter laboratory switched to ketamine-xylazine, glyburide was then found to abolish preconditioning.[60] The authors did not have an explanation for these findings. In another study in rabbit papillary muscles, Tan et al[61] determined the effect of preconditioning on the rate of cellular electrical uncoupling during ischemia. They preconditioned papillary muscles with 10 minutes of ischemia followed by a prolonged (40 minutes) period of ischemia. Uncoupling occurred at 15 ± 0.7 minutes in sham tissue and at 22.8 ± 1.5 minutes in preconditioned tissue. This protective effect was completely abolished by glyburide. Interestingly, action potential shortening during the prolonged ischemia was enhanced in the preconditioned papillary muscles, suggesting an increased outward potassium current, an effect which was also abolished by glyburide.

Several studies have suggested that preconditioning can be observed in man. Since glyburide is an approved drug, it was logical to determine the effect of this K_{ATP} blocker on preconditioning in man. Tomai et al[62] preconditioned patients by inflating an intracoronary balloon for 2 minutes followed by recovery for approximately 5 minutes. ST-segment shifts were recorded during the brief ischemic episode. A second 2 minute inflation was then performed and ST-segment shifts were once again determined. The second occlusion resulted in a significantly attenuated ST-shift, suggesting a preconditioning effect. In a second group of patients glyburide did not affect ST-segment deviation during the first balloon occlusion, but abolished the protective effect of the second occlusion. This study suggested that preconditioning in man is mediated by K_{ATP}, although further proof is needed as ST-segment shifts may not be predictive. Studies from Yellon's laboratory[63] showed that human atrial trabecula could be preconditioned using hypoxia and reoxygenation. In this model, a short period of hypoxia was followed by a prolonged period of hypoxia. Preconditioned tissue showed a significantly enhanced recovery of contractile function compared to sham. This protective effect was completely abolished by glyburide. Interestingly, cromakalim was cardioprotective in this model at similar concentrations to that observed in rat hearts.

Since K_{ATP} modulation can have effects on electrical uncoupling and refractory period, it may also affect arrhythmogenesis. Preconditioning has been shown by several investigators to reduce arrhythmias in different models.[64,65] Several studies have shown however, that K_{ATP} blockers do not have any effect on the antiarrhythmic action of preconditioning. In one study, meclofenamate was found to block the antiarrhythmic effect of preconditioning in dogs,[66] and recently reported data from our laboratory indicates meclofenamate to be a potent K_{ATP} blocker,[30] although the significance of this is presently unknown.

The specter of species differences in preconditioning has presented itself with K_{ATP} as it has in other areas (reviewed in chapter 12). In studies conducted thus far, preconditioning in rat hearts appears to occur via a different mechanism compared to other species such as dogs and pigs.[67,68] Studies from several laboratories have shown that glyburide does not block preconditioning in rat isolated heart models of ischemia and reperfusion.[68,69] Rat isolated hearts can be preconditioned, and it is presently unknown why glyburide is not effective in attenuating preconditioning in this species. One hypothesis is that glyburide loses some of its K_{ATP} blocking efficacy under ischemic conditions[14] and therefore the negative results could not be used to completely exclude the involvement of K_{ATP} in rat hearts. To address this question, we determined the effect of sodium 5-hydroxydecanoate on preconditioning in rat isolated hearts. Sodium 5-hydroxydecanoate, an ischemia-selective K_{ATP} blocker, also did not affect preconditioning in the rat heart.[70] It is of interest that adenosine receptor antagonists are also without effect on preconditioning in rat hearts.[67] While the mechanism of preconditioning in rat hearts is presently unclear, studies have suggested the potential importance of α-adrenoceptor involvement in some aspects of the cardioprotection.[71] Another interesting study was published from Lazdunski's laboratory showing that rat cerebrum could be preconditioned and that this protective effect was abolished by glyburide.[72] These data suggest the possibility that K_{ATP} is involved with preconditioning in rats, but there may be tissue differences or that we are not performing cardiac studies under appropriate conditions to show K_{ATP} involvement. Further work is warranted in this area of study.

8.5. HOW DOES K_{ATP} FIT INTO THE CASCADE OF EVENTS IN PRECONDITIONING?

8.5.1. Interaction of Adenosine and KATP

Data from numerous laboratories show that the activation of A_1-adenosine receptors is involved in the protective effects of preconditioning.[73] Studies have suggested that adenosine receptor activation can open K_{ATP} and it is logical to hypothesize that adenosine release during preconditioning could open K_{ATP} and mediate preconditioning. This was first suggested by Kirsch et al[74] in rat cardiomyocytes in which adenosine A_1 receptor stimulation increased K_{ATP} open probability via a G_i protein coupled mechanism. We then showed that the cardioprotective effects of the A_1-adenosine receptor agonist R-PIA was abolished by glyburide in a canine model of infarction.[56] Van Winkle et al[75] also showed that the protective effect of R-PIA was abolished by sodium 5-hydroxydecanoate in a porcine model of infarction. Adenosine has also been shown to protect ischemic dog and rabbit hearts and this protective effect was abolished by glyburide.[76,77] These results strongly suggest a link between adenosine agonist release and K_{ATP}. They also suggest that adenosine release stimulates K_{ATP}, which then protects the heart.

There have been several studies suggesting that K_{ATP} activation causes adenosine receptor stimulation. A study by Tsuchida et al[78] showed that pinacidil can reduce infarct size in ketamine-xylazine anaesthetized rabbits and this effect was blocked by the adenosine receptor antagonist sulphophenyltheophylline (SPT). Similar results were observed in rabbit ventricular myocytes using SPT, 3,7-dimethyl-1-propargylxanthine (DPCPX) or adenosine deaminase,[79] suggesting that K_{ATP} openers were protecting ischemic myocytes by enhancing the release of adenosine. Kitakaze et al[80] proposed that K_{ATP} openers protect ischemic myocardium by increased adenosine formation through activation of 5'-nucleotidase, a reverse sequence of events to that outlined in the previous paragraph. Other studies have shown that the cardioprotective effects of bimakalim are not abolished by the selective A_1-adenosine receptor antagonist DPCPX.[81] In the same study the cardioprotective effect of bimakalim in rat isolated hearts was not abolished by DPCPX.[81] While there is still some controversy over this issue, most investigators feel that K_{ATP} is involved in preconditioning, but whether K_{ATP} is an "end-effector" has yet to be unequivocally shown.

8.5.2. Mechanism of K_{ATP}-Induced Cardioprotection in Preconditioning: Is It the "End-Effector?"

Determination of the end effector for preconditioning has been an elusive target, although some progress has been made. While activation of protein kinase C has been strongly suggested by several investigators to be involved with preconditioning, it is not clear what protein or ion channel is phosphorylated.[71] Recent studies from Yellon's laboratory suggest that protein kinase C may phosphorylate a protein which is directly or indirectly involved with K_{ATP}.[63] These data are exciting not only because of their mechanistic implications, but because they were collected in human tissue. While these data indicate an important role for K_{ATP} in preconditioning, they still do not prove that K_{ATP} is the ultimate end-effector because we do not know what biochemical cascade of events follows K_{ATP} opening. Until we have a better understanding of how K_{ATP} activation causes cardioprotection, it will be difficult to make significant advances in our understanding of the role of K_{ATP} in preconditioning. Probably the best means at our disposal is to use pharmacologic K_{ATP} openers to determine mechanisms. I will briefly review what is known and suggest future directions.

As described above, the cardioprotective effects of K_{ATP} openers can be dissociated from action potential shortening.[45,46] For example, the cardioprotective effects of cromakalim were preserved even when the action potential shortening effects of cromakalim were abolished by the delayed rectifier blocker dofetilide.[46] Unpublished data from our laboratories have also shown that dofetilide has no effect on preconditioning, despite complete inhibition of action potential shortening. Cromakalim analogs which are virtually devoid of vasodilator and action potential shortening activity retain glyburide-reversible cardioprotective activity.[42] Our hypothesis is that sarcolemmal K_{ATP} may not be involved with preconditioning but that mitochondrial K_{ATP} might well be. It would be worthwhile to determine whether those glyburide-reversible agents which do not open sarcolemmal K_{ATP}, still protect myocardial tissue by opening mitochondrial K_{ATP}.

An important tool for determining the site of action of K_{ATP} openers would be to find their binding site. Presently we do not know the cardioprotective binding site of K_{ATP} openers, although it is known that the sulphonylurea binding site is distinct and allosterically linked. Since glyburide is the most commonly used tool for determining the role of K_{ATP} in preconditioning, it would be useful to determine its relevant binding site in terms of abolishing cardioprotection. A sulphonylurea binding site (SUR) has been described to be a member of the family of ATP binding cassette proteins.[31,32] This protein is 140-170 kDa. No ion permeant properties have been found with the expressed protein. Much of this work has been done in relation to insulin secreting cells and it remains to be determined if this site is relevant to ischemia or preconditioning in the myocardium. Glyburide is known to have other activities which may be involved with myocardial ischemia, such as metabolism of fatty acids and glycolysis, and it is unknown if any of these activities can influence preconditioning.[82]

8.6. CONCLUSIONS

The overwhelming majority of the data collected suggest a role for K_{ATP} in preconditioning. While some species and model differences need to be explained, K_{ATP} cannot be excluded as an important part of the mechanism of preconditioning. In the future it is critical to understand how K_{ATP} mediates preconditioning; only in this way can we understand its role in preconditioning. An important issue is the design of potential therapeutic agents based on the mechanism of preconditioning. The design of such agents will depend on the discovery of the enzyme or receptor important in preconditioning which is the most specific or will lend itself to the development of potent chemotypes. Therefore the end effector may not necessarily be the best target for pharmacological manipulation. At the present time, specific K_{ATP} openers may represent the best opportunity to pharmacologically exploit an endogenous protective mechanism.

REFERENCES

1. Gross GJ, Auchampach JA. Blockade of ATP-sensitive potassium channels prevents myocardial preconditioning in dogs. Circ Res 1992; 70:223-233.
2. Noma A. ATP-regulated K^+ channels in cardiac muscle. Nature 1983; 305:147-148.
3. Spruce AE, Standen NB, Stanfield PR. Voltage-dependent ATP-sensitive potassium channels of skeletal muscle membrane. Nature 1985; 316:736-738.

4. de Weille J, Schmid-Antomarchi H, Fosset M et al. ATP-senstive K^+ channels that are blocked by hypoglycemia-inducing sulfonylureas in insulin-secreting cells are activated by galanin, a hyperglycemia-inducing hormone. Proc Natl Acad Sci USA 1988; 85:1312-1316.
5. Treherne JM, Ashford MLJ. The regional distribution of sulfphonylurea binding sites in rat brain. Neuroscience 1991; 40:523-531.
6. Edwards G, Weston AH. The pharmacology of ATP-sensitive potassium channels. Ann Rev Pharmacol Toxicol 1993; 33:597-637.
7. Light PE, Allen BG, Walsh MP et al. Regulation of adenosine triphosphate-sensitive potassium channels from rabbit ventricular myocytes by protein kinase C and type 2A protein phosphatase. Biochemistry 1995; 34:7252-7257.
8. Ribalet B, Ciani S, Eddlestone GT. ATP mediates both activation and inhibition of K_{ATP} channel activity via cAMP-dependent protein kinase in insulin-secreting cell lines. J Gen Physiol 1989; 94:693-717.
9. de Weille J, Schmid-Antomarchi H, Fosset M et al. Regulation of ATP-sensitive K^+ channels in insulinoma cells: activation by somatostatin and protein kinase C and the role of cAMP. Proc Natl Acad Sci USA 1989; 86:2971-2975.
10. Kirsch CE, Codina J, Birnbaumer L et al. Coupling of ATP-sensitive K^+ channels to A_1 receptors by G proteins in rat ventricular myocytes. Am J Physiol 1990; 259:H820-826.
11. Ito K, Kanno T, Suzuki K et al. Effects of cromkalim on the contraction and the membrane potential of the circular smooth muscle of guinea-pig stomach. Br J Pharmacol 1992; 105:335-340.
12. Quast U, Cook NS. Moving together: K^+ channel openers and ATP-sensitive K^+ channels. Trends Pharmacol Sci 1989; 10:431-435.
13. Cook NS. The pharmacology of potassium channels and their therapeutic potential. Trends Pharmacol 1988; 9:21-28.
14. Venkatesh N, Lamp ST, Weiss JN. Sulfonylureas, ATP-senstive K^+ channels, and cellular K^+ loss during hypoxia, ischemia, and metabolic inhibition in mammalian ventricle. Circ Res 1991; 69:623-367.
15. Kantor PF, Coetzee WA, Carmeliet EE et al. Reduction of ischemic K^+ loss and arrhythmias in rat hearts. Effect of glybenclamide, a sulfonylurea. Circ Res 1990; 66:478-485.
16. Wilde AAM, Escande D, Schumacher CA et al. Potassium accumulation in the globally ischemic mammalian heart. A role for the ATP-sensitive potassium channel. Circ Res 1990; 67:835-843.
17. Kubota I, Yamaki M, Shibata T et al. Role of ATP-sensitive K^+ channel on ECG ST segment elevation during a bout of myocardia ischemia. A study on epicardial mapping in dogs. Circulation 1993; 88:1845-1851.
18. Gross GJ, Peiper GM, Warltier DC. Comparative effects of nicorandil, nitroglycerin, nicotinic acid, and SG-86 on the metabolic status and functional recovery of the ischemic-reperfused myocardium. J Cardiovasc Pharmacol 1987; 10(Suppl 8):S76-84.
19. Pieper GM, Gross GJ. Salutary action of nicorandil, a new antianginal

drug, on myocardial metabolism during ischemia and on postischemic function in a canine preparation of brief, repetitive coronary artery occlusion: comparison with isosorbide dinitrate. Circulation 1987; 76:916-928.
20. Grover GJ, McCullough JR, Henry DE et al. Anti-ischemic effects of the potassium channel activators pinacidil and cromakalim and the reversal of these effects with the potassium channel blocker glyburide. J Pharmacol Exp Ther 1989; 251:98-104.
21. Grover GJ, Sleph PG, Dzwonczyk S. Pharmacologic profile of cromakalim in the treatment of myocardial ischemia in isolated rat hearts and anesthetized dogs. J Cardiovasc Pharmacol 1990; 16:853-864.
22. Sargent CA, Dzwonczyk S,Sleph PG et al. Cardioprotective effects of the cyanoguanidine potassium channel opener P-1075. J Cardiovasc Pharmacol 1993; 22:564-570.
23. Grover GJ, Dzwonczyk S, Sleph PG. Reduction of ischemic damage in isolated rat hearts by the potassium channel opener RP 52891. Eur J Pharmacol 1990; 191:11-19.
24. Sargent CA, Dzwonczyk S, Grover GJ. Effect of the potassium channel opener EMD 56431 on globally ischemic rat hearts. Pharmacology 1992; 45:260-268.
25. Grover GJ, Dzwonczyk S, Parham CS et al. The protective effects of cromaklim and pinacidil on reperfusion function and infarct size in isolated perfused rat hearts and anesthetized dogs. Cardiovasc Drugs Ther 1990; 4:465-74.
26. Cole WC, McPherson CD, Sontag D. ATP-regulated channels protect the myocardium against ischemia/reperfusion damage. Circ Res 1991; 69:571-581.
27. Galiñanes M, Shattock MJ, Hearse DJ. Effects of potassium channel modulation during global ischaemia in isolated rat heart with and without cardioplegia. Cardiovasc Res 1992; 26:1063-1068.
28. Ohta H, Jinno Y, Harada K et al. Cardioprotective effects of KRN2391 and nicorandil on ischemic dysfunction in perfused rat heart. Eur J Pharmacol 1991; 204:171-177.
29. McCullough JR, Normandin DE, Conder ML et al. Specific block of the anti-ischemic actions of cromakalim by sodium 5-hydroxydecanoate. Circ Res 1991; 69:949-958.
30. Grover GJ, D'Alonzo J, Sleph PG et al. The cardioprotective and electrophysiologic effects of cromakalim are attenuated by meclofenamate through a cyclooxygenase-independent mechanism. J Pharmacol Exp Ther 1994; 269:536-540.
31. Philipsone LH, Steiner DF. Pas de deux or more: the sulfonylurea receptor and K^+ channels. Science 1995; 268:372-373.
32. Aguilar-Bryan L, Nichols CG, Wechsler SW et al. Cloning of the beta-cell high-affinity sulfonylurea receptor: A regulator of insulin secretion. Science 1995; 268:423-425.
33. Venkatesh N, Lamp ST, Weiss JN. Sulfonylureas, ATP-sensitive K^+ loss during hypoxia, ischemia, and metabolic inhibition in mammalian ven-

tricle. Circ Res 1991; 623-629.
34. Mizumura T, Gross GJ. The cardioprotective effect of nicorandil, a K_{ATP} channel opener nitrate, is blocked by glyburide in dogs. J Mol Cell Cardiol 1995; 27:A24.
35. Grover GJ, Sleph PG, Parham CS. Nicorandil improves postischemic contractile function independently of direct myocardial effects. J Cardiovasc Pharmacol 1990; 15:698-705.
36. Auchampach JA, Maruyama M, Cavero I et al. The new K^+ channel opener aprikalim (RP 52891) reduces experimental infarct size in dogs in the absence of hemodynamic changes. J Pharmacol Exp Ther 1991; 259:961-967.
37. Auchampach JA, Maruyama M, Cavero I et al. Pharmacologic evidence for a role of ATP-dependent potassium channels in myocardial stunning. J Pharmacol Exp Ther 1992; 86:311-319.
38. Rohmann S, Weygandt H, Schelling P et al. Involvement of ATP-sensitive potassium channels in preconditioning protection. Basic Res Cardiol 1994; 89:563-576.
39. Imai N, Liang C, Stone CK et al. Comparative effects of nitroprusside and pinacidil on myocardial blood flow and infarct size in awake dogs with acute myocardial infarction. Circulation 1988; 77:705-711.
40. Thornton JD, Thornton CS, Sterling KL et al. Blockade of ATP-sensitive channels increases infarct size but does not prevent preconditioning in rabbit hearts. Circ Res 1993; 72:44-49.
41. Kitzen JM, McCallum JD, Harvey C et al. Potassium channel activators cromakalim and celikalim (WAY-120,491) fail to decrease myocardial infarct size in the anesthetized canine. Pharmacology 1992; 45:71-82.
42. Grover GJ, Newburger J, Sleph PG et al. Cardioprotective effects of the potassium channel opener cromakalim: stereoselectivity and effects on myocardial adenine nucleotides. J Pharmacol Exp Ther 1991; 257:156-162.
43. McPherson CD, Pierce GN, Cole WC. Ischemic cardioprotection by ATP-sensitive potassium channels involves high-energy phosphate preservation. Am J Physiol 1993; 265:H1809-H1818.
44. Behling RW, Malone HJ. K_{ATP}-channel openers protect against increaed cytosolic calcium during ischemia and reperfusion. J Mol Cell Cardiol 1995; 27:1804-1817.
45. Yao Z, Gross GJ. Effects of the K_{ATP} opener bimakalim on coronary blood flow, monophasic action potential duration, and infarct size in dogs. Circulation 1994; 89:1769-1775.
46. Grover GJ, D'Alonzo AJ, Parham CS. Cardioprotection with the K_{ATP} opener cromakalim is not correlated with ischemic myocardial action potential duration. J Cardiovasc Pharmacol 1995; 26:145-152.
47. Grover GJ, D'Alonzo AJ, Hess TA et al. Glyburide-reversible cardioprotective effect of BMS-180448 is independent of action potential shortening in guinea pig hearts. Cardiovasc Res (In Press).
48. Atwal KS, Grover GJ, Ahmed S et al. Cardioselective anti-ischemic ATP-sensitive potassium channel openers. J Med Chem 1993; 36:3971-3974.

49. Inoue I, Nagase H, Kishi K et al. ATP-sensitive K^+ channel in the mitochondrial inner membrane. Nature 1991; 352:244-7.
50. Paucek P, Mironova G, Mahdi F et al. Reconstitution and partial purification of the glibenclamide-sensitive, ATP-dependent K^+ channel from rat liver and beef heart mitochondria. J Biol Chem 1992; 36:26062-26069.
51. Paucek P, Yarov-Yarovoy V, Sun X et al. Physiological and pharmacological activators of the mitochondrial K_{ATP} channel. Biophys J 1995; 68:A145 (Abstract).
52. Jennings RB, Murry CE, Reimer KA. Preconditioning myocardium with ischemia. Cardiovasc Drug Ther 1991; 5:933-938.
53. Escande D, Thuringer D, Le Guern S et al. Potassium channel openers act through an activation of ATP-sensitive K^+ channels in guinea-pig cardiac myocytes. Plugers Arch 1989; 414:669-675.
54. Yao Z, Gross GJ. Activation of ATP-sensitive potassium channels lowers threshold for ischemic preconditioning in dogs. Am J Physiol 1994; 267:H1888-H1894.
55. Auchampach JA, Grover GJ, Gross GJ. Blockade of ischemic preconditioning in dogs by the novel ATP dependent potassium channel antagonist sodium 5-hydroxydecanoate. Cardiovasc Res 1992; 26:1054-1062.
56. Grover GJ, Sleph PG, Dzwonczyk S. Role of myocardial ATP-sensitive potassium channels in mediating preconditioning in the dog heart and their possible interaction with adenosine A_1-receptors. Circulation 1992; 86:1310-1316.
57. Schulz R, Rose J, Heusch G. Involvement of activation of ATP-dependent potassium channels in ischemic preconditioning in swine. Am J Physiol 1994; 267:H1341-H1352.
58. Toombs CF, Moore TL, Shebuski RJ. Limitation of infarct size in the rabbit by ischemic preconditioning is reversible with glibenclamide. Cardiovasc Res 1993; 27:617-622.
59. Thornton JD, Thornton CS, Sterlin DL et al. Blockade of ATP-sensitive potassium channels increases infarct size but does not prevent preconditioning in rabbit hearts. Circ Res 1993; 72:44-49.
60. Walsh RS, Tsuchida A, Daly JJF et al. Ketamine-xylazine anesthesia permits a K_{ATP} channel antagonist to attenuate preconditiong in rabbit myocardium. Cardiovasc Res 1994; 28:1337-1341.
61. Tan HL, Mazon P, Verberne HJ et al. Ischemic preconditioning delays ischemia induced cellular uncoupling in rabbit myocardium by activation of ATP-sensitive potassium channels. Cardiovasc Res 1993; 27:644-651.
62. Tomai F, Crea F, Gaspardone A et al. Ischemic preconditioning during coronary angioplasy is prevented by glibenclamide, a selective ATP-sensitive K^+ channel blocker. Circulation 1994; 90:700-705.
63. Speechly-Dick ME, Grover GJ, Yellon DM. Does ischemic preconditioning in the human involve protein kinase C and the ATP-dependent potassium channel? Studies of contractile function in an in vitro model. Circ Res (In Press).

64. Lu H Remeysen P, De Clerck F. The protection by ischemic preconditioning against myocardial ischemia and reperfusion induced arrhythmias is not mediated by ATP-sensitive potassium channels in rats. Coron Artery Dis 1993; 4:649-654.
65. Vegh A, Papp JG, Szekeres L et al. Are ATP sensitive potassium channels involved in the pronounced antiarrhythmic effects of preconditioning? Cardiovasc Res 1993; 27:638-643.
66. Vegh A, Szekeres L, Parratt JR. Protective effects of preconditioning of the ischaemic myocardium involve cyclo-oxygenase products. Cardiovasc Res 1990; 24:1020-1023.
67. Li Y, Kloner RA. The cardioprotective effects of ischemic 'preconditioning' are not mediated by adenosine receptors in rat hearts. Circulation 1993; 87:1642-1648.
68. Fralix TA, Steenbergen C, London RE et al. Glibenclamide does not abolish the protective effect of preconditioning on stunning in the isolated perfused rat heart. Cardiovasc Res 1993; 27:630-637.
69. Grover GJ, Dzwonczyk S, Sleph PG et al. The ATP-sensitive potassium channel blocker glibenclamide (glyburide) does not abolish preconditioning in isolated ischemic rat hearts. J Pharmacol Exp Ther 1993; 265:559-564.
70. Grover, GJ, Murray HN, Baird A et al. The K_{ATP} blocker sodium 5-hydroxydecanoate does not abolish preconditioning in isolated rat hearts. Eur J Pharmacol. 1995; 277:271-274.
71. Banerjee A, Locke-Winter C, Rogers KB et al. Preconditioning against myocardial dysfunction after ischemia and reperfusion by an $alpha_1$-adrenergic mechanism. Circ Res 1993; 73:656-670.
72. Heurteaux C, Lauritzen I, Widmann C et al. Essential role of adenosine, adenosine A_1 receptors, and ATP-sensitive K^+ channels in cerebral ischemic preconditioning. Proc Natl Acad Sci USA 1995; 92:4666-4670.
73. Downey JM, Liu GS, Thornton JD. Adenosine and the anti-infarct effects of preconditioning. Cardiovasc Res 1993; 27:3-8.
74. Kirsch CE, Codina J, Birnbaumer L et al. Coupling of ATP-sensitive K^+ channels to A_1 receptors by G proteins in rat ventricular myocytes. Am J Physiol 1990; 258:H820-H826.
75. Van Winkle DM, Chien GL, Wolff RA et al. Cardioprotection provided by adenosine receptor activation is abolished by blockade of the K_{ATP} channel. Am J Physiol 1994; 266:H829-H839.
76. Toombs CF, McGee DS, Johnston WE et al. Protection from ischaemic-reperfusion injury with adenosine treatment is reversed by inhibition of ATP-sensitive potassium channels. Cardiovasc Res 1993; 27:623-629.
77. Auchampach JA, Gross GJ. Adenosine A_1 receptors, K_{ATP} channels and ischemic preconditioning in dogs. Am J Physiol 1993; 264:H1327-1336.
78. Tsuchida A, Walsh RS, Downey J. Protection by the ATP-sensitive K^+ opener pinacidil can be blocked with an adenosine receptor antagonist. Circulation 1993; 88:I632 (Abstract).

79. Armstrong SC, Liu G, Downcy J ct al. K_{ATP} channcls and prccondition-ing of rabbit cardiomyocytes. J Mol Cell Cardiol 1995; 27:A23 (Abstract).
80. Kitakaze M, Minamino T, Node K et al. Opening of K^+ channels mimics the infarct size limiting effect of ischaemic preconditioning: role of activation of ectosolic 5'-nucleotidase. Eur Heart J 1994; 15(Suppl):482 (Abstract).
81. Gross GJ, Sleph PG, Grover PG. Cardioprotective effects of K_{ATP} openers occur independently of adenosine A_1 receptor activation. J Mol Cell Cardiol 1995; 27:A24 (Abstract).
82. Freyss Beguin M, Simon J et al. Effect of glibenclamide on the metabolism of fatty acids in cultures of new born rat heart cells under normoxic and hypoxic conditions. Prostaglandins Leukotrienes Essent Fatty Acids 1995; 52:325-331.

CHAPTER 9

The Role of G Proteins in Myocardial Preconditioning

Lucia Piacentini and Nigel J. Pyne

9.1. INTRODUCTION

Several comprehensive reviews have recently been published examining the role of G proteins in cardiovascular tissue.[1-3] This chapter will focus on modifications of G protein function under conditions of myocardial ischemia, reperfusion and ischemic preconditioning and wille examine the role of G proteins in receptor-mediated cardioprotection. Firstly, as an introduction to outlining their regulation under ischemic conditions, a description of the fundamental studies of G protein function in many cell types is given.

9.1.1. What are G Proteins?

G proteins, or guanine nucleotide-binding regulatory proteins, serve to transduce information from agonist-bound receptors to effector enzymes or ion channels. The heterotrimeric G proteins exist as a large family of isoforms which are encoded by distinct genes and are also processed from the differential splicing of pre-mRNA strands.[4] Their function is mediated by guanine nucleotide releasing proteins, the binding of guanosine triphosphate (GTP) and subsequent binding and activation of effector proteins.[4,5] The effector protein may also act as GTPase activating proteins in order to promote the termination of the action of the G protein.[6]

Heterotrimeric G proteins are composed of three nonidentical subunits:[4,5] an α subunit (39-52 kDa), a β subunit (35-36 kDa) and a γ subunit (8-10 kDa). In unstimulated cells, the G protein is inactive with guanosine diphosphate (GDP) bound to the α subunit. However, dependent upon the guanine-nucleotide binding equilibrium, GDP is

Myocardial Preconditioning, edited by Cherry L. Wainwright and James R. Parratt.

slowly released from the α subunit and can be replaced by GTP with the subsequent activation of effector enzymes.

9.1.2 G Protein Family

Approximately 20 α subunit isoforms have been identified and these can be grouped into four major classes: $G_{s\alpha}$, $G_{i\alpha}$, $G_{q\alpha}$ and $G_{12\alpha}$.[4,5] In addition to stimulating adenylyl cyclase, $G_{s\alpha}$ directly opens Ca^{2+} channels and closes Na^+ channels in cell membranes, including those from cardiac tissue.[7,8] Cholera toxin catalyses the transfer of the ADP-ribose moiety of NAD to Arg-201 of $G_{s\alpha}$. This modification leads to inhibition of GTPase activity and, thus, constitutive activation of the α subunit.[9] There are at least four cDNA clones of $G_{s\alpha}$ and two of these ($G_{s\alpha\text{-}1}$ and $G_{s\alpha\text{-}4}$) are expressed as single polypeptide chains and are derived from the differential splicing of pre-mRNA from a single gene.[10] The expression of $G_{s\alpha}$ cDNAs yield 52 and 45 kDa isoforms.[11] The difference between these is the inclusion or absence of 14 amino acids at the splice junction. Both forms appear to be capable of activating adenylyl cyclase and Ca^{2+} channels to an equal extent[12] despite the rate of release of GDP from the α subunit being faster for the 52 kDa isoform.[13] The 45 kDa isoform displays an increased susceptibility to in vitro phosphorylation by protein kinase C[14] and has been shown to be the predominant $G_{s\alpha}$ subunit in canine ventricular tissue.[15,16]

Members of the G_i family can undergo ADP-ribosylation at a cysteine residue near the C-terminal end of the α subunit in the presence of pertussis toxin.[17] $G_{i\alpha}$ exists as three isoforms, each having a molecular mass of approximately 41 kDa.[18] Recombinant forms of $G_{i\alpha 1/2/3}$ have all been shown to regulate K^+ conductance in atrial cells.[19] Similarly, myristoylated recombinant forms of all three can inhibit adenylyl cyclase.[20] However, only mRNA for $G_{i\alpha 2}$ and $G_{i\alpha 3}$ are expressed in cardiac tissue,[21-23] with $G_{i\alpha 2}$ being the predominant form.[24] The localization of mRNA for $G_{i\alpha 2}$ with mRNA for $G_{s\alpha}$[25] suggests that this G_i isoform is likely to be functionally associated with G_s, participating in the dual regulation of adenylyl cyclase. $G_{o\alpha}$ exists as two isoforms, each with a molecular mass of 39 kDa. In noncardiac tissue, $G_{o\alpha}$ has been shown to couple to muscarinic and $GABA_B$ receptors. Although present in myocardial tissue,[23] the function of $G_{o\alpha}$ has yet to be identified. A role for the modulation of phospholipase C by a pertussis toxin-sensitive substrate has been suggested for a number of cell types including those of cardiac origin.

The G_q class of G proteins are characterized by their insensitivity to either cholera or pertussis toxins.[26] All five members ($G_{q\alpha}$, $G_{11\alpha}$, $G_{14\text{-}16\alpha}$) can activate phospholipase C-β isozymes in vitro.[27] At least two isoforms, the G_q and G_{11} α subunits, have been identified in cardiac tissue.[28] $G_{z\alpha}$ is a recently described 42 kDa G protein predominantly found in platelets.[29] Although observed in neonatal cardiac tissue, this G protein does not appear to be present in adult cardiac tissue.[28]

There are five isoforms of β subunit (35-36 kDa) and six isoforms of γ subunit (8-10 kDa).[5] βγ subunits appear to have diverse functions. They may serve to anchor the α subunit to the lipid milieu of the plasma membrane, but may also serve to modulate effector function. For instance, βγ subunits have been shown to:

1. inhibit Type 1 adenylyl cyclase (Ca^{2+}/calmodulin dependent) by binding directly to it and inhibiting activity;[30]
2. bind to $G_{s\alpha}$-GTP and thereby inhibit agonist-stimulated adenylyl cyclase activity;[31]
3. activate Type II and Type IV adenylyl cyclase; this is conditional upon G_s activation;[32,33]
4. stimulate phospholipase C.[34,35] The α subunits are required in pmol amounts to elicit regulation, whereas nmol amounts of βγ subunits are required to stimulate phospholipase C. This is an important point since the differential effects of α versus βγ subunits will, therefore, be dependent upon the density of receptor occupancy;
5. stimulate phospholipase A_2 in retinal rod cells;[36] and
6. bind to β-adrenoceptor kinase (βARK) and muscarinic receptor kinase and facilitate homologous desensitization.[37,38] Desensitization is, therefore, dependent upon agonist occupancy of the receptor, dissociation of G_s into α and βγ subunits and the rapid translocation of βARK to specific free βγ subunit binding sites.

9.1.3. Function of G Proteins

At saturating concentrations of GTP and at physiological concentrations of Mg^{2+} (mM), agonist-receptor complexes markedly enhance the rate of GDP/GTP exchange in the G protein α subunit.[39] In this regard, the βγ subunits perform an important role, since at mM concentrations of Mg^{2+} they inhibit GDP release from the α subunit.[39] This permits the receptor to recycle and, therefore, allows it to activate many molecules of G protein sequentially. Thus, G proteins enable single agonist-occupied receptors to regulate several effector molecules and, therefore, provide the cell with a means of amplification of the initial hormonal signal. GTP also elicits dissociation of the heterotrimeric G protein into α-GTP and βγ subunits. Termination of the transmembrane signaling event results from the hydrolysis of GTP by the intrinsic GTPase activity of the α subunit.

9.2. MODIFICATION OF G PROTEIN FUNCTION AS A CONSEQUENCE OF MYOCARDIAL ISCHEMIA

As discussed above, G proteins in cardiovascular tissue modify several effectors including adenylyl cyclase[40-42] and certain ion channels.[43-45] The function of G proteins under ischemic conditions is of importance since modification at this level in the signal transduction pathway could alter the responsiveness of effectors to G protein-coupled receptors.

Recent evidence indicates that ischemia induces changes in both G protein levels and function in plasma membranes. Consequently, the effectors regulated by affected G proteins will also be modified by ischemia. The extent of these changes appears to depend on the species studied and type of ischemia employed and can be evaluated by both direct (e.g., measurement of the G proteins' intrinsic GTPase activity) and indirect (e.g., measurement of effector response) methods.

9.2.1. G_s Proteins

The dual regulation of adenylyl cyclase by G_i and G_s proteins is a frequently used model for the study of G protein function. After an initial period during which adenylyl cyclase is sensitized,[46] ischemia induces an overall decline in activity.[46-48] While some ischemia-induced changes occur at the level of the enzyme, modifications can occur upstream in the signaling pathway and subsequently affect the overall adenylyl cyclase activity. The deterioration in adenylyl cyclase activity is due, at least in part, to both the loss of $G_{s\alpha}$ subunits from the plasma membrane[49] and decreased activity of the G_s protein. Using a rabbit model of regional ischemia, Iwase et al[50] found that 10 minutes of ischemia induced a reduction in the relative stimulation of adenylyl cyclase by isoprenaline and the nonhydrolysable GTP analog Gpp(NH)p. This could be correlated with a reduction in β-adrenoceptor numbers and decreased activity of G_s proteins in the ischemic region compared to the nonischemic region, where G_s activity was assessed by the ability of isolated G_s protein to activate adenylyl cyclase when reconstituted in cyc-cells (a mutant cell line which lacks endogenous G_s).

9.2.2. G_i Proteins

In canine hearts subjected to coronary artery occlusion, an early sensitization of adenylyl cyclase occurs.[51] However, in contrast to rat myocardium where sensitization persists for 15 minutes,[52] after 5 minutes of in vivo ischemia, sensitization is no longer observed and both basal adenylyl cyclase activity and activity in the presence of the directly acting agent forskolin are reduced.[52] The relative isoprenaline-induced stimulation of adenylyl cyclase is similar in both ischemic and nonischemic tissue, implicating an intact β-adrenoceptor-linked signal transduction pathway in this model. However, at this point an ischemia-induced reduction in carbachol-mediated inhibition of both forskolin- and isoprenaline-stimulated adenylyl cyclase is apparent,[52] which suggests an impairment of muscarinic receptor-mediated control of adenylyl cyclase. Furthermore, the loss of Gpp(NH)p-mediated inhibition of forskolin-stimulated adenylyl cyclase indicates that the modification lies at the level of the G_i protein. These ischemic changes are still apparent when ischemia duration is extended to 15 minutes. Similar findings in porcine tissue after global ischemia have been shown to coincide with a loss of $G_{i\alpha}$ subunits from the plasma membrane,[53] but only when the duration of ischemia was greater than 30 minutes. However, studies utilizing membranes prepared from canine hearts demonstrate that a loss of $G_{i\alpha}$ is not apparent when measured by either pertussis toxin catalyzed ADP-ribosylation[52] or quantitive immunoblotting.[54]

Muscarinic inhibition of adenylyl cyclase is intrinsically linked to the GTPase activity of G_i proteins in canine sarcolemmal membranes[55] and the reduction in muscarinic inhibition of adenylyl cyclase activity is similarly associated with a reduction in carbachol-stimulated GTPase activity.[54] A number of observations indicate that this effect is a result of ischemia-induced modification of G_i protein:

1. the relative increase in GTPase response to carbachol is similar whether tissue is ischemic or nonischemic;[56]
2. both the amount of receptors and affinity of muscarinic agonists to receptors remains unchanged;[54]
3. as indicated above, the actual amount of $G_{i\alpha}$ subunits remains constant.[52,54]

Furthermore, since the carbachol stimulated binding of a nonhydrolysable analog of GTP is not modified by ischemia, then modification of G_i protein is suggested to be at a site distal to the guanine nucleotide-binding site.[54]

9.3. MODIFICATION OF G PROTEIN FUNCTION AS A CONSEQUENCE OF PRECONDITIONING

9.3.1. G_i PROTEINS

When ischemic canine myocardium is reperfused, the functional modification of G_i protein observed under ischemic conditions is reversed, such that after 15 minutes of reperfusion the carbachol-mediated increase in the GTPase activity of G_i protein is greater than in nonischemic tissue.[52,54] An improvement in carbachol-mediated inhibition of isoprenaline-stimulated adenylyl cyclase is also observed.[52,54] Moreover, when the myocardium is subjected to a second ischemic episode of 5 minutes, the depression of G_i protein function observed during a single ischemic episode does not occur.[54] Thus, in preconditioned canine myocardium the muscarinic receptor-mediated inhibitory control of adenylyl cyclase is preserved. These effects are not associated with increased amounts of $G_{i\alpha}$ subunits after either reperfusion[56,57] or a second period of ischemia,[56] but would appear to represent a reversal of the functional modification of G_i proteins which occurs during ischemia.

There is also evidence that the activity of G_i protein under ischemic conditions is modified by ischemic preconditioning in other species. Iwase et al[50] examined the effects of ischemia on the activity of Na^+/K^+ ATPase, assessed by measurement of oubain-sensitive ATP hydrolysis, in sarcolemmal vesicles prepared from rabbit myocardium. Preconditioning prevented the ischemia-induced reduction in Na^+/K^+ ATPase activity and this protection did not occur if rabbits were pretreated with pertussis toxin. Since pertussis toxin uncouples receptors from G_i and G_o proteins, preservation of Na^+/K^+ ATPase activity as a consequence of preconditioning would appear to require intact receptor G_i/G_o protein coupling.

9.3.2. G_s Protein

As discussed previously, ischemia induced a reduction in β-adrenoceptor mediated stimulation of adenylyl cyclase activity in membranes prepared from rabbit hearts which was associated with a reduction in the density of β-adrenoceptors at the plasma membrane and a decrease in G_s protein activity.[50] Biopsies taken from hearts which had undergone ischemic preconditioning and then a second period of ischemia indicate that in these tissues the ischemia-induced reductions in isoprenaline and Gpp(NH)p stimulation of adenylyl cyclase were abolished. Interestingly, this preservation was not associated with prevention of the ischemia-induced reductions in β-adrenoceptors at the plasma membrane but was concomitant with attenuation of the deleterious effects of ischemia on G_s protein activity.[50] One possibility, therefore, is that the ischemia-induced loss of the stimulatory effect of isoprenaline on adenylyl cyclase activity was through a reduction in G_s activity while preconditioning prevented this loss. However, the apparent improvement in isoprenaline-mediated stimulation observed in preconditioned tissue is, at least in part, a consequence of lower levels of response in the nonischemic myocardium from these hearts. This could be indicative of greater functional activity of G_i protein in these tissues when the adjacent area is ischemic. Restoration of adenylyl cyclase responsiveness to isoprenaline after pertussis toxin pretreatment in membranes from this region further supports this suggestion. Modification of 'virgin' myocardium as a consequence of the preconditioning occlusion has been noted previously[58] and could confound a direct comparison between ischemic and nonischemic tissues from the same heart. These results would suggest that pertussis toxin-sensitive G proteins modulate the response of preconditioned hearts to ischemia. Thus, although the protective effect on isoprenaline stimulation of G_s is concomitant with preservation of G_s activity, a role for improved G_i protein function, as has been suggested,[54] cannot be ruled out in this species.

9.4. ACTIVATION OF G PROTEIN COUPLED RECEPTORS AS A MECHANISM OF PRECONDITIONING

9.4.1. Ischemic Preconditioning

Several investigators have demonstrated that the cardioprotective effect of preconditioning in rabbit, dog and pig models is triggered by endogenous adenosine, released during the preconditioning occlusion, and mediated via the A_1-adenosine receptor.[59-61] Furthermore, adenosine release during the sustained ischemic episode is also crucial for cardioprotection.[62] The effectiveness of K_{ATP} channel blockers in preventing the infarct-size limiting effects of both ischemic preconditioning[63-65] and pretreatment with adenosine[64-68] or adenosine A_1-receptor agonists[63] suggests a possible effector for this phenomenon (reviewed in chapter 1).

A_1-adenosine receptors are linked to associated effectors, including ventricular K_{ATP} potassium channels, by pertussis toxin-sensitive G proteins.[69-72] Thornton et al[73] found that pretreating rabbits with pertussis toxin, thus preventing G_i/G_o protein interaction with associated receptors, attenuated the cardioprotective effect of ischemic preconditioning. Employing an in vivo rabbit model, it was shown that whilst pertussis toxin pretreatment did not affect the size of infarct measured in control animals, preconditioning was no longer effective (Fig. 9.1).

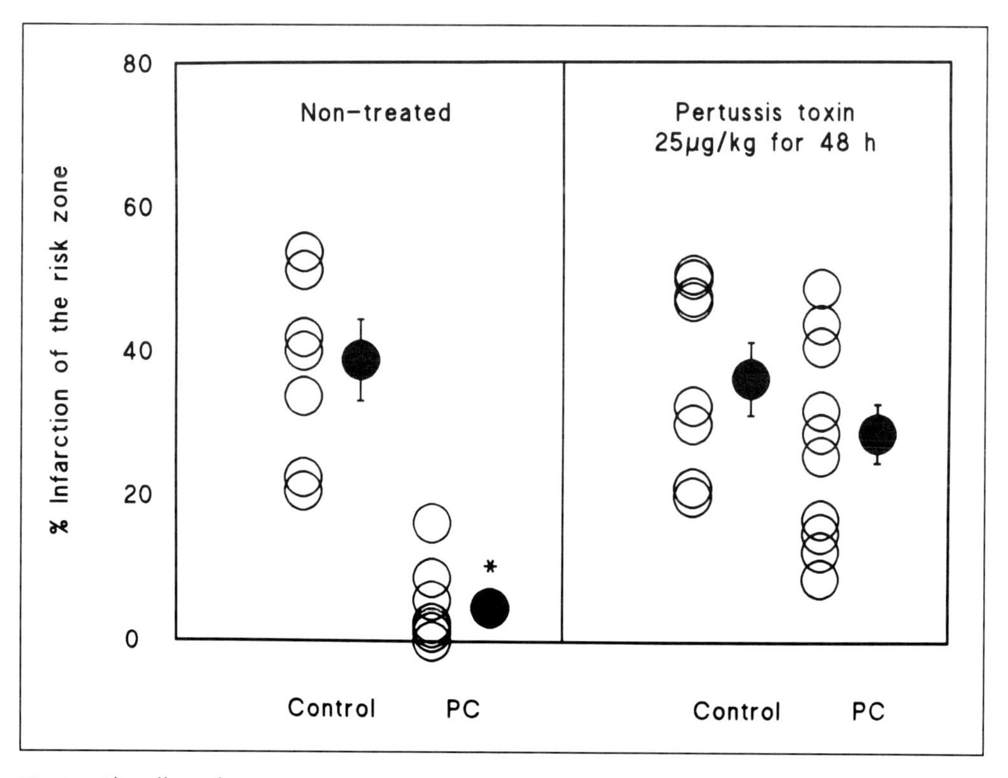

*Fig. 9.1. The effect of pertussis toxin on preconditioning against infarct size in rabbits. Infarct sizes are shown for control and preconditioned rabbits in both pertussis toxin-treated and nontreated groups. The closed symbol in each group indicates the group mean (with SEM) while open symbols indicate infarct sizes from individual hearts. * p<0.05 compared to appropriate control. Reproduced with permission from Thornton J et al, J Mol Cell Cardiol 1993; 25:311-320.*

In this species, therefore, intact coupling between receptors and pertussis toxin-sensitive G proteins seems necessary for cardioprotection by ischemic preconditioning.

In rat models however, there is evidence that adenosine is not involved in the improvement of postischemic functional recovery[74] nor the infarct size-limiting[75,76] and antiarrhythmic[77] effects of preconditioning (reviewed in chapter 12). Nonetheless, the preconditioning effect observed in rat models has characteristics similar to those seen in other species, i.e., the transient nature and type of protection. Despite the initial 'trigger' for protection appearing at variance with cardioprotection observed in other species, a universal underlying signal transduction system may be operative.

Several studies have been carried out examining this possibility with somewhat conflicting results.[78-82] Using rat isolated crystalloid-perfused hearts, the antiarrhythmic effect of preconditioning was examined in detail[78] and was found to be lost when hearts were taken from pertussis toxin-treated rats (Table 9.1). Furthermore, a recent study, again employing rat isolated crystalloid-perfused hearts, has shown that the improvement in postischemic function after global ischemia, associated with preconditioning, is also pertussis toxin-sensitive.[81] In contrast, when pertussis toxin-treated rats were subjected to in vivo myocardial ischemia, both the infarct size-limiting[80] (Fig. 9.2) and the antiarrhythmic[82] effects (Fig. 9.3) remained intact. Additionally, when rat hearts are isolated from pertussis toxin-treated animals and perfused, not with physiological salt solution but with blood from untreated donor rats, preconditioning against ischemia-induced arrhythmias

Table 9.1. Effect of pretreatment with pertussis toxin or of its vehicle on ventricular arrhythmias occurring during a 30 min occlusion of the left coronary artery in rat isolated hearts and on the effects of preconditioning (3 min ischemia and 10 min reperfusion prior to the 30 min occlusion)

Group	n	Total No of VPB	Incidence (%) VT	Incidence (%) VF	Sustained VF
Vehicle-treated					
Control	16	240 ± 90	88	63	43
Preconditioned	10	64 ± 8	70	0**	0*
Pertussis toxin-treated					
Control	9	226 ±120	89	33	11
Preconditioned	9	241 ± 93	89	33	11

VPB, ventricular premature beats, with values expressed as mean ± SEM, in hearts without sustained VF during the 30 min occlusion; VT, ventricular tachycardia; VF, ventricular fibrillation; *p < 0.05, ** p < 0.01 when compared to appropriate control group. Reproduced with permission from Piacentini L et al, Cardiovasc Res 1993; 27: 674-680.

is still apparent.[79] These conflicting results are perhaps indicative of a multifactorial basis of preconditioning, at least in this species; the pertussis toxin-sensitive protective effect, apparent in rat isolated crystalliod-perfused hearts, can be readily overwhelmed by a pertussis toxin-insensitive cardioprotection observed in vivo.

Adenosine released during ischemia has a role as an endogenous cardioprotectant (reviewed in refs. 83 and 84). However, the mechanism underlying its protective effect in ischemic preconditioning is still uncertain. As indicated above, convincing evidence in canine models indicates that it mediates its effect through the activation of K_{ATP} channels. During ischemia, whether tissue is preconditioned or not, mM amounts of adenosine are released by the myocardium.[85] Since maximal stimulation of adenosine receptors occurs at nM concentrations,[86] it seems unlikely that increased production of adenosine results in enhanced K_{ATP} channel opening during ischemia. Several studies have indicated that adenosine release is in fact decreased during ischemia in preconditioned hearts.[87-89] Furthermore, it can be demonstrated that the potassium channel blocker glibenclamide, at a concentration which

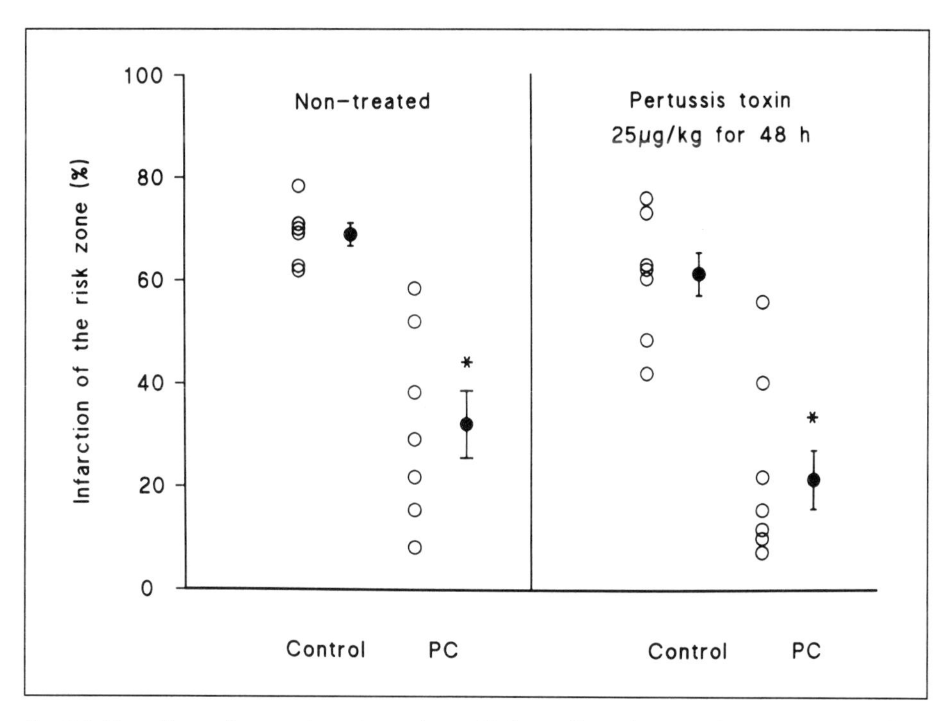

*Fig. 9.2. The effects of pertussis toxin on the anti-infarct effect of preconditioning in rats. Infarct sizes are shown for the control and preconditioned rats in both pertussis toxin-treated and nontreated groups. The closed symbol in each group indicates the group mean (with SEM) while open symbols indicate infarct sizes in individual hearts. * p<0.001 compared to appropriate control. Reproduced with permission from Liu Y et al, Cardiovasc Res 1993; 27:608-611.*

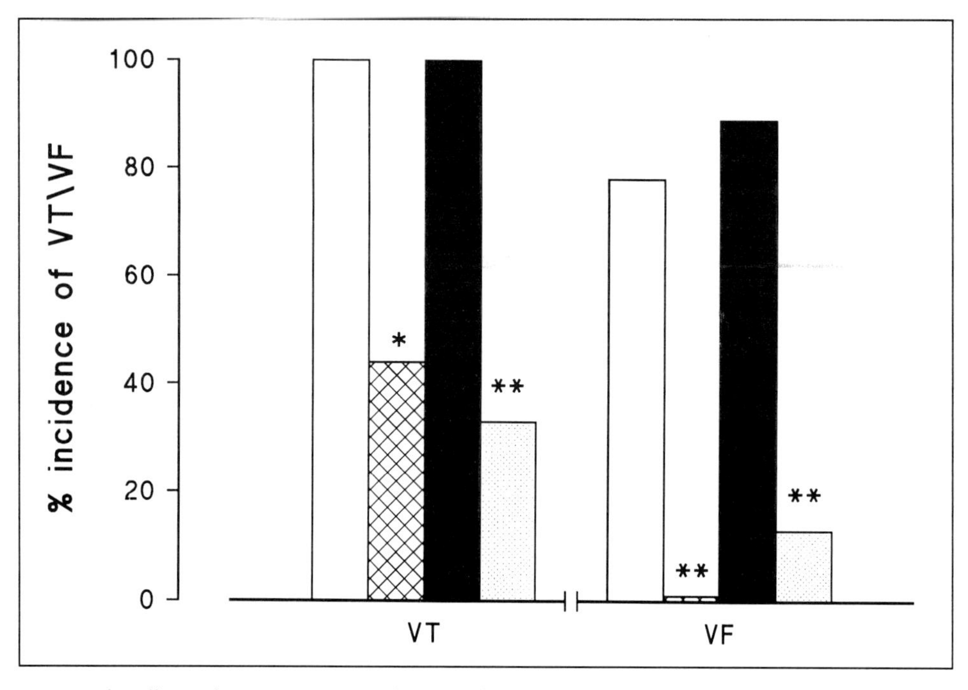

Fig. 9.3. The effects of pertussis toxin on the antiarrhythmic effects of preconditioning in anesthetized rats. The incidences of ventricular tachycardia (VT) and ventricular fibrillation (VF) occurring in anesthetized rats during a 30 minute occlusion of the left coronary artery. Control rats were subjected only to a 30 minute occlusion (open and solid columns for vehicle- and pertussis *toxin-treated rats, respectively). In preconditioned animals, the preconditioning cycle consisted of a 3 minute occlusion and 10 minute reperfusion (cross-hatched and stippled columns for vehicle- and* pertussis *toxin-treated rats, respectively). For each of the four groups, n = 9. * $p<0.05$, ** $p<0.001$ compared to the same treatment control. Reproduced with permission from Piacentini L et al, Br J Pharmacol 1995; 114:755-760.*

prevents ischemic preconditioning, does not increase infarct size as a consequence of ischemia in hearts which have not first been preconditioned.[65] Taken together, these findings suggest that in the preconditioned heart, adenosine is more effective at activating K_{ATP} channels. The prevention of ischemia-induced impairment of G_i protein function by preconditioning[54] could be the basis for this increased responsiveness.

9.4.2. Receptor-Mediated Preconditioning

G proteins are linked to multiple receptors which, in turn, share some of the same effectors.[90] Thus, it could be predicted that stimulation of receptors other than A_1 adenosine receptors would result in cardioprotection during subsequent periods of ischemia and reperfusion. Receptors which can be coupled to pertussis toxin-sensitive G proteins in ventricular tissue include M_2-muscarinic receptors,[91] angiotensin II receptors[92] and α_1-adrenoceptors.[93-95] It can be shown that protection against ischemia-induced damage is prevented by pretreatment with acetylcholine[96] or carbachol, a muscarinic receptor agonist.[73] Treatment with noradrenaline or α_1-adrenoceptors agonists[97-100] results in similar cardioprotection. Preliminary results indicate that pretreatment with angiotensin II[101] and opiate receptor agonists[102] also has a cardioprotective effect.

Caution should be applied in the interpretation of such results since, rather than mediating cardioprotection by activating the same signal transduction pathway as ischemic preconditioning, drug interventions could simply be reproducing the metabolic changes (e.g., increased metabolic demand) associated with the ischemic preconditioning stimulus. This has recently been suggested to be the mechanism of noradrenaline-mediated cardioprotection in isolated rat hearts.[100] Furthermore, not only can G proteins be coupled to multiple receptors, receptors can also be linked to more than one G protein and can thus have differential responses.[103,104] Comparison between the cardiac M_2-muscarinic and A_1-adenosine receptors exemplifies this last point; both can decrease hormonally-stimulated adenylyl cyclase[72,90,105] and open K_{ATP} channels,[43] but whilst there is much evidence for a cardiac M_2-muscarinic receptor coupled to phosphoinositide metabolism,[106-108] the evidence of a similar coupling of A_1-adenosine receptors in cardiac tissue is conflicting.[108,109]

9.5. POSSIBLE INTERACTION WITH PROTEIN KINASE C

Activation of protein kinase C has been implicated in the cardioprotective effect of preconditioning in rabbit and rat models (see chapter 11). Most of the agents mentioned above which can induce cardioprotection, stimulate phosphoinositide hydrolysis[106,110,111] and consequently protein kinase C activation, via the pertussis toxin-insensitive G_q protein. Whether or not protein kinase C activation, which has been shown to be important in the cardioprotective effect of preconditioning,[112,113] is a direct consequence of receptor activation or the acidic conditions associated with ischemia[114] is unclear and could depend on the species studied. Protein kinase C can interact with other signaling pathways by 'cross-talk'[115] and several plasma membrane components have been shown to be substrates for PKC-mediated phosphorylation including adenylyl cyclase[116] and G proteins[117-121] with a consequential modification of the affected protein's activity. Phosphorylation of G_i protein[117] and adenylyl cyclase[116] results in a decrease and increase, respectively, of functional activity. The physiological significance of this is not yet clear, but since protein kinase C activation is an early consequence of acute myocardial ischemia,[122] its interaction with other signaling pathways should be considered. In addition, some aspects of phosphoinositide metabolism have been shown to be modified by a pertussis toxin-sensitive G protein.[123]

9.6. CONCLUSIONS AND FUTURE DIRECTIONS FOR RESEARCH

It is clear that G proteins, especially pertussis toxin-sensitive G proteins, play a role in ischemic preconditioning by transduction of the receptor-mediated effect. Furthermore, the activity of G_i protein function under ischemic conditions is modified by ischemic preconditioning. Whether or not this modification is a manifestation of the overall cardioprotective effect of preconditioning or actively contributes to cardioprotection is not known. The difficulty in establishing the importance of G protein modification lies in the endpoints measured; tissue biopsies must be taken during the early phase of ischemia for biochemical assessment, consequently endpoints such as infarct size-limitation and recovery of left ventricular function cannot be measured concomitantly. What remains to be clarified is whether the modification of G protein function observed in canine myocardium can be reproduced with agents known to induce cardioprotection. It would be of special interest to correlate these changes in G_i protein function with activation or antagonism of adenosine receptors.

ACKNOWLEDGMENTS

Lucia Piacentini is a recipient of a Wellcome Travelling Research Fellowship. Thanks is given to Dr. Feraydoon Niroomand for allowing us to use reference 54 prior to publication.

REFERENCES

1. Holmer SR, Homcy CJ. G proteins in the heart. A redundant and diverse transmembrane signaling network. Circulation 1991; 84:1891-1902.
2. Fleming JW, Wisler PL, Watanabe AM. Signal transduction by G proteins in cardiac tissues. Circulation 1992; 85:420-433.
3. Johnson MD, Friedman E. G proteins in cardiovascular function and dysfunction. Biochem Pharmacol 1993; 45:2365-2372.
4. Hepler JR, Gilman AG. G proteins. TIBS 1992; 17:383-387.
5. Neer EJ. Heterotrimeric G proteins: organizers of transmembrane signals. Cell 1995; 80:249-257.
6. Berstein G, Blank JL, Jhon DY et al. Phospholipase C-β1 is a GTPase-activating protein for $G_{q/11}$, its physiological regulator. Cell 1992; 70:411-418.
7. Yatani A, Codina J, Imoto Y et al. A G protein directly regulates mammalian cardiac calcium channels. Science 1987; 238:1288-1293.
8. Schubert B, Van Dongen AMJ, Kirsch GE et al. β-adrenergic inhibition of cardiac sodium channels by dual G protein pathways. Science 1989; 245:516-519.
9. Kaziro Y, Itoh H, Nakafuku M. Organization of genes coding for G-protein α subunits in higher and lower eukaryotes. In: Iyengar R and Birnbaumer L eds. G-Proteins. San Diego: Academic Press 1990.
10. Bray P, Carter A, Guo V et al. Human cDNA clones of four species of

$G_{s\alpha}$ signal transduction protein. Proc Natl Acad Sci 1986; 83:8893-8897.
11. Robishaw JD, Smigel MD, Gilman AG. Molecular basis for two forms of the G-protein that stimulates adenylate cyclase J Biol Chem 1986; 261:9587-9590.
12. Graziano MP, Freissmuth M, Gilman AG. Expression of G_{sa} in *E. coli*: purification and properties of two forms of the protein. J Biol Chem 1989; 264:409-418.
13. Mattera AM, Graziano MP, Yatani A et al. Bacterially synthesised splice variants of the α subunit of the G-protein G_s activate both adenylyl cyclase and dihydropyridine-sensitive calcium channels. Science 1989; 243:804-807.
14. Pyne NJ, Freissmuth M, Palmer S. Phosphorylation of the spliced variant forms of the recombinant stimulatory guanine-nucleotide-binding regulatory protein (G_{sa}) by protein kinase C. Biochem J 1992; 285:333-338.
15. Longabaugh JP, Vatner DE, Graham DE et al. NADP improves the efficiency of cholera toxin catalysed ADP-ribosylation in liver and heart membranes. Biochem Biophys Res Comm 1986; 137:328-333.
16. Longabaugh JP, Vatner DE, Vatner SF et al. Decreased stimulatory guanosine triphosphate binding protein in dogs with pressure-overload left ventricular failure. J Clin Invest 1988; 81:420-424.
17. West RE, Moss J, Vaughan M et al. Pertussis toxin-catalysed ADP-ribosylation of transducin. J Biol Chem 1985; 260:14428-14430.
18. Suki W, Abramowitz J, Mattera R et al. The human genome encodes at least three non-allelic G proteins with α_i-type subunits. FEBS Lett 1987; 220:187-192.
19. Yatani A, Mattera R, Codina J et al. The G protein-gated atrial K^+ channel is stimulated by three distinct G_i α-subunits. Nature 1988; 336:680-682.
20. Taussig R, Iniguez-Lluhi JA, Gilman AG. Inhibition of adenylyl cyclase by $G_{i\alpha}$. Science 1993; 261:218-221.
21. Jones DJ, Randall RR. Molecular cloning of five GTP-binding protein cDNA species from rat olafactory neuroepithelium. J Biol Chem 1987; 262:14241-14249.
22. Katoh Y, Komuro I, Takaku F et al. Messenger RNA levels of guanine nucleotide-binding proteins are reduced in the ventricle of cardiomyopathic hamsters. Circ Res 1990; 67:235-239.
23. Luetje CW, Tietje KM, Christian JL et al. Differential tissue expression and developmental regulation of guanine nucleotide binding regulatory proteins and their messenger RNAs in rat heart. J Biol Chem 1988; 263:13357-13365.
24. Holmer SR, Stevens S, Homcy CJ. Tissue- and species-specific expression of inhibitory guanine nucleotide-binding proteins. Circ Res 1989; 65:1136-1140.
25. Brann MR, Collins RM, Spiegel A. Localization of mRNAs encoding the α-subunits of signal transducing G-proteins within rat brain and among peripheral tissues. FEBS Lett 1987; 222:191-198.

26. Strathman M, Simon M. G-protein diversity: a distinct class of α subunit is present in vertebrates and invertebrates. Proc Natl Acad Sci USA 1990; 87:9113-9117.
27. Kozasa T, Hepler JR, Smrka AV et al. Purification and a characterization of recombinant $G_{16\alpha}$ from Sf9 cells: activation of purified phospholipase C isozymes by G-protein a subunits. Proc Natl Acad Sci USA 1993; 90:9176-9180.
28. Hansen CA, Schroering AG, Robishaw JD. Subunit expression of signal transducing G proteins in cardiac tissue: implications for phospholipase C-β regulation. J Mol Cell Cardiol 1995; 27:471-484.
29. Carlson KE, Brass LF, Manning DR. Thrombin and phorbol esters cause the selective phosphorylation of G-protein other than G_i in human platelets. J Biol Chem 1989; 264:13298-13305.
30. Katada T, Kusakabe K, Oinuma M et al. A novel mechanism for the inhibition of adenylate cyclase via the inhibitory GTP-binding protein. Calmodulin-dependent inhibition of the cyclase catalyst by the βγ subunits of the GTP-binding proteins. J Biol Chem 1987; 262:11897-11900.
31. Hildebrandt JD, Kohnken RE. Hormone inhibition of adenylyl cyclase. Differences in the mechanisms for inhibition by hormones and G protein βγ. J Biol Chem 1990; 265:9825-983.
32. Tang WJ, Gilman A. Type specific regulation of adenylyl cyclase by G-protein βγ subunits. Science 1991; 254:1500-1503.
33. Federman AD, Conklin BR, Schrader KA et al. Hormonal stimulation of adenylyl cyclase through G_i-protein βγ subunits. Nature 1992; 356: 159-161.
34. Camps M, Hou C, Sidiropoulos D et al. Stimulation of phospholipase C by guanine nucleotide-binding protein βγ subunits. Eur J Biochem 1992; 206:821-831.
35. Boyer JL, Waldo GL, Harden TK. βγ-subunit activation of G-protein-regulated phospholipase C. J Biol Chem 1992; 267:25454-25456.
36. Jeshma CL, Axelrod J. Stimulation of phospholipase A_2 in bovine rod outer segments by the βγ sub-units of transducin and the inhibition by the α sub-unit. Proc Natl Acad Sci USA 1987; 80:3899-3902.
37. Pitcher JA, Inglese J, Higgins JB et al. Role of βγ subunits of G proteins in targeting the β-adrenergic receptor kinase to membrane-bound receptors. Science 1992; 257:1264-1267.
38. Kameyama K, Haga K, Haga T et al. Activation of G-protein βγ subunits of β-adrenergic and muscarinic receptor kinase. J Biol Chem 1993; 268:7753-7758.
39. Gilman AG. G proteins: transducers of receptor-generated signals. Ann Rev Biochem 1987; 56:615-49.
40. Birnbaumer L, Codina J, Mattera R et al. Structural basis of adenylate cyclase stimulation and inhibition by distinct guanine nucleotide regulatory proteins. In: Cohen P, Houslay MD, eds. Molecular mechanisms of transmembrane signalling. Elsevier Science Publishers B.V. (Biomedical Division) 1985.

41. Gilman AG. Regulation of adenylyl cyclase by G proteins. In: Nishizuka Y, ed. The biology and medicine of signal transduction. New York: Raven Press 1990.
42. Levitzki A, Bar-Sinai A. The regulation of adenylyl cyclase by receptor-operated G proteins. In: Taylor CW, ed. Intracellular Messengers. Oxford: Pergamon Press 1993.
43. Szabo G, Otero AS. G protein mediated regulation of K^+ channels in heart. Ann Rev Physiol 1990; 52:293-305.
44. Brown AM, Birnbaumer L. Direct G protein gating of ion channels. Am J Physiol 1988; 254:H401-H410.
45. Robishaw JD, Foster KA. Role of G proteins in the regulation of the cardiovascular system. Ann Rev Physiol 1989; 51:229-244.
46. Strasser RH, Krimmer J, Braun Dullaeus R et al. Dual sensitization of the adrenergic system in early myocardial ischemia: independent regulation of the β-adrenergic receptors and the adenylyl cyclase. J Mol Cell Cardiol 1990; 22:1405-1423.
47. Drummond RW, Sordahl LA. Temporal changes in adenylate cyclase activity in acutely ischemic dog heart: evidence of functional subunit damage. J Mol Cell Cardiol 1981; 13:323-330.
48. Will-Shahab L, Krause EG, Bartel S et al. Reversible inhibition of adenylate cyclase activity in the ischemic myocardium. J Cardiovasc Pharmacol 1985; 7(Suppl. 5):S23-S27.
49. Maisel AS, Ransnäs LA, Insel PA. β-adrenergic receptors and the G_S protein in myocardial ischemia and injury. Bas Res Cardiol 1990; 85(Suppl. 1):47-56.
50. Iwase T, Murakami T, Tomita T et al. Ischemic preconditioning is associated with a delay in ischemia-induced reduction of β-adrenergic signal transduction in rabbit hearts. Circulation 1993; 88:2827-2837.
51. Wollenberger A, Krause EG, Heier G. Stimulation of 3'5'-cyclic AMP formation in dog myocardium following arrest of blood flow. Biochem Biophys Res Comm 1969; 36:664-670.
52. Niroomand F, Bangert M, Beyer T et al. Reduced adenylyl cyclase inhibition by carbachol and GTP during acute myocardial ischemia. J Mol Cell Cardiol 1992; 24:471-475.
53. Will-Shahab L, Rosenthal W, Schulze W et al. G protein function in the ischemic myocardium. Eur Heart J 1991; 12 (Suppl F):135-138.
54. Niroomand F, Weinbrenner C, Weis A et al. Impaired function of inhibitory G proteins during acute myocardial ischemia of canine hearts and its reversal during reperfusion and a second period of ischemia: possible implications for the protective mechanism of ischemic preconditioning. Circ Res 1995; 76:861-870.
55. Fleming JW, Watanabe AM. Muscarinic cholinergic-receptor stimulation of specific GTP hydrolysis related to adenylate cyclase activity in canine cardiac sarcolemma. Circ Res 1988; 64:340-350.
56. Weis A, Zeifang F, Rauch B et al. A functional modification of G_i proteins as the underlying mechanism of ischemic preconditioning. Circulation 1993; 88:3406 (Abstract).

57. Fu LX, Kirkeboen KA, Liang QM et al. Free radical scavenging enzymes and G protein mediated receptors signalling systems in ischemically preconditioned porcine myocardium. Cardiovasc Res 1993; 27:612-616.
58. Przyklenk K, Bauer B, Ovize M et al. Regional ischemic 'preconditioning' protects remote virgin myocardium from subsequent sustained coronary occlusion. Circulation 1993; 87:893-899.
59. Liu GS, Thornton J, Van Winkle DM et al. Protection against infarction afforded by preconditioning is mediated by A_1 adenosine receptors in rabbit heart. Circulation 1991; 84:350-356.
60. Auchampach JA, Gross GJ. Adenosine A_1 receptors, K_{ATP} channels, and ischemic preconditioning. Am J Physiol 1993; 264:H1327-H1336.
61. Schwarz ER, Mohri M, Sack S et al. Preconditioning by ischemia lasts for only 30 minutes and is inhibited by adenosine antagonists. J Mol Cell Cardiol 1992; 24 (Suppl.I):S.93 (Abstract).
62. Thornton JD, Thornton CS, Downey JM. Effect of adenosine receptor blockade: preventing protective preconditioning depends on time of initiation. Am J Physiol 1993; 265:H504-H508.
63. Grover GJ, Sleph PG, Dzwonczyk S. Role of myocardial ATP sensitive potassium channels in mediating preconditioning in the dog heart and their possible interaction with adenosine A_1 receptors. Circulation 1992; 86:1310-1316.
64. Gross GJ, Auchampach JA. Blockade of ATP sensitive potassium channels prevents myocardial preconditioning in dogs. Circ Res 1992; 70:223-233.
65. Toombs CF, Moore TL, Shebuski RJ. Limitation of infarct size in the rabbit by ischemic preconditioning is reversible with glibenclamide. Cardiovasc Res 1993; 27:617-622.
66. Toombs CF, McGee DS, Johnston WE et al. Protection from ischemic-reperfusion injury with adenosine pretreatment is reversed by inhibition of ATP-sensitive potassium channels. Cardiovasc Res 1993; 27:623-629.
67. Van Winkle DM, Chien GL, Wolff RA et al. Cardioprotection provided by adenosine receptor activation is abolished by blockade of the K_{ATP} channel. Am J Physiol 1994; 266:H829-H839.
68. Yao Z, Gross GJ. A comparison of adenosine-induced cardioprotection and ischemic preconditioning in dogs. Circulation 1994; 89:1229-1236.
69. Kirsch GE, Condina J, Birnbaumer L et al. Coupling of ATP-sensitive K^+ channels to A_1 receptors by G proteins in rat ventricular myocytes. Am J Physiol 1990; 259:H820-H826.
70. Endoh M, Masahiko M, Taira N. Modification by islet-activating protein of direct and indirect inhibitory actions of adenosine on rat atrial contraction in relation to cyclic nucleotide metabolism. J Cardiovasc Res 1983; 5:131-142.
71. Böhm M, Brückner R, Neumann R et al. Role of guanine nucleotide-binding protein in the regulation by adenosine of cardiac potassium conductance and force of contraction. Evaluation with pertussis toxin. Naunyn-Schmiedeberg's Arch Pharmacol 1986; 332:406-405.

72. Kubalak SW, Newman WH, Webb JG. Differential effect of pertussis toxin on adenosine and muscarinic inhibition of cyclic AMP accumulation in canine ventricular myocytes. J Mol Cell Cardiol 1991; 23:199-205.
73. Thornton J, Liu GS, Downey JM. Pretreatment with pertussis toxin blocks the protective effects of preconditioning: evidence for a G protein mechanism. J Mol Cell Cardiol 1993; 25:311-320.
74. Asimakis GA, Inners-McBride K, Conti VR. Attenuation of postischemic dysfunction by ischemic preconditioning is not mediated by adenosine in the isolated rat heart. Cardiovasc Res 1993; 27:1522-1530.
75. Liu Y, Downey JM. Ischemic preconditioning protects against infarction in rat heart. Am J Physiol 1992; 263:H1107-H1112.
76. Li Y, Kloner R. The cardioprotective effects of ischemic 'preconditioning' are not mediated by adenosine receptors in rat hearts. Circulation 1993; 87:1642-1648.
77. Piacentini L, Wainwright CL, Parratt JR. The antarrhythmic effect of preconditioning, in rat isolated hearts, does not involve A_1 adenosine receptors. Br J Pharmacol 1992; 107:137P (Abstract).
78. Piacentini L, Wainwright CL, Parratt JR. The antiarrhythmic effect of ischemic preconditioning in isolated rat heart involves a pertussis toxin-sensitive mechanism. Cardiovasc Res 1993; 27:674-680.
79. Lawson CS, Coltart DJ, Hearse DJ. The antiarrhythmic action of ischemic preconditioning in rat hearts does not involve functional G_i proteins. Cardiovasc Res 1993; 27:681-687.
80. Liu Y, Downey JM. Preconditioning against infarction in the rat heart does not involve a pertussis toxin-sensitive G protein. Cardiovasc Res 1993; 27:608-611.
81. Hu K, Nattel S. Signal transduction systems underlying ischemic preconditioning in rat hearts. Circulation 1994; 90 (Suppl.):0578 (abstract).
82. Piacentini L, Wainwright CL, Parratt JR. Effects of *Bordetella* pertussis toxin pretreatment on the antiarrhythmic action of ischemic preconditioning in anesthetized rats. Br J Pharmacol 1995; 114:755-760.
83. Boachie-Ansah G, Kane KA, Parratt JR. Is adenosine an endogenous myocardial protective (antiarrhythmic) substance under conditions of ischemia? Cardiovasc Res 1993; 27:77-83.
84. Ely SW, Berne RM. Protective effects of adenosine in myocardial ischemia. Circulation 1992; 85:893-904.
85. Olsson RA. Changes in the content of purine nucleosides in canine myocardium during coronary occlusion. Circ Res 1970; 26:310-316.
86. Londos C, Cooper DM, Schlegel W et al. Adenosine analogs inhibit adipocyte adenylate cyclase by a GTP-dependent process: basis for actions of adenosine and methylxanthines on cyclic AMP production and lipolysis. Proc Natl Acad Sci 1978; 75:5362-5366.
87. Wikstrom G, Waldenstrom A, Ronquist G. Adenosine release is decreased by ischemic preconditioning. A study with microdialysis in thoracotamised pigs. J Mol Cell Cardiol 1993; 25 (Suppl I):S.35 (Abstract).

88. Dorheim TA, Mentzer RM, Van Wylen DGL. Preconditioning reduces interstitial fluid purine metabolites during prolonged myocardial ischemia. Circulation 1991; 84:0760 (Abstract).
89. Van Wylen DGL. Effect of ischemic preconditioning on interstitial purine metabolite and lactate accumulation during myocardial ischemia. Circulation 1994; 89:2253-2289.
90. Birnbaumer L. G proteins in signal transduction. Ann Rev Pharmacol Toxicol 1990; 30:675-705.
91. Kurose H, Ui M. Functional uncoupling of muscarinic receptors from adenylate cyclase in rat cardiac membranes by the active component of islet-activating protein, pertussis toxin. J Cyc Nuc Prot Phos Res 1983; 9:305-318.
92. Anand-Srivastava MB. Angiotensin II receptors negatively coupled to adenylate cyclase in rat myocardial sarcolemma. Biochem Pharmacol 1989; 38:489-469.
93. Barrett S, Honbo N, Karliner JS. α_1-adrenoceptor-mediated inhibition of cellular cAMP accumulation in neonatal rat ventricular myocytes. Naunyn Schmeidberg Arch Pharmacol 1993; 347:384-393.
94. Shah A, Cohen IS, Rosen MR. Stimulation of cardiac α-receptors increases Na/K pump current and decreases gK via a pertussis toxin-sensitive pathway Biophys J 1988; 54:219-225.
95. Vulliemoz Y, Verosky M, Horn EM et al. A pertussis toxin substrate regulates the cGMP response to α-adrenergic agonists in mouse heart. Circulation 1987; 76 (Suppl. IV):0245 (Abstract).
96. Yao Z, Gross GJ. Role of nitric oxide, muscarinic receptors, and the ATP-sensitive K^+ channel in mediating the effects of acetylcholine to mimic preconditioning in dogs. Circ Res 1993; 73:1193-1201.
97. Banerjee A, Locke Winter C, Rogers KB et al. Preconditioning against myocardial dysfunction after ischemia and reperfusion by an α_1-adrenergic mechanism. Circ Res 1993; 73:656-670.
98. Bankwala Z, Hale SL, Kloner RA. α-adrenoceptor stimulation with exogenous norepinephrine or release of endogenous catecholamines mimics ischemic preconditioning. Circulation 1994; 90:1023-1028.
99. Tsuchida A, Liu Y, Liu GS et al. $\alpha 1$-adrenergic agonists precondition rabbit ischemic myocardium independent of adenosine by direct activation of protein kinase C. Circ Res 1994; 75:576-585.
100. Asimakis GK, Inners-McBride K, Conti VR et al. Transient β-adrenergic stimulation can precondition the rat heart against postischemic contractile dysfunction. Cardiovasc Res 1994; 28:1726-1734.
101. Liu Y, Tsuchida A, Cohen MV et al. Pretreatment with angiotensin II limits infarction in rabbit heart. J Mol Cell Cardiol 1994; 26 : CLV (Abstract).
102. Schultz JE, Yoa Z, Gross GJ. Evidence for the involvement of opioid receptors in ischemic preconditioning in the rat heart. Circulation 1994; 90 (Suppl.):2556 (Abstract).
103. Ashkenzai A, Winslow JW, Peralta EG et al. An M_2 muscarinic receptor

subtype coupled to both adenylyl cyclase and phosphoinositide turnover. Science 1987; 238:672-673.
104. Peralta EG, Ashkenazai A, Winslow JW et al. Differential regulation of PI hydrolysis and adenylyl cylase by muscarinic receptor subtypes. Nature 1988; 334:434-437.
105. Hershberger RE, Feldman AM, Bristow MR. A_1 adenosine receptor inhibition of adenylate cyclase in failing and nonfailing human ventricular myocardium. Circulation 1991; 83:1343-1351.
106. Brown Masters S, Martin MW, Harden TK et al. Pertussis toxin does not inhibit muscarinic-receptor-mediated phosphoinositide hydrolysis or calcium mobilization. Biochem J 1985; 227:933-937.
107. Tajima T, Tsuji Y, Heller Brown J et al. Pertussis toxin-insensitive phosphoinositide hydrolysis, membrane depolarization, and positive inotropic effect of carbachol in chick atria. Circ Res 1987; 61:436-445.
108. Leung E, Johnston CI, Woodcock EA. Stimulation of phosphatidylinositol metabolism in atrial and ventricular myocytes. Life Sci 1986; 39: 2215-2220.
109. Kohl C, Linck B, Schmitz W et al. Effects of carbachol and (-)-N6-phenylisopropyladenosine on myocardial inosital phosphate content and force of contraction. Br J Pharmacol 1990; 101:829-834.
110. Endoh M, Hiramoto T, Ishihata A et al. Myocardial α_1-adrenoceptors mediate positive inotropic effect and changes in phosphatidylinositol metabolism. Species differences in receptor distribution and the intracellular coupling process in mammalian ventricular myocardium. Circ Res 1991; 68:1179-1190.
111. Meggs LG, Coupet J, Huang H et al. Regulation of angiotensin II receptors on ventricular myocytes after myocardial infarction in rats. Circ Res 1993; 72:1149-1162.
112. Ytrehus K, Liu Y, Downey JM. Preconditioning protects ischemic rabbit heart by protein kinase C activation. Am J Physiol 1994; 266: H1145-H1152.
113. Speechly-Dick ME, Mocanu MM, Yellon DM. Protein kinase C. Its role in ischemic preconditioning. Circ Res 1994; 75:586-590.
114. Strasser RH, Simonis G, Weinbrenner C et al. Ischemia-induced activation of protein kinase C: evidence for an acidosis-dependent, but energy-independent mechanism. Circulation 1994; 90:1117 (Abstract).
115. Houslay MD. 'Crosstalk': a pivotal role for protein kinase C in modulating relationships between signal transduction pathways. Eur J Biochem 1991; 195:9-27.
116. Yoshimasa T, Sibley DR, Bouvier M et al. Cross-talk between cellular signalling pathways suggested by phorbol-ester-induced adenylate cyclase phosphorylation. Nature 1987; 327:67-70.
117. Katada T, Gilman AG, Watanabe Y et al. Protein kinase C phosphorylates the inhibitory guanine-nucleotide-binding regulatory component and apparently suppresses its function in hormonal inhibition of adenylate cyclase. Eur J Bioch 1985; 1985:431-437.

118. Bell JD, Brunton LL. Enhancement of adenylate cyclase activity in S49 lymphona cells by phorbal esters. J Biol Chem 1986; 261:12036-12041.
119. Pyne NJ, Murphy GJ, Milligan G et al. Treatment of intact hepatocytes with either the phorbal ester TPA or glucagon elicits the phosphorylation and functional inactivation of the inhibitory guanine nucleotide regulatory protein G_i. FEBS Lett 1989; 243:77-82.
120. Gordeladze JO, Bjoro T, Torjesen PA et al. Protein kinase C stimulates adenylyl cyclase activity in prolactin-secreting rat adenoma (GH4C1) pituicytes by inactivation of the inhibitory GTP-binding protein G_i. Eur J Biochem 1989; 183:397-406.
121. Bushfield M, Murphy GJ, Lavan BE et al. Hormonal regulation of $G_{i2\alpha}$-subunit phosphorylation in intact hepatocytes. Biochem J 1990; 268:449-457.
122. Strasser RH, Braun-Dullaeus R, Walendzik H et al. α_1-receptor independent action of protein kinase C in acute myocardial ischemia. Circ Res 1992; 70:1304-1312.
123. Traynor-Kaplan AE, Harris AL, Thompson BL et al. Transient increase in phosphatidylinisitol 3,4-biphosphate and phosphatidylinisotol triphosphate during activation of human neutrophils. J Biol Chem 1989; 264:15668-15673.

CHAPTER 10

Mimicking Preconditioning with Catecholamines

Tanya Ravingerová

10.1. INTRODUCTION

The last decade has witnessed the development of a novel approach to myocardial protection against ischemia that exploits the heart's own endogenous protective mechanisms. The concept of ischemic preconditioning has offered powerful new tools to combat the deleterious effects of long-lasting ischemia through adaptation of the heart by means of preceding, short episodes of the same ischemic stress. This short-term adaptive phenomenon is mediated by cell signaling mechanisms which open possibilities for pharmacological modulation at different levels of signal transduction (receptors, second messengers, effectors).[1,2]

A variety of substances are known to be released from the myocardium in the early period of ischemia that may modulate the severity of ischemic injury. These biochemical factors include both potentially protective substances (like bradykinin, adenosine and nitric oxide) and those exacerbating cellular damage (for example, catecholamines, oxygen-derived free radicals and thromboxane).[3,4] A general approach towards the pharmacological induction of preconditioning has been represented by attempts either to enhance the effects of protective substances, or to antagonize the detrimental effects of toxic substances. An alternative approach could be mimicking the stressful situation caused by a brief episode of ischemia, by exogenous application of potentially deleterious substances or by potentiation of their release from the myocardium. It has been recently found that, somewhat unexpectedly, a stressful stimulus such as administration of catecholamines can elicit similar protective effects to ischemic preconditioning. This chapter deals with the pharmacological induction of preconditioning

Myocardial Preconditioning, edited by Cherry L. Wainwright and James R. Parratt.

by catecholamines in different experimental settings and discusses their possible mechanisms.

10.2. MYOCARDIAL ISCHEMIA AND THE RELEASE OF CATECHOLAMINES

It is well established, on the basis of experimental data and clinical observations, that both plasma and cardiac catecholamines are increased early in the course of myocardial ischemia due to activation of the sympathetic nervous system.[5-9] In addition to the reflex release of catecholamines into the systemic circulation, which is a calcium-dependent exocytotic process,[10] several mechanisms contribute to this enhanced release of catecholamines from sympathetic nerve endings in the myocardium. In the normal myocardium, sympathetic nerve stimulation increases the exocytotic release of this neurotransmitter leading to the stimulation of myocardial cells.[11] This process has been found to be modulated by different mechanisms such as presynaptic inhibition via α–adrenoceptors, adenosine, prostanoids,[12-14] clearance from the synaptic cleft by a combination of neuronal reuptake (uptake$_1$) and washout of the neurotransmitter.[11] In the first few minutes of ischemia, enhanced neuronal uptake of noradrenaline, presynaptic inhibition by adenosine, as well as ATP and substrate dependency of exocytosis, all contribute to effective prevention of excess extracellular accumulation of noradrenaline.[15] With longer durations of ischemia (10-40 minutes), the heart is no longer protected against excess adrenergic stimulation, since mechanisms of local metabolic release become more important, whilst central stimulation-induced exocytotic release progressively fails. In this phase massive accumulation of noradrenaline in the extracellular space occurs. Studies in isolated hearts have revealed the time course and the mechanisms of this nonexocytotic calcium independent release. Under conditions of energy depletion, the release of noradrenaline is a two-step process, starting with loss from the storage vesicles into the neuron axoplasm followed by transport across the cell membrane into the extracellular space by the uptake$_1$ carrier operating against its normal direction.[16] Enhanced axoplasmic concentrations of noradrenaline, due to its loss from storage vesicles and to suppressed inactivation by monoamine oxidase (facilitated by exhaustion of energy metabolism and inhibition of vesicular H^+ ATPase), explain the reversal in the direction of transport by the uptake carrier.[15] However, this mechanism starts operating only under conditions of impaired homeostasis of sodium.[17] An increase in intracellular sodium concentration, an important prerequisite for the outward cotransport of catecholamines across the cell membrane, occurs during early myocardial ischemia due to inhibition of Na^+/K^+ ATPase activity and to enhanced sodium influx across the cell membrane via Na^+/H^+ exchange mechanism.[16] Nonexocytotic noradrenaline release is not modulated by presynaptic inhibition nor by calcium entry blockers effective in blocking exocytotic mechanisms. On the other hand, blockade of the uptake carrier by tricyclic antidepressants, such as desipramine, inhibits both the inward and outward catecholamine transport and thus, under conditions of ischemia up to 30 minutes duration, effectively suppresses noradrenaline release.[15] Thus, the carrier-mediated efflux of noradrenaline represents the crucial mechanism of enhanced release of catecholamines for up to 40 minutes of ischemia. In periods of ischemia of longer duration, with resultant and progressive energy deficiency, the storage vesicles become progressively depleted of noradrenaline, and release is then mediated by passive diffusion through leaky membranes. In this phase, neuronal reuptake blockade can no longer reduce the accumulation of catecholamines in the extracellular space.

Although this time course of catecholamine release has been studied mainly in isolated hearts, the above mechanisms apply in general to conditions in vivo. In humans, however, the situation can be even more complicated, since different mechanisms of catecholamine release can operate in parallel in the ischemic and nonischemic regions of the myocardium. Accordingly, one must bear in mind that these local inhomogeneities within the ischemic myocardium can be manifested by opposing responses in different regions of the heart to agents designed to inhibit the release and accumulation of catecholamines and consequently to suppress the detrimental effects of excess adrenergic stimulation of the heart.

10.3. CATECHOLAMINES AND MYOCARDIAL INJURY

It is well known that catecholamines affect both α- and β-adrenoceptors modulating heart rate, contractility and peripheral resistance. Their positive inotropic and lusitropic effects are mediated via stimulation of adenylyl cyclase activity (coupled via stimulatory G_s proteins to β-receptors) and consequent elevation of cAMP leading to activation of protein kinase A (PKA).[18] This cAMP-dependent enzyme is known to induce metabolic effects such as acceleration of lipolysis and glycogenolysis and stimulation of trans-sarcolemmal and sarcoplasmic Ca^{2+} transport via cAMP-dependent phosphorylation of the cell membrane proteins.[19,20] Stimulation of α_1-adrenoceptors is associated with an increased formation of two second messengers, inositol triphosphate (IP_3) and diacylglycerol (DAG). The former is responsible for the stimulation of Ca^{2+} release from the sarcoplasmic reticulum, while the latter has been found to activate protein kinase C (PKC).[18,21] The role of this enzyme in the α_1-adrenoceptor-mediated positive inotropy is not well understood, although some studies have demonstrated their role in increasing the slow inward Ca^{2+} current by means of a cAMP-independent phosphorylation of major cardiac proteins.[22,23] An additional mechanism contributing to the maintenance of the α_1-adrenoceptor-mediated inotropic effect may be intracellular alkalinization due to stimulation of Na^+/H^+ exchange.[24] The final response of the heart is manifested by an increase in heart rate and contractile function, thus increasing myocardial oxygen demand. High concentrations of catecholamines can cause damage to the myocardium by leukocyte infiltration, myofibrillar degeneration, and myofiber necrosis.[25]

The electrophysiological responses of the heart to α- and β-adrenoceptor stimulation differ with regard to species, location of the receptors in the myocardium, and whether the myocardium is normal. Whilst in ventricular myocardium the main response to β-adrenoceptor stimulation under physiological conditions is refractory period shortening (a proarrhythmic effect), the contribution of α-adrenoceptors under normal conditions is negligible.[26] This is in contrast to their important role under pathological conditions.[18,27] In Purkinje tissue in normal hearts, α_1-adrenoceptor stimulation prolongs refractoriness, increases action potential duration and decreases automaticity.[28,29] In general, these electrophysiological changes are considered to be antiarrhythmic. In addition, in intact anesthetized dogs Purkinje refractoriness can also be prolonged by pre- and postsynaptic α_2-adrenoceptor stimulation.[28] However, α- and β-receptor-mediated effects can occur simultaneously in different areas of the myocardium and are a source of spatial dispersion of refractoriness, predisposing the heart to re-entrant arrhythmias.[27,30]

Under conditions of myocardial ischemia, elevated catecholamine concentrations and adrenergic stimulation are believed to aggravate myocardial injury. Catecholamines accelerate cell damage in the border zones of infarction by increasing energy demand and facilitating calcium influx into the cells.[31] There is also little doubt that increased sympathetic activity plays an important role in exacerbating arrhythmias by enhancing automaticity and triggered activity.[32] Furthermore, a large number of studies have shown the protective effect in myocardial ischemia of α_1- and β-adrenoceptor blockade, as well as of blockade of the uptake$_1$ carrier, thus inhibiting noradrenaline release.[27,30,33-35]

There is, however, evidence suggesting that the role of catecholamines in myocardial ischemia is not unequivocally deleterious. First of all, there is no clear evidence about the relationship between increased catecholamine concentrations and the incidence of arrhythmias in the clinical setting.[7] Furthermore, in experimental studies it has been demonstrated that neither an increased central sympathetic activity, nor elevated circulating catecholamines are essential for the occurrence of arrhythmias.[36,37] Moreover, a great number of interventions aimed at adrenergic/sympathetic modulations have little or no effect on arrhythmias.

In the failing heart adrenoceptor stimulation can certainly be beneficial by maintaining a sufficient cardiac output. In addition, it can lead to attenuation of the arrhythmogenic effects in heart failure induced by rapid pacing in dogs.[38] Electrophysiological effects of catecholamines on partially depolarized myocardium may counteract reentry by prolonging action potential duration and refractoriness and by increasing conduction velocity.[39] Variability in the results obtained can be explained, to a large extent, by species differences, variability in experimental protocols (in vivo versus in vitro), in clinical settings by the choice of the investigated end-point, as well as by the multifactorial nature of the ischemic injury. The effects of catecholamines have been also shown to depend on their concentrations, being cardiotoxic in a higher concentration range.

Several lines of evidence indicate that catecholamines, under certain conditions, can exert a beneficial effect on the myocardium and their possible role in myocardial preconditioning will now be discussed.

10.4. MIMICKING PRECONDITIONING WITH CATECHOLAMINES

10.4.1. Protection Against Contractile Dysfunction

One of the first attempts to establish a role of catecholamines as mediators of myocardial preconditioning was the study of Locke-Winter et al.[40] In this study using rat isolated hearts, a preconditioning-like effect on myocardial dysfunction after ischemia/reperfusion was induced by a short infusion of β-adrenoceptor agonists prior to the onset of global ischemia. Heart function was significantly better preserved in the hearts pretreated with isoprenaline or forskolin, pointing to the role of cAMP stimulation. Protection was abolished by reserpine and α_1-adrenoceptor blockade, indicating a role for the neuronal release of noradrenaline and subsequent α_1-adrenoceptor stimulation. Later, using the same model of postischemic myocardial dysfunction in rat isolated heart, Banerjee et al[41] confirmed that transient ischemic preconditioning is mediated by the sympathetic neurotransmitter noradrenaline since its release was markedly increased after transient ischemia and functional protection was lost after depletion of endogenous stores by pretreatment with reserpine. The authors clearly demonstrated that preconditioning can be also induced pharmacologically by 2 minutes of perfusion with low concentrations of exogenous noradrenaline or the α_1-adrenoceptor agonist phenylephrine, followed by a 10 minute recovery period before the ischemic insult. This short pretreatment facilitated better postischemic recovery of contractile function which was accompanied by higher levels of tissue ATP at the end of reperfusion. The crucial role of α_1-adrenoceptor stimulation in the preconditioning cascade has been documented by inhibition of this protection by selective α_1-adrenoceptor blockade.[41] In contrast, Asimakis et al[42] have shown that the protective effect of noradrenaline pretreatment on postischemic contractile dysfunction in rat isolated hearts (Fig. 10.1) is mediated by β-adrenoceptor stimulation that inducing a 'transient demand' ischemia. These investigators were unable to achieve any protection with an α_1-adrenoceptor agonist; furthermore, β-adrenoceptor antagonists blocked the protective effect of noradrenaline. Protection

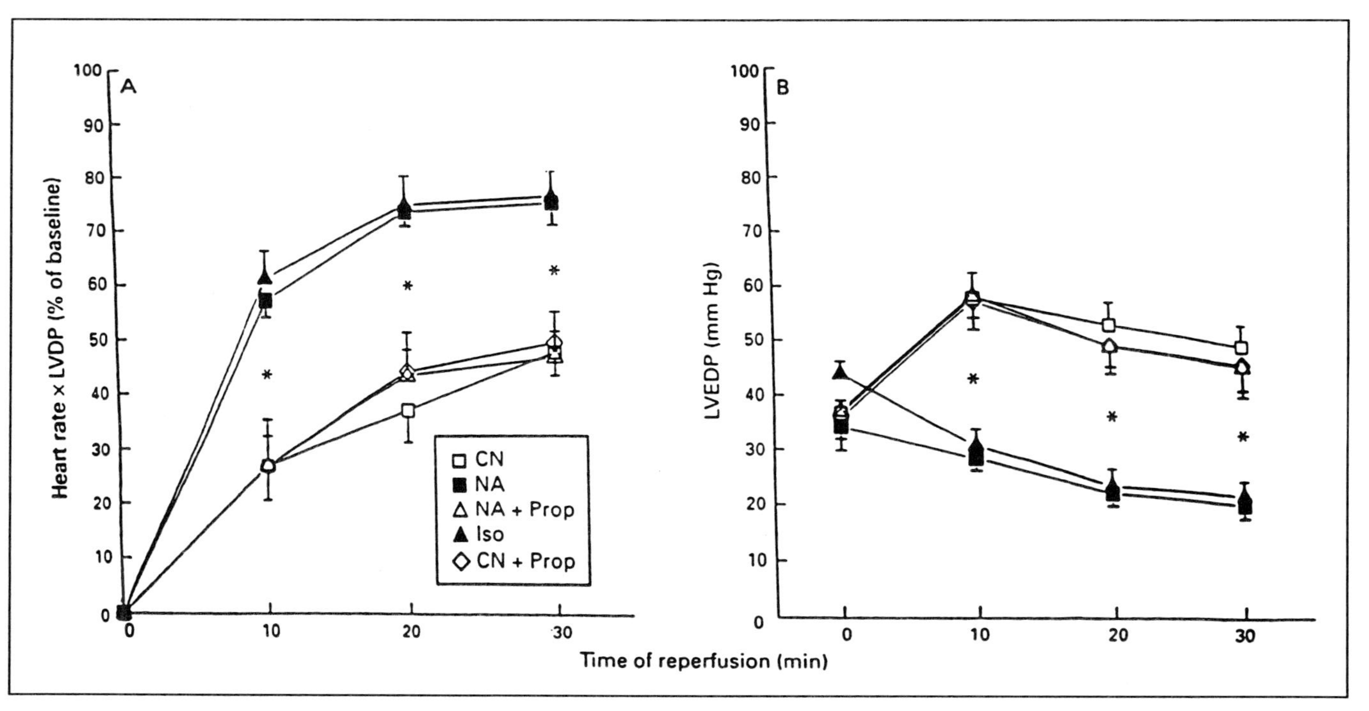

*Fig. 10.1. Postischemic recovery of left ventricular function in control hearts and in noradrenaline and isoprenaline treated hearts after 30 minute ischemia. (A) Postischemic recovery of heart rate x left ventricular developed pressure (LVDP). (B) Postischemic left ventricular end diastolic pressure (LVEDP) during reperfusion. NA, noradrenaline; Iso, isoprenaline; Prop, propranolol; CN, control; n = 10-12 hearts for each group. Error bars - SEM. *p<0.05, noradrenaline and isoprenaline values (filled symbols) v control. Reproduced with permission from Asimakis GK et al, Cardiovasc Res 1994; 28:1726-1734.*

against contractile dysfunction by low doses of intracoronary infused noradrenaline has been also achieved in the ischemic canine heart.[43] The protective effect of α_1-adrenoceptor stimulation was found to be linked to the enhanced production of adenosine due to stimulation of 5'-nucleotidase activity. Subsequent activation of A_1-adenosine receptors coupled to G_i proteins attenuated β–adrenoceptor-mediated alterations in the myocardium, as well as inhibiting presynaptic noradrenaline release.[43]

However, the most recent paper of Weselcouch et al[44] is in strong disagreement with the studies discussed which suggest that endogenous catecholamines are important mediators of ischemic preconditioning and that exogenous catecholamines precondition the heart against contractile dysfunction. These authors utilized a similar model of myocardial dysfunction in rat isolated hearts to that of Banerjee et al,[41] but with a more aggressive ischemic insult (30 minutes global ischemia) and a different preconditioning protocol (four episodes of 5 minute global ischemia and 5 minute reperfusion). In contrast to the results of Banerjee et al,[41] Weselcouch et al[44] found that depletion of endogenous catecholamines stores did not modify the improvement in postischemic cardiac function recovery afforded by preconditioning. Furthermore, it had no effect on either preischemic or postischemic parameters of heart function. Pretreatment with exogenous noradrenaline to mimic preconditioning failed to attenuate the severity of ischemic injury. The authors concluded that endogenous catechololamines play little or no role in the response to ischemia in this preparation.

At first glance it is not clear why these results are so disparate. However, these discrepancies may be related to differences in experimental settings, variabilities in the preconditioning protocols utilized and in the multiple mechanisms of protection operating in each case.

10.4.2. Reduction of Myocardial Necrosis

Apart from the improvement in myocardial dysfunction, catecholamines have been shown to play a role in reducing myocardial necrosis. Thornton et al[45] have recently shown that infusion of tyramine, an agent which induces the release of endogenous noradrenaline from its stores, when given shortly before coronary artery occlusion, pharmacologically preconditions rabbit hearts and significantly reduces infarct size. Infarct size limitation by endogenous catecholamines appeared to be mediated by α-adrenoceptors, since the protective effect of tyramine could be blocked by α_1-, but not by β-adrenoceptor antagonists. However, since α-adrenoceptor blockade did not influence classical preconditioning, the authors concluded that α-adrenoceptor activation is not a mechanism of ischemic preconditioning, at least in the rabbit model. Neither does preconditioning in rabbits require the presence of tissue catecholamines[46] since marked depletion of myocardial noradrenaline stores by surgical sympathectomy did not attenuate the protective effect of preconditioning on infarct size. Similarly, depletion of catecholamine stores by reserpine (5 mg/kg, 18 hours prior to preconditioning) only partially blocked protection. This is in contrast with the data of Toombs et al[47] who pretreated rabbits with reserpine (5 mg/kg) 24 hours before the induction of preconditioning and showed that reserpinized preconditioned animals had infarct sizes that did not differ significantly from those in the control nonpreconditioned group, suggesting the loss of protection in noradrenaline-depleted myocardium. The results of Toombs et al[47] are consistent with the results obtained by Bankwala et al[48] in an in vivo rabbit model of occlusion and reperfusion. These investigators demonstrated that the release of endogenous catecholamines induced by tyramine is essential in triggering protection against infarct size in rabbits. In addition, these authors observed similar protection after infusion of exogenous noradrenaline (5 minutes; 0.25 μg/kg/min), which was blocked by prazosin treatment (Fig. 10.2). In other words, preconditioning in rabbits can be simulated by either endogenously released or exogenously applied catecholamines and α-adrenoceptor mechanisms participate in this protection. The authors

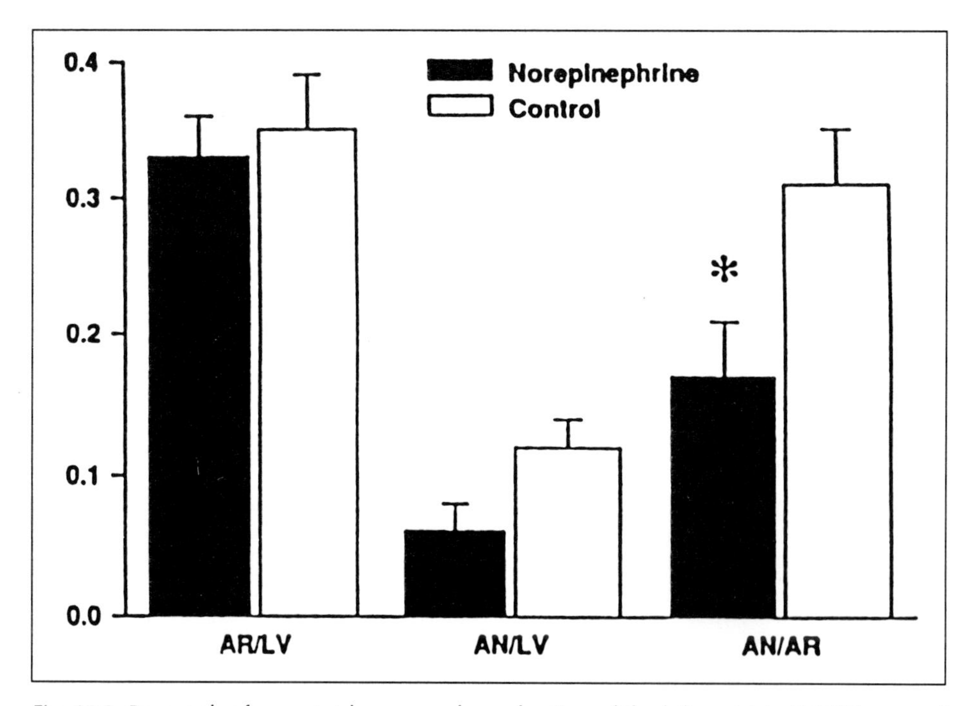

*Fig. 10.2. Bar graph of area at risk expressed as a fraction of the left ventricle (AR/LV), area of necrosis expressed as a fraction of the left ventricle (AN/LV), and area of necrosis expressed as a fraction of the area at risk (AN/AR) of animals randomized to noradrenaline (norepinephrine) or control groups. *p<.02, control vs. norepinephrine. Reproduced with permission from Bankwala Z et al, Circulation 1994; 90:1023-1028.*

explained the observed protective effect by the increase in oxygen demand due to the increased systemic arterial pressure and heart rate, and thus a relative myocardial ischemia. Another possible explanation[45] is that catecholamines stimulate the release of adenosine with a subsequent activation of A_1-adenosine receptors.[43]

On the other hand, Tsuchida et al[49] have recently demonstrated that pharmacological preconditioning of isolated rabbit hearts with phenylephrine (5 minute infusion, 10 minutes prior to prolonged ischemia and reperfusion) was independent of adenosine and rather involved a direct activation of PKC. This conclusion was based upon the fact that an adenosine A_1-receptor blocker, applied at the same time as phenylephrine, did not abolish the infarct size-reducing effect of α-agonist pretreatment, whereas the inhibitor of PKC, polymyxin B, blocked this protection. However, when applied during prolonged ischemia, the A_1 blocker abolished protection afforded by preconditioning, suggesting that A_1-receptor occupancy is necessary during the prolonged ischemia to reactivate PKC and mediate protection.

These discrepancies indicate that both α_1-adrenoceptor and A_1-adenosine receptor stimulation can activate parallel pathways in the myocardium leading to a common effector.

10.4.3. Protection Against Arrhythmias

The effect of ischemic preconditioning on arrhythmias in vivo and in vitro has been well documented.[50,51] To investigate whether adrenoceptor stimulation by exogenous noradrenaline can mimic the antiarrhythmic effects of preconditioning we have used rat isolated Langendorff-perfused hearts.[52] In this study we compared the effect of classical ischemic preconditioning (induced by a single 5 minute coronary artery occlusion and a subsequent 10 minute reperfusion before a long occlusion) with the effect of perfusion with exogenous noradrenaline (50 nmol). This was introduced in a manner similar to preconditioning; a 5 minute infusion at constant flow rate (10 ml/min) followed by 10 minutes of normal perfusion before the onset of ischemia. Both interventions significantly reduced the incidence of ventricular tachycardia (VT) and the total number of ventricular premature beats (VPB) and abolished ventricular fibrillation (VF) during 30 minutes of subsequent ischemia (Fig. 10.3). In addition, in noradrenaline-pretreated hearts, sustained VF induced by reperfusion was also suppressed, in contrast to hearts subjected to ischemic preconditioning, which never restored their normal sinus rhythm (Fig. 10.4). Infusions of noradrenaline caused transient increases in heart rate and perfusion pressure. However, these hemodynamic effects completely waned during the 10 minute recovery period before the onset of ischemia (Fig. 10.5). Thus, this study demonstrated that a short-term exposure of rat hearts to exogenous noradrenaline renders it more resistant to subsequent ischemia- and reperfusion-induced arrhythmias, in a manner similar

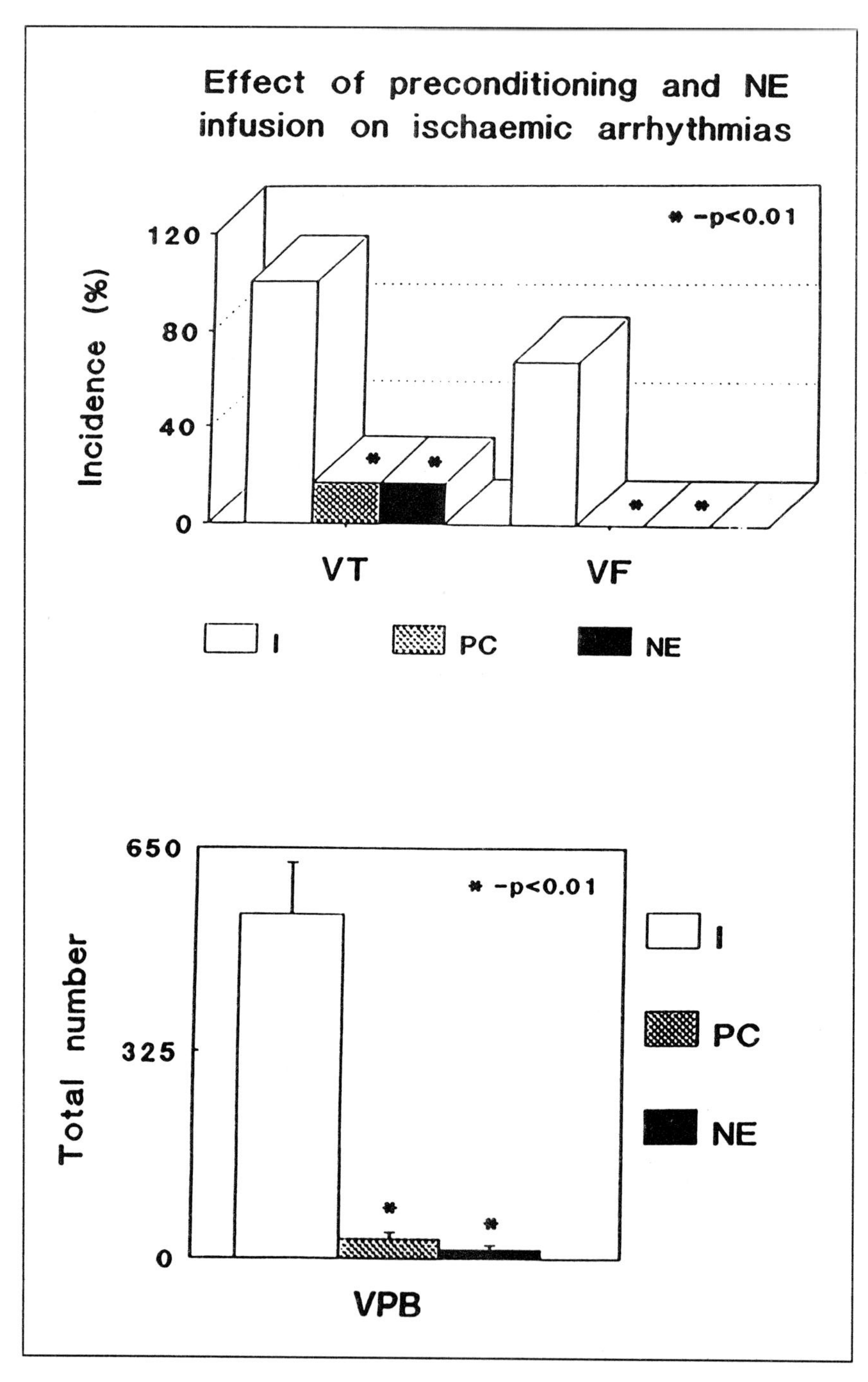

*Fig. 10.3. Effect of ischemic preconditioning and noradrenaline infusion on ischemic arrhythmias in rat isolated heart. VT, ventricular tachycardia; VF, ventricular fibrillation; VPB, ventricular premature beats; I, ischemia; PC, preconditioning; NE, noradrenaline. Error bars - SEM.*p<0.01, noradenaline vs. ischemic controls.*

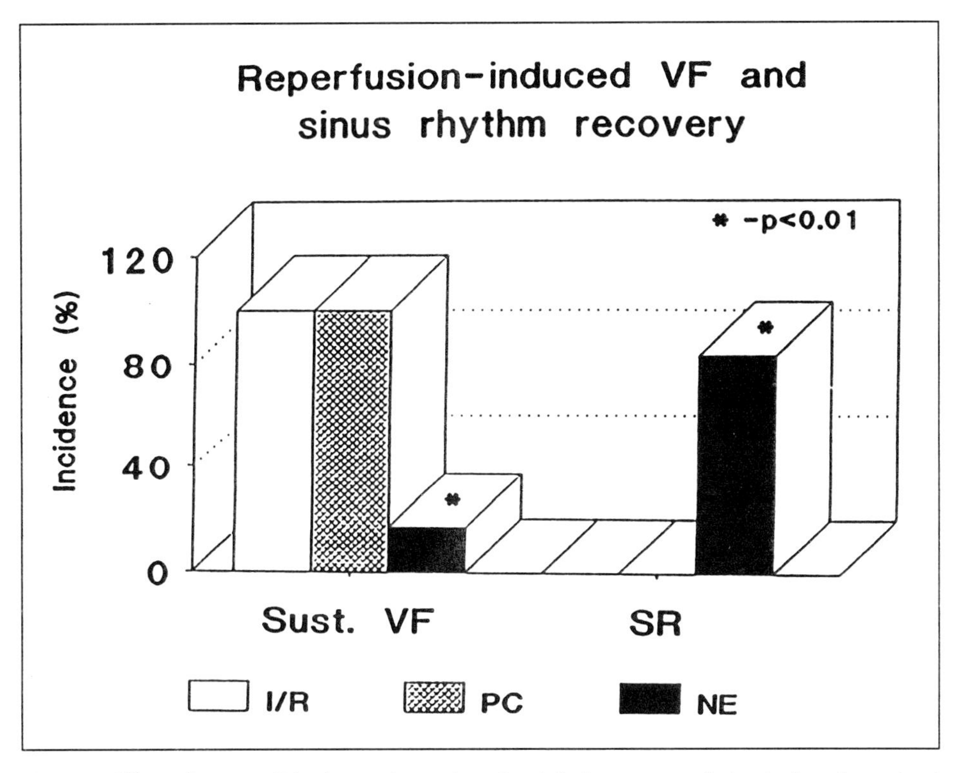

*Fig. 10.4. Effect of preconditioning and noradrenaline infusion on reperfusion-induced sustained ventricular fibrillation and sinus rhythm recovery in rat isolated heart. Abbreviations as in Fig. 10.3.; SR = sinus rhythm. *p<0.01, noradrenaline vs. ischemic controls.*

to a brief episode of ischemia. This is in agreement with the finding of Parratt et al[53] that the infusion of exogenous catecholamines can protect rat hearts against ischemia-induced arrhythmias in vivo. Further evidence that catecholamines can mimic the effect of preconditioning on ischemia-induced ventricular arrhythmias has been recently provided in a study by Vegh et al[54] in anesthetized dogs; the intracoronary administration of a low dose of noradrenaline significantly reduced the severity of subsequent ischemia-induced arrhythmias.

In our study on isolated hearts we did not explore the relationship between the preconditioning-like antiarrhythmic effect and changes in hemodynamics. These transient changes in heart rate and perfusion pressure (due to coronary vasoconstriction) could probably be sufficient to produce a certain degree of 'demand ischemia' leading to preconditioning. On the other hand, we cannot rule out a direct adrenergic effect on the myocardium triggering adaptive processes mediated via signal transduction systems.

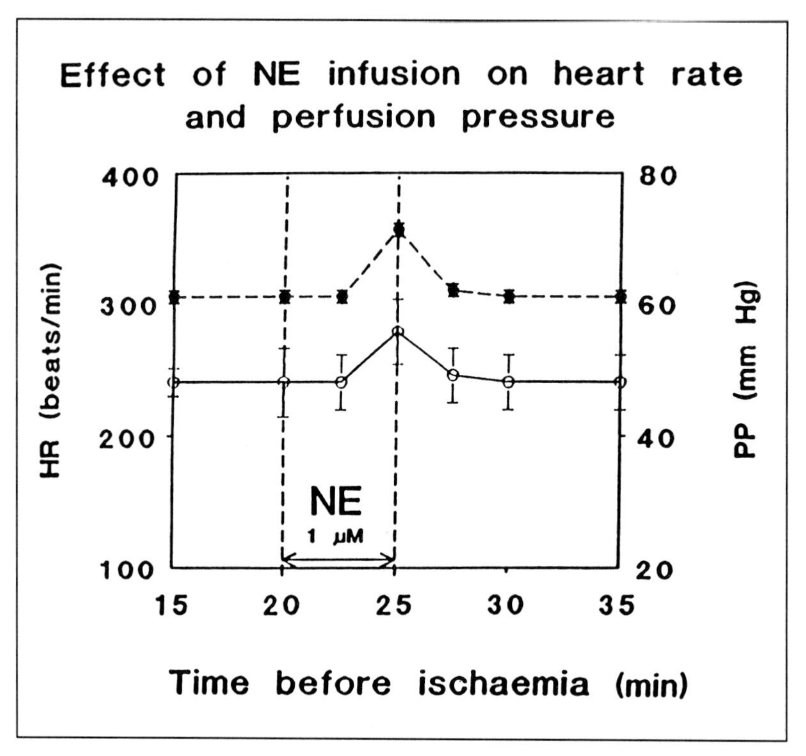

Fig. 10.5. Effect of noradrenaline infusion on heart rate and perfusion pressure in rat isolated heart. Noradrenaline (NE, 10^{-6} M) was infused for 5 minutes at a rate of 10 ml/min. PP, perfusion pressure (solid line), HR, heart rate (dashed line). Error bars - SEM.

It has been recently demonstrated by Richard et al[55] that adrenergic stimulation of isolated rat hearts by normoxic perfusion with exogenous noradrenaline stimulates adenosine formation in the heart, in a manner similar to ischemia-induced nonexocytotic noradrenaline release. The effect of noradrenaline on the release of adenosine has been shown to be mediated via β-adrenoceptor stimulation. It can thus be inferred that activation of adenosine receptors may occur before the onset of ischemia and trigger a further cascade of events. On the other hand, Winter et al[56] have demonstrated that in rats, α-adrenoceptor-mediated preconditioning does not involve adenosine receptors. Thus, the role of adenosine (or A_1 agonists) in myocardial preconditioning in rats has to be discounted, as has also been confirmed by Asimakis et al.[57]

To gain a further insight into postreceptor mechanisms in the preconditioning cascade, we focused on the role of G proteins as main transducers of adrenergic signals from receptors to effectors in myocardial cells.[18] The potential role of G_i proteins in the mechanism(s) of preconditioning has been suggested on the basis of abolition of protection by pertussis toxin, at least in some models.[51,58] However, until now G proteins have not been directly estimated in the myocardium under conditions of brief ischemia, or in preconditioned hearts. We used rat isolated hearts with a high incidence of ischemic arrhythmias, and demonstrated an increase in the inhibitory G_i proteins and a reduction in the stimulatory G_s proteins in heart tissue (as revealed by Western blot analysis) immediately after a single 5 minute episode of regional ischemia followed by 10 minutes of reperfusion.[59,60] These changes persisted in the preconditioned hearts during the subsequent long-lasting ischemia and corresponded to the time interval between 10 and 20 minutes of ischemia, when the antiarrhythmic effect of preconditioning was markedly expressed. This finding points indirectly to a role of α-adrenoceptor stimulation by endogenous noradrenaline in the mechanisms of preconditioning. Since G_s and G_i proteins are coupled to different effectors, the most important of them being adenylyl cyclase and phospholipase C (PLC), two potential G-protein-mediated pathways of signal transduction leading to a final protective response can be considered:

1. inhibition of the activity of adenylate cyclase with a consequent reduction in cAMP production, and
2. activation of PLC leading via formation of DAG to translocation of protein kinase C from the cytosol to the cell membrane and its activation.

Subsequently, phosphorylation of different membrane proteins occurs. Evidence that in rat isolated hearts signal transduction mechanisms underlying ischemic preconditioning involve stimulation of α_1-adrenoceptors by endogenous catecholamines and G_i protein-mediated activation of PKC has been recently provided by Hu and Nattel.[61]

An alternative mechanism might be stimulation of Na^+/K^+ ATPase activity by catecholamines. There is indirect evidence that β-adrenoceptor activation by endogenous catecholamines during ischemia stimulates the activity of this enzyme resulting in hyperpolarization of heart cell membrane.[62] In addition, Haag et al[63] found that Na^+/K^+ ATPase activity in rat cerebral cortex plasma membranes could be directly stimulated by noradrenaline, which efficiently reversed the inhibition of this enzyme by calcium. It is known that the disturbed homeostasis of ions due to the alteration of membrane cation systems (Na^+/K^+ ATPase, Na^+/Ca^{2+} and Na^+/H^+ exchangers) during ischemia and reperfusion plays a crucial role in the electrical instability of cell membranes.[64] These changes not only predispose the heart to arrhythmias, but also cause myocardial dysfunction and additional ultrastructural deterioration upon reperfusion.[65-67] Concomitantly, the membrane-stabilizing effect of any intervention is capable of improving the function of the Na^+ pump; this may contribute to suppression of ischemia- and reperfusion-induced arrhythmias and/or attenuate myocardial dysfunction.

10.5. CONCLUSION

Recent findings indicate that catecholamines are now recognized as important mediators of ischemic preconditioning. These include attempts to simulate ischemic preconditioning with catecholamines; impressive results have been achieved in different species (dog, rat, rabbit) with protection against various consequences of ischemia, such as infarction, stunning and arrhythmias. The experimental diversity arises from the variety of interactive factors involved in signal transduction of preconditioning at different levels. On the other hand, stimulation of different receptors is believed to activate several pathways leading through central mediators (G proteins) to a common effector. In some species and in some experimental settings protein kinase C has been identified as this common effector. Therefore, a cascade of events starting with activation of adrenoceptors by catecholamines, and transduction of the signal through G proteins to protein kinase C, seems to be one of the most probable mechanisms of the preconditioning-like effect of catecholamines. Hence, this recognition of what appeared paradoxical may offer a new approach towards the pharmacological induction of preconditioning and may lead to development of new strategies for the treatment of myocardial ischemia. However, the exact mechanisms of the protection afforded by catecholamines require further exploration.

REFERENCES

1. Das DK. Ischemic preconditioning and myocardial adaptation to ischemia. Cardiovasc Res 1993; 27:2077-2079.
2. Banerjee A. Could the "real" preconditioning mechanism please stand up? (or, a polyplanic platitude to preconditioning). Cardiovasc Res 1994; 28:1872-1873.
3. Parratt JR. Endogenous myocardial protective (antiarrhythmic) substances. Cardiovasc Res 1993; 27:693-702.
4. Curtis MJ, Pugsley MK, Walker MJA. Endogenous chemical mediators of arrhythmogenesis in ischemic heart disease. Cardiovasc Res 1993; 27:703-719.
5. Karlsberg RP, Cryer PE, Roberts R. Serial plasma carecholamine response early in the course of clinical acute myocardial infarction: relationship to infarct extent and mortality. Am Heart J 1988; 102:24-29.
6. Yamaguchi M, de Champlain J, Nadeau R. Correlation between the response of the heart to sympathetic stimulation and the release of endogenous catecholamines into the coronary sinus of the dog. Circ Res 1975; 36:662-628.
7. Bertel O, Buhler FR, Baitsch G et al. Plasma adrenaline and noradrenaline in patients with acute myocardial infarction. Relationship to ventricular arrhythmias of varying sensitivity. Chest 1982; 82:64-68.
8. Dart AM, Riemersma RA. Neurally mediated and spontaneous release of noradrenaline in the ischemic and reperfused rat heart. J Cardiovasc Pharmacol 1985; 7(Suppl 5):545-549.

9. Karlsberg RP, Penkoske PA, Cryer PE et al. Rapid activation of the sympathetic nervous system following coronary occlusion: relationship to infarct size site and hemodynamic impact. Cardiovasc Res 1979; 13:523-531.
10. Knight DE. Calcium and exocytosis. Ciba Found Symp 1986; 122: 250-270.
11. Dart AM. Influence of myocardial ischemia on exocytotic noradrenaline release. In: Brachman J, Schömig A, eds. Adrenergic system and ventricular arrhythmias in myocardial infarction. Berlin: Springer-Verlag 1989:34-44.
12. Dart AM, Dietz R, Hieronymus K et al. Effects of alpha and beta adrenoceptor blockade on the neurally evoked overflow of endogenous noradrenaline from the rat isolated heart. Br J Pharmacol 1984; 81:475-478.
13. Forfar JC, Riemersma RA. Metabolic modulation of cardiac neurosympathetic activity in vivo: effects of potassium and adenosine. Cardiovasc Res 1987; 21:821-829.
14. Hedqvist P. Basic mechanisms of prostaglandin action on autonomic neurotransmission. Ann Rev Pharmacol Toxicol 1977; 17:259-279.
15. Schömig A. Increase of cardiac and systemic catecholamines in myocardial ischemia. In: Brachman J, Schömig A, eds. Adrenergic system and ventricular arrhythmias in myocardial infarction. Berlin: Springer-Verlag 1989:61-77.
16. Schömig A, Fischer S, Kurz T et al. Nonexocytotic release of endogenous noradrenaline in the ischemic and anoxic rat heart: mechanism and metabolic requirements. Circ Res 1987; 60:194-205.
17. Graefe K-H. On the mechanism of nonexocytotic release of noradrenaline from noradrenergic neurons. In: Brachman J, Schömig A, eds. Adrenergic system and ventricular arrhythmias in myocardial infarction. Berlin: Springer-Verlag 1989:44-53.
18. Fleming JW, Wisler PL, Watanabe AM. Signal transduction by G proteins in cardiac tissues. Circulation 1992; 85:419-433.
19. Tada M, Katz AM. Phosphorylation of the sarcoplasmatic reticulum and sarcolemma. Ann Rev Physiol 1982; 44:401-423.
20. Fedelesova M, Ziegelhoeffer A. Enhanced calcium accumulation related to increased protein phosphorylation in cardiac sarcoplasmic reticulum induced by cyclic 3',5'-AMP or isoproterenol. Experientia 1975; 31:518-520.
21. Kohl C, Schmitz W, Scholz H et al. Evidence for α_1-adrenoceptor-mediated increase of inositol triphosphate in the human heart. J Cardiovasc Pharmacol 1989; 13:324-327.
22. Presti CF, Scott BT, Jones LR. Identification of an endogenous protein kinase C activity and its intrinsic 15-kdalton substrate in purified canine cardiac sarcolemmal membrane. J Biol Chem 1985; 260:13879-13889.
23. Lindemann JP. α-Adrenegic stimulation of sarcolemmal protein phosphorylation and slow responses in intact myocardium. J Biol Chem 1986; 261:4860-4867.

24. Hanem S, Gronas T, Holten T et al. Intracellular pH measured by ^{31}P-NMR in isolated rat heart during alpha-1-adrenoceptor stimulation. J Mol Cell Cardiol 1994; 26:669-674.
25. Jiang JP, Downing SE. Catecholamine cardiomyopathy: review and analysis of pathogenetic mechanisms. Yale J Biol Med 1990; 63:581-591.
26. Martins JB, Zipes DP. Effects of sympathetic and vagal nerves on recovery properties of the endocardium and epicardium of the canine left ventricle. Circ Res 1980; 46:100-110.
27. Tanabe T, Takahashi K, Kitada M et al. Effects of sympathetic stimulation, with and without previous α_1 and β adrenoceptor blockade, on refractoriness dispersion in canine heart. Cardiovasc Res 1994; 28:1787-1793.
28. Cable DG, Rath TE, Dreyer ER et al. Refractory period response of cardiac Purkinje tissue to α_1- and α_2-adrenergic influences. Am J Physiol 1994; 267:H376-H382.
29. Wendt DJ, Martins JB. Autonomic neural regulation of intact Purkinje system of dogs. Am J Physiol 1990; 258:H1420-H1426.
30. Benfey BG. Antifibrillatory effects of α_1-adrenoceptor blocking drugs in experimental coronary artery occlusion and reperfusion. Can J Physiol Pharmacol 1993; 71:103-111.
31. Rona G. Catecholamine cardiotoxicity. J Mol Cell Cardiol 1985; 17:291-306.
32. Penny WJ. The deleterious effects of myocardial catecholamines on cellular electrophysiology and arrhythmias during ischemia and reperfusion. Eur Heart J 1984; 5:960-973.
33. Sheridan DJ, Penkoske PA, Sobel BE et al. Alpha adrenergic contributions to dysrhythmia during myocardial ischemia and reperfusion in cats. J Clin Invest 1988; 65:161-171.
34. Yusuf S, Petro R, Lewis J et al. Beta blockade during and after myocardial infarction: an overview of the randomized trials. Prog Cardiovasc Dis 1985; 27:335-371.
35. Bril A, Rochette L. Comparison of the effect of antidepressant drugs on arrhythmias in the isolated rat heart subjected to myocardial ischemia and reperfusion. Pharmacol Toxicol 1987; 60:249-254.
36. Dietz R, Offner B, Dart AM et al. Ischemia-induced noradrenaline release mediates ventricular arrhythmias. In: Brachman J, Schömig A, eds. Adrenergic system and ventricular arrhythmias in myocardial infarction. Berlin: Springer-Verlag 1989:313-321.
37. Curtis MJ, Botting JH, Hearse DJ et al. The sympathetic nervous system, catecholamines and ischemia-induced arrhythmias: dependence upon serum potassium concentration. In: Brachman J, Schömig A, eds. Adrenergic system and ventricular arrhythmias in myocardial infarction. Berlin: Springer-Verlag 1989:205-219.
38. Li HG, Jones DL, Yee R et al. Arrhythmogenic effects of catecholamines are decreased in heart failure induced by rapid pacing in dogs. Am J Physiol 1993; 265(Heart Circ Physiol 34):H1654-H1662.
39. Janse MJ, Schwartz PJ, Wilms-Schopman F et al. Effects of unilateral

stellate ganglion stimulation and ablation on electrophysiologic changes induced by acute myocardial ischemia in dogs. Circulation 1985; 72:585-595.

40. Locke-Winter CR, Winter CB, Nelson DW et al. cAMP stimulation facilitates preconditioning against ischemia-reperfusion through norepinephrine and alpha1 mechanisms. Circulation 1991; 84(Suppl II):II-433 (Abstract).
41. Banerjee A, Locke-Winter C, Rogers KB et al. Preconditioning against myocardial dysfunction after ischemia and reperfusion by an α_1-adrenergic mechanism. Circ Res 1993; 73:656-670.
42. Asimakis GK, Inners-McBride K, Conti VR et al. Transient β adrenergic stimulation can precondition the rat heart against postischemic contractile dysfunction. Cardiovasc Res 1994; 28:1726-1734.
43. Kitakaze M, Hori M, Kamada T. Role of adenosine and its interaction with α adrenoceptor activity in ischemic and reperfusion injury of the myocardium. Cardiovasc Res 1993; 27:18-27.
44. Weselcouch EO, Baird AJ, Sleph PG et al. Endogenous catecholamines are not necessary for ischemic preconditioning in the isolated perfused rat heart. Cardiovasc Res 1995; 29:126-132.
45. Thornton JD, Daly JF, Cohen MV et al. Catecholamines can induce adenosine receptor-mediated protection of the myocardium but do not participate in ischemic preconditioning in the rabbit. Circ Res 1993; 73:649-655.
46. Ardell JL, Yang Xi-M, Thornton JD et al. Depletion of norepinephrine by chronic surgical sympathectomy does not block protection from ischemic preconditioning. Circulation 1994; 90:I-108 (Abstract).
47. Toombs CF, Wiltse AL, Shebuski RJ. Ischemic preconditioning fails to limit infarct size in reserpinized rabbit myocardium: implication of norepinephrine release in the preconditioning effect. Circulation 1993; 88:2351-2358.
48. Bankwala Z, Hale SL, Kloner RA. α-Adrenoceptor stimulation with exogenous norepinephrine or release of endogenous catecholamines mimics ischemic preconditioning. Circulation 1994; 90:1023-1028.
49. Tsuchida A, Liu Y, Liu GS et al. Alpha 1-adrenergic agonists precondition rabbit ischemic myocardium independent of adenosine by direct activation of protin kinase C. Circ Res 1994; 75:576-585.
50. Vegh A, Komori S, Szekeres L et al. Antiarrhythmic effects of preconditioning in anesthetized dogs and rats. Cardiovasc Res 1992; 26:487-495.
51. Piacentini L, Wainwright CL, Parratt JR. The antiarrhythmic effect of ischemic preconditioning in isolated rat heart involves a pertussis toxin sensitive mechanism. Cardiovasc Res 1993; 27:674-680.
52. Ravingerova T, Pyne NJ, Parratt JR. Adrenergic stimulation protects rat hearts against severe arrhythmias: relevance to ischemic preconditioning. J Mol Cell Cardiol 1994; 26:CXII (Abstract).
53. Parratt JR, Campbell C, Fagbemi O. Catecholamines and early postinfarction arrhythmias: the effects of α- and β-adrenoceptor blockade. In:

Delius W, Gerlach E, Grobecker H, Kübler W, eds. Catecholamines and the heart. Berlin: Springer-Verlag 1981:269-284.
54. Vegh A, Papp J Gy, Parratt JR. Intracoronary noradrenaline suppresses ischemia-induced ventricular arrhythmias in anesthetized dogs. J Mol Cell Cardiol 1994; 26:LXXXVII (Abstract).
55. Richard G, Blessing R, Schömig A. Cardiac noradrenaline release accelerates adenosine formation in the ischemic rat heart: role of neuronal noradrenaline carrier and adrenergic receptors. J Mol Cell Cardiol 1994; 26:1321-1328.
56. Winter CB, Mitchell MB, Locke-Winter CR et al. Adenosine induced preconditioning is dependent upon α_1-adrenoreceptor activation. Circulation 1992; 86(Suppl I):I-25 (Abstract).
57. Asimakis GK, Inners-McBride K, Conti VR. Attenuation of postischemic dysfunction by ischemic preconditioning is not mediated by adenosine in the isolated rat heart. Cardiovasc Res 1993; 27:1522-1530.
58. Lasley RD, Mentzer RM. Pertussis toxin blocks adenosine A_1-receptor mediated protection of the ischemic heart. J Mol Cell Cardiol 1993; 25:815-821.
59. Ravingerova T, Pyne NJ, Parratt JR. Ischemic preconditioning in the rat heart: the role of cell signaling. Physiol Res 1994; 43:10P (Abstract).
60. Ravingerova T, Pyne NJ, Parratt JR. Ischemic preconditioning in the rat heart: the role of G-proteins and adrenergic stimulation. Mol Cell Biochemistry 1995; 147:123-128.
61. Hu K, Nattel S. Signal transduction systems underlying ischemic preconditioning in rat hearts. Circulation 1994; 90:I-108 (Abstract).
62. Wilde AAM, Peters RJG, Janse MJ. Catecholamine release and potassium accumulation in the isolated globally ischemic rabbit heart. J Mol Cell Cardiol 1988; 20:887-896.
63. Haag M, Gevers W, Bohmer RG. The interaction between calcium and the activation of Na^+,K^+-ATPase by noradrenaline. Mol Cell Biochemistry 1985; 66:111-116.
64. Bers DM, Lederer WJ, Berlin JR. Intracellular Ca transients in rat cardiac myocytes: role of Na-Ca exchange in excitation-contraction coupling. Am J Physiol; 258:C944-C954.
65. Avkiran M, Haddock P, Ibuki C. Antifibrillatory effect of transient acidic reperfusion: role of Na^+/K^+ ATPase activity. J Mol Cell Cardiol 1994; 26:CVII (Abstract).
66. Janse MJ. The premature beat. Cardiovasc Res 1992; 26:89-100.
67. Nayler WG, Panagiotopoulos S, Elz JS et al. Calcium mediated damage during post-ischemic reperfusion. J Mol Cell Cardiol 1988; 20(suppl II):41-54.

CHAPTER 11

Activation of Protein Kinase C is Critical to the Protection of Preconditioning

Michael V. Cohen, Yongge Liu and James M. Downey

11.1. INTRODUCTION

The phenomenon of cardiac preconditioning was first clearly described in 1986 by Murry et al[1] when they announced that brief ischemia could actually make the heart more tolerant to further ischemia. In dogs infarct size following a 40 minute coronary occlusion was reduced by 75% if four cycles of 5 minute occlusion/5 minute reperfusion preceded the 40 minute ischemia. Perhaps not unpredictably these data received little attention until a series of reports appeared in 1990-1991 confirming and extending Murry's observation.[2-4] In the succeeding 3-4 years the preconditioning phenomenon has been well characterized and many laboratories have been attempting to determine the mechanism. The latter has not yet been fully characterized, but efforts to date have strongly suggested the participation of protein kinase C (PKC).[5] In this review the evidence for involvement of this important cell enzyme will be outlined, and our current hypothesis of the mechanism of preconditioning will be detailed.

11.2. PROTEIN KINASE C AND ITS ACTIVATION PATHWAYS

The cellular protein kinases are important enzymes which are responsible for phosphorylating many proteins including other enzymes

Myocardial Preconditioning, edited by Cherry L. Wainwright and James R. Parratt.

and ion channels. Protein kinase A is perhaps best known because it is activated by cyclic AMP which itself is a product of adenylyl cyclase activation. Protein kinase A can then phosphorylate troponin I, myofibrillar C protein, phospholambam, and calcium channels,[6] and in so doing can modulate their function. Protein kinase C has been implicated in the phosphorylation of potassium channels,[7] cardiac myosin light chains,[8] and sarcolemmal Ca^{2+} ATPase,[9] and it is this kinase which will concern us in this review. Protein kinase C is usually present in the cell cytosol in its inactive form. One of the peculiar properties of PKC is that activation of the enzyme is also accompanied by its physical translocation into the lipid bilayer of the cell membrane. For that reason many investigators have used this translocation as an index of its activation.

G proteins are trimeric proteins which bind GTP following receptor occupancy. Many receptors including α- and β-adrenergic, muscarinic, adenosine, angiotensin, bradykinin, and endothelin are coupled to their effectors by these intracellular messengers.[6,10] A number of isoforms of α, β, and γ subunit proteins exist and these may combine to form different G proteins, each of which is activated by a specific receptor. For example, adenosine A_1 and muscarinic M_2 receptors activate G_i which can be ADP-ribosylated and therefore inactivated by pertussis toxin,[6] while α_1-adrenergic and angiotensin receptors activate G_q which is not sensitive to pertussis toxin.[10] The $G_{q\alpha}$.GTP complex can activate phospholipase C (PLC) which metabolizes phosphatidylinositol-4,5 bisphosphate in the cell membrane yielding diacylglycerol (DAG) and an intermediate which is quickly phosphorylated to inositol 1,4,5-trisphosphate (IP_3). The latter releases Ca^{2+} from nonmitochondrial intracellular stores increasing contractility, whereas the former activates PKC. Some have proposed that the βγ complex of G_i can also stimulate PLC,[10] although this view is not universal.[6]

Catecholamines[6,11-13] and angiotensin[14,15] are known to activate cardiac PKC while at the same time causing enhanced myocardial contractility because of the increased production of IP_3. Henrich and Simpson[16] as well as Bogoyevitch et al[17] have demonstrated that catecholamines cause translocation of PKC from the cytosol into the membrane. There is evidence that G_i activated by adenosine A_1- and muscarinic M_2-agonist-receptor interactions also activates phospholipase C and thus PKC.[18] However, the absence of an increase in myocardial contractility following A_1-adenosine-receptor stimulation has caused some to doubt this pathway.

11.3. ADENOSINE TRIGGERS ISCHEMIC PRECONDITIONING

As demonstrated in Figure 11.1, ischemic preconditioning in the rabbit with a single 5 minute coronary occlusion prior to a 30 minute period of ischemia caused approximately 10% infarction of the risk area, significantly smaller than the 35-40% infarction in nonpreconditioned animals. These data are very similar to those reported by Murry in his seminal observations in the dog.[1] Early investigations into the preconditioning phenomenon in our laboratory identified the involvement of endogenous adenosine. Pretreatment with the nonspecific adenosine antagonists 8-(p-sulfophenyl)theophylline (SPT) or PD115,199 blocked ischemic preconditioning's protection.[4] Further experiments demonstrated that exogenous adenosine, and more specifically selective A_1-adenosine agonists, in lieu of the 5 minute coronary occlusion could also protect the ischemic heart.[4] Therefore, it was proposed that adenosine receptors, and probably the A_1-adenosine receptor subtype, were initiating preconditioning. Additional observations have identified possible involvement of the A_3-adenosine receptor as well, which is similar in many respects to the A_1-adenosine receptor.[19]

As noted above the A_1-adenosine receptor is thought to be coupled exclusively to G_i. To determine whether preconditioning might also be

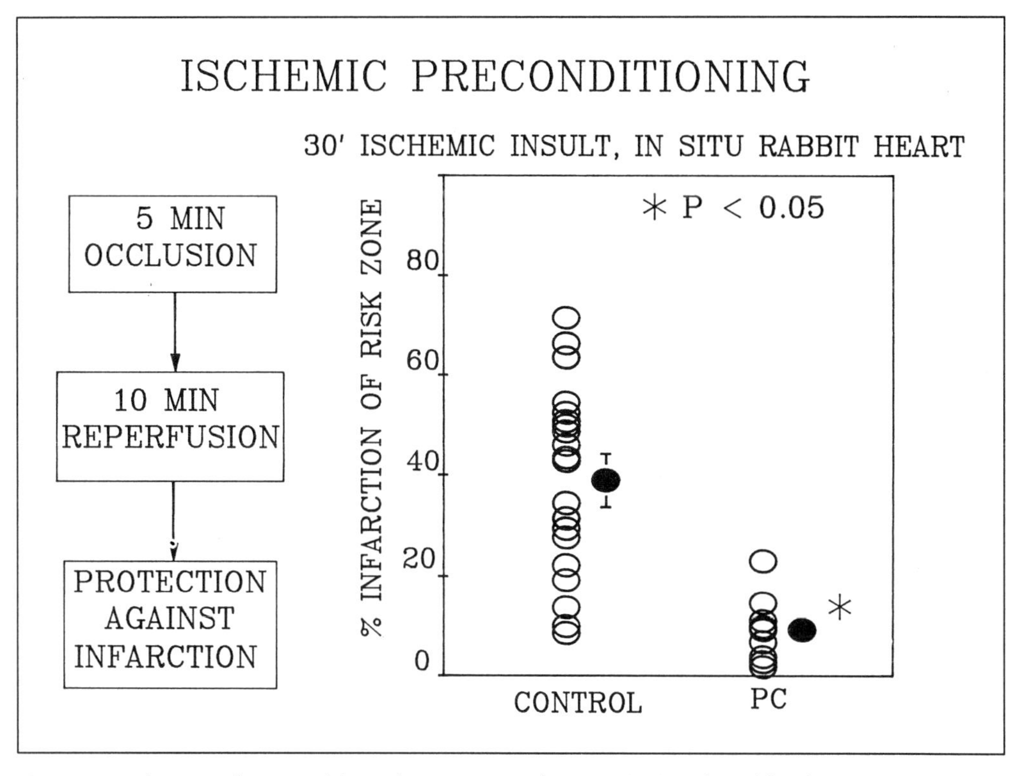

Fig. 11.1. In the typical protocol for ischemic preconditioning (PC) in the rabbit the coronary artery is occluded for 5 minutes and then released for 10 minutes prior to a more prolonged 30 minute ischemic insult. This early 5 minute occlusion reduces infarction from approximately 40% of the risk zone in control animals to about 10% (p<0.05). Open symbols represent data from individual animals, while the closed symbols are the group averages. The vertical bars represent SEM.

dependent upon this coupling, rabbits were pretreated with pertussis toxin which ADP-ribosylates G_i, and, therefore, effectively blocks activation.[20] Forty-eight hours later attempts were made to precondition rabbit myocardium with the standard 5 minute myocardial ischemia. Successful blockade of G_i was indicated by the inability of acetylcholine, also coupled to G_i, to produce its expected bradycardic effect. As anticipated, myocardium pretreated with pertussis toxin could not be protected by ischemic preconditioning.

11.4. ADENOSINE RECEPTORS TRIGGER AS WELL AS MEDIATE PROTECTION

Although it had been known that adenosine blockade could abort the protection of ischemic preconditioning, it was uncertain when the blockade had to occur. The adenosine receptor antagonist SPT is a polar compound which stays in the extracellular space and does not enter cells. In isolated, buffer-perfused rabbit hearts it is possible to administer SPT for a specified period and then withdraw it with subsequent restoration of receptor function. Blocking adenosine receptors during only the preconditioning ischemia completely blocked protection, indicating that adenosine acts as a trigger for the protection.[19] In an in vivo rabbit model Thornton et al[21] delayed SPT administration to preconditioned hearts until shortly before the long ischemic period. Protection was equally blocked revealing that renewed adenosine receptor occupation during the long ischemia was also needed for protection. When SPT administration was further delayed until the end of the 30 minute coronary occlusion, protection was no longer blocked. Hence activation of adenosine receptors during preconditioning triggers the protective changes, but adenosine receptors must be reactivated during the subsequent ischemia to maintain the protected state. A possible explanation of this dual role of adenosine is discussed below.

11.5. THE ROLE OF PROTEIN KINASE C IN PRECONDITIONING

Since G_i can activate the PKC pathway we investigated whether PKC might be involved in preconditioning. Direct activation of PKC with phorbol myristate acetate was as effective as ischemic preconditioning in protecting the heart.[22] Following phorbol ester treatment a 30 minute coronary occlusion resulted in only $6.4 \pm 1.4\%$ infarction of the risk zone as opposed to $28.0 \pm 4.5\%$ in untreated control hearts. Likewise, pretreatment with oleoyl acetyl glycerol, a water-soluble form of DAG, also protected the heart with only $11.7 \pm 3.3\%$ infarction.[22]

Further evidence that the PKC cascade was operational in preconditioning was obtained from experiments in which the activity of PKC was blocked. Staurosporine, a potent kinase inhibitor, blocks the ATP-binding site of PKC.[23] Pretreatment of hearts with this agent prevented the protection normally seen with ischemic preconditioning

(36.2 ± 2.7% infarction of risk zone).[22] However, staurosporine is not very specific (10:1) for PKC among kinases,[23] and, therefore, we could not unequivocally conclude that PKC blockade accounted for the loss of protection. Polymyxin B, which prevents phospholipid cofactors from binding to the enzyme, is highly specific for PKC among the various kinases,[23] and likewise blocked the protective effect of ischemic preconditioning (40.9 ± 2.5% infarction).[22] Unfortunately, a recent preliminary report has asserted that polymyxin B is also capable of blocking the ATP-dependent K^+ channel,[24] another proposed mediator of preconditioning.[25-27] Chelerythrine is a highly selective antagonist of PKC which competitively inhibits the phosphate acceptor of the enzyme[23] with few other known biological effects. It also blocked preconditioning's protection (31.5 ± 2.8% infarction).[28] The only property which staurosporine, polymyxin B, and chelerythrine appear to have in common is blockade of PKC. Hence, these studies strongly implicate PKC as an important part of the protection pathway.

The above observations serve as the foundation of our hypothesis that PKC plays a central role in the preconditioning phenomenon. Other laboratories have now examined this issue and many have confirmed our data.[29-35] Additionally, calphostin C, another highly specific and potent PKC antagonist, can block the salutary effects of simulated ischemic preconditioning in isolated myocytes.[36] However, it is important to realize that several investigators have been unable to block the protection of ischemic preconditioning with PKC antagonists.[37-40] That failure may, in part, reflect an inability to completely block PKC in large animal models, or it may reflect true species differences. Others have failed to mimic preconditioning with activators of PKC.[39,41,42] In the studies by Vogt and his colleagues[38,42] the dose of PMA injected into the ischemic myocardium (1 μM) may have been too high. As noted in Figure 11.2 PMA perfusate concentrations of 0.2 and 0.4 nM in the isolated rabbit model in our laboratory provided protection, but a dose of 2 nM no longer salvaged ischemic myocardium. We suspect that a single PKC isoform is responsible for the protection, while other PKC isoforms may actually be injurious. A low dose of PMA may selectively stimulate only the protective isoform. The results of studies examining the importance of PKC to the preconditioning phenomenon are summarized in Table 11.1.

11.6. WHAT IS PRECONDITIONING'S MEMORY?

Consideration of PKC involvement in protection of ischemic myocardium helped to explain a baffling observation related to preconditioning made several years ago. As originally performed, ischemic preconditioning in rabbits involved a 5 minute period of ischemia followed by 10 minutes of reperfusion prior to the 30 minute ischemic insult. As shown in Figure 11.3 this reperfusion in the anesthetized animal could be prolonged to approximately 60 minutes before protection was

Table 11.1.* *Studies supporting or denying the involvement of protein kinase C in the preconditioning phenomenon

Investigator	PKC involvement	Species	Model	End point	PC modality	PKC antagonist	PKC activator	Other agents
Ytrehus et al[22]	+	Rabbit	i.s.	Infarct size	Ischemia	Staurosporine		
	+	Rabbit	i.s./i.v.	Infarct size	Ischemia	Polymyxin B		
	+	Rabbit	i.v.	Infarct size			PMA	
	+	Rabbit	i.v.	Infarct size			OAG	
Liu et al[28]	+	Rabbit	i.s.	Infarct size	Ischemia	Chelerythrine		
Liu et al[45]	+	Rabbit	i.s.	Infarct size	Ischemia	Staurosporine		
	+	Rabbit	i.s.	Infarct size	Ischemia			Colchicine
Goto et al[29]	+	Rabbit	i.s.	Infarct size	PIA	Staurosporine		
Tsuchida et al[52]	+	Rabbit	i.v.	Infarct size	PE	Polymyxin B		
Liu et al[54]	+	Rabbit	i.v.	Infarct size	Angiotensin II	Polymyxin B		
Goto et al[60]	+	Rabbit	i.v.	Infarct size	Bradykinin	Polymyxin B		
Armstrong et al[36]	+	Rabbit	myocytes	cell survival	Substrate deprivation	Calphostin C		
Kitakaze et al[32]	+	Dog	i.s.	Infarct size	Ischemia	Polymyxin B		
Speechly-Dick et al[30]	+	Rat	i.s.	Infarct size	Ischemia	Chelerythrine		
	+	Rat	i.s.	Infarct size			DOG	
Li and Kloner[35]	+	Rat	i.s.	Infarct size	Ischemia	Calphostin C		

Table 11.1 (continued)

Mitchell et al[34]	+	Rat	i.v.	LV function	Ischemia or PE	Staurosporine	
	+	Rat	i.v.	LV function	Ischemia or PE	Chelerythrine	
	+	Rat	i.v.	LV function			SAG
Hu and Nattel[31]	+	Rat	i.v.	LV function/ CK release	Ischemia	H-7	
	+	Rat	i.v.	LV function/ CK release			PMA
Cave and Apstein[33]	+	Rat	i.v.	LV function	Ischemia	Polymyxin B	
Brew et al[59]	+	Rat	i.v.	LV function	Bradykinin	Chelerythrine	
Mitchell et al[34]	+						
Przyklenk and Kloner[40]	–	Dog	i.s.	Infarct size	Ischemia	H-7	
	–	Dog	i.s.	Infarct size	Ischemia	Polymyxin B	
Vogt et al[39]	–	Pig	i.s.	Infarct size	Ischemia	Staurosporine	
	–	Pig	i.s.	Infarct size	Ischemia	Bisindolylmaleimide	
	–	Pig	i.s.	Infarct size			PMA
Vogt et al[42]	–	Pig	i.s.	Infarct size			PMA
Tsuchida et al[38]	–	Rat	i.s.	Infarct size	Ischemia	Polymyxin B	
Kolocassides and Galiñanes[37]	–	Rat	i.v.	LV function	Ischemia	Chelerythrine	
Mitchell et al[41]	–	Rat	i.v.	LV function			PdBu

Abbreviations: CK, creatinine kinase; DOG, 1,2-dioctanoyl-sn-glycerol; H-7, 1-[5-isoquinolinylsulfonyl]-2-methylpiperazine; i.s., in situ; i.v., in vitro; LV, left ventricular; OAG, 1-oleoyl-2-acetyl-sn-glycerol; PdBu, phorbol 12,13 dibutyrate; PE, phenylephrine; PIA, R(-)N^6-(2-phenylisopropyl)adenosine; PMA, phorbol 12-myristate 13-acetate; SAG, 1-stearoyl-2-arachidonyl-sn-glycerol.

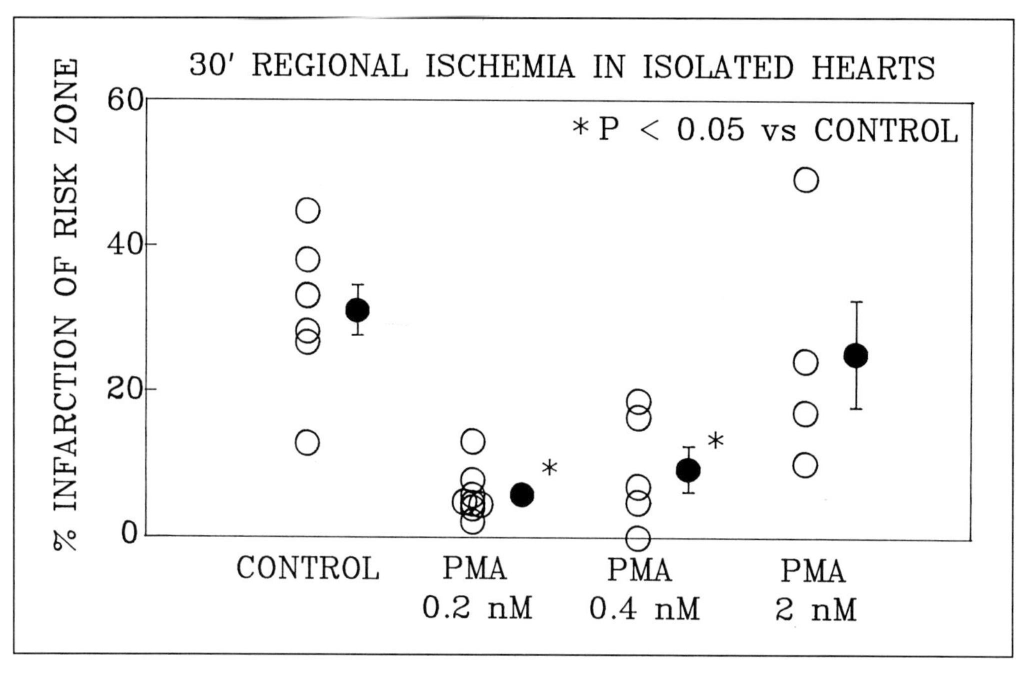

Fig. 11.2. The concentration of the protein kinase C (PKC) activator phorbol 12-myristate 13-acetate (PMA) in the perfusate of isolated rabbit hearts suspended in a Langendorff apparatus affects the degree of protection of ischemic myocardium. At a concentration of PMA of 0.2 nM the protection offered by infused PMA mimicked that observed after early brief ischemia. Doubling the concentration of PMA to 0.4 nM produced little difference. But a much higher concentration of 2 nM caused the protection to be lost. Hence only some agonist concentrations will selectively stimulate and trigger protection PKC, while higher concentrations may have no or deleterious effects. Open symbols represent data from individual animals, while the closed symbols are the group averages. The vertical bars represent SEM.

lost.[43,44] Similarly, when preconditioning is pharmacologically induced with adenosine, the latter can be washed out for ten minutes prior to the onset of ischemia without loss of protection.[4] These curious observations raised an obvious question regarding the mechanism by which the myocardial cell could "remember" that it had been preconditioned 10-60 minutes earlier.

We tested whether the memory could be the result of phosphorylation of a key protein by PKC. It was hypothesized that as long as this key protein was phosphorylated, the heart would be in a preconditioned state; protein dephosphorylation would end the protection. Timing experiments with staurosporine were used to test the phosphorylation theory of the memory.[22,45] When the kinase blocker was administered 5 minutes before the long occlusion, protection was successfully aborted thus revealing that kinase activity was a requirement at this stage of PKC activation.[22] To determine whether staurosporine could also prevent protection if the kinase activity were blocked during

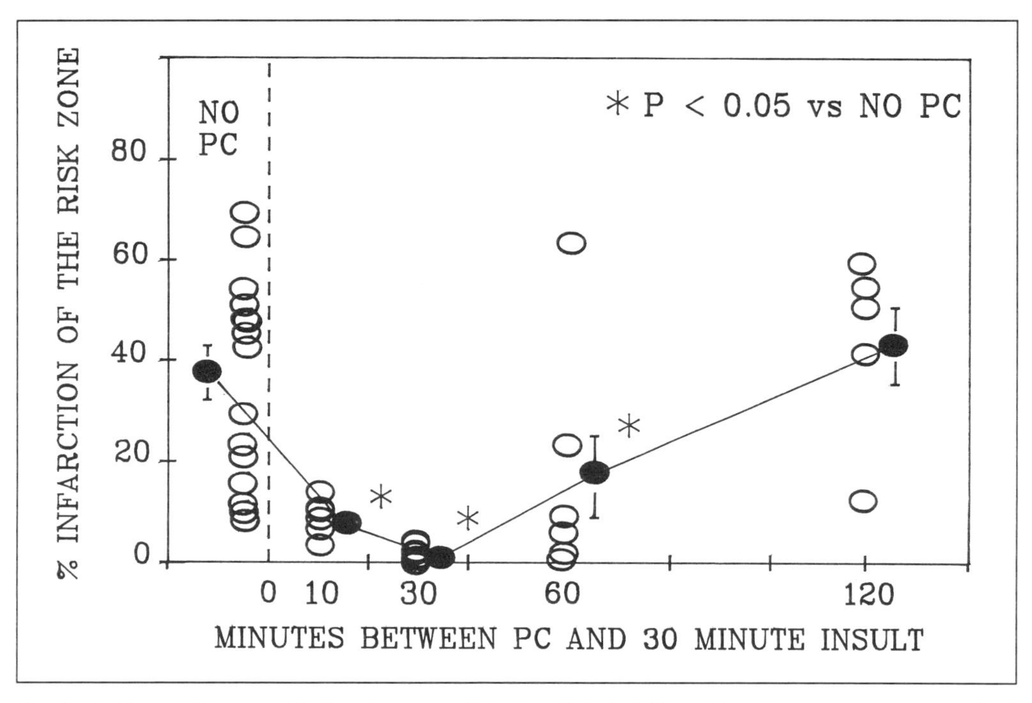

Fig. 11.3. Natural history of ischemic preconditioning (PC). Rabbits with only a 30 minute period of myocardial ischemia have approximately 40% infarction of the jeopardized region. If a 5 minute period of preconditioning ischemia precedes the 30 minute ischemic insult, the degree of myocardial salvage is very dependent on the interval between the two ischemias. Ten and 30 minute intervals resulted in equally small infarcts of 5-10% of the risk zone. A 60 minute reflow period caused 20% infarction. Although the latter was still significantly less than in control animals (p<0.05), it was also higher than levels of protection seen in the 10 and 30 minute groups. All protection was gone with 120 minute reflow periods. Open symbols represent data from individual animals, while the closed symbols are the group averages. The vertical bars represent SEM. (Adapted with permission from Van Winkle et al.[43])

only the preconditioning ischemia, the drug was administered 5 minutes before a 5 minute preconditioning ischemia.[45] Under these conditions staurosporine did not block protection. Staurosporine does not affect the translocation of PKC into the cell membrane, but rather only blocks the enzyme's ability to phosphorylate. Hence it appeared that kinase activity was necessary during only the mediation, but not the trigger, phase of ischemic preconditioning. The failure to abort protection by blockade of PKC's ability to phosphorylate during the preconditioning phase implies that protein phosphorylation cannot be involved in the "memory" function of preconditioning.

11.7. THE PROTEIN KINASE C TRANSLOCATION THEORY OF PRECONDITIONING

The above data with adenosine and staurosporine demonstrating that there are two distinct phases of preconditioning, the trigger and mediation phases, suggested that PKC was indeed being activated twice, but that the characteristics of each activation differed. The translocation theory of preconditioning is outlined in Figure 11.4. The first ischemic episode results in net formation of adenosine from ATP breakdown. Adenosine and A_1-receptor occupancy activate G_i, which in turn stimulates a phospholipase to produce DAG which acts as a cofactor for PKC. Normally, adenosine receptors are poorly coupled to PKC, but activation of the PKC pathway appears to upregulate that coupling, perhaps through translocation of PKC. Whatever the mechanism, the coupling between adenosine receptors and PKC appears to be established for approximately 1 hour. Adenosine released during the subsequent episode of ischemia results in immediate kinase activity of PKC as long as the latter is in an upregulated state. Ultimately, some effector protein must be phosphorylated early during this second ischemic period if protection is to occur.

11.8. DOES TRANSLOCATION OF PKC ACCOUNT FOR THE UPREGULATION?

As demonstrated by Strasser et al[46] ischemia is capable of activating PKC, and in so doing causes its translocation into the cell membrane. It takes approximately 10 minutes for this translocation process to occur. Since kinase activity will parallel the translocation, there will be a 10 minute delay in phosphorylation of substrate. Furthermore, Bogoyevitch et al[17] have noted that, following exposure of ventricular myocytes to phorbol esters, translocated PKC remains in the cell membrane for approximately 1 hour. The delay associated with translocation could explain why adenosine receptor stimulation during a single, prolonged episode of myocardial ischemia fails to be protective. Although adenosine will be produced by the ischemic cells and PKC will be activated, up to 10 minutes will have elapsed before a steady-state level of activated PKC is present in the membrane. Therefore, by the time phosphorylation of the protective protein can begin, high-energy phosphate stores may be sufficiently depleted to limit this reaction, thus minimizing any protective effect. However, as long as PKC is already in the membrane because of a prior episode of ischemia, reoccupation of adenosine receptors would result in immediate phosphorylation of the protective protein and subsequent protection of the cell. That PKC translocation is critical to preconditioning is supported by experiments with colchicine.[45] This anti-inflammatory drug disrupts microtubules which are required for PKC to be physically pulled from the cytosol into the membrane. As expected, pretreatment with colchicine prevented ischemic preconditioning from protecting the heart ($38.3 \pm 1.9\%$ infarction). It should be noted, however, that some investigators have been unable to detect PKC translocation in preconditioned canine hearts.[47,48]

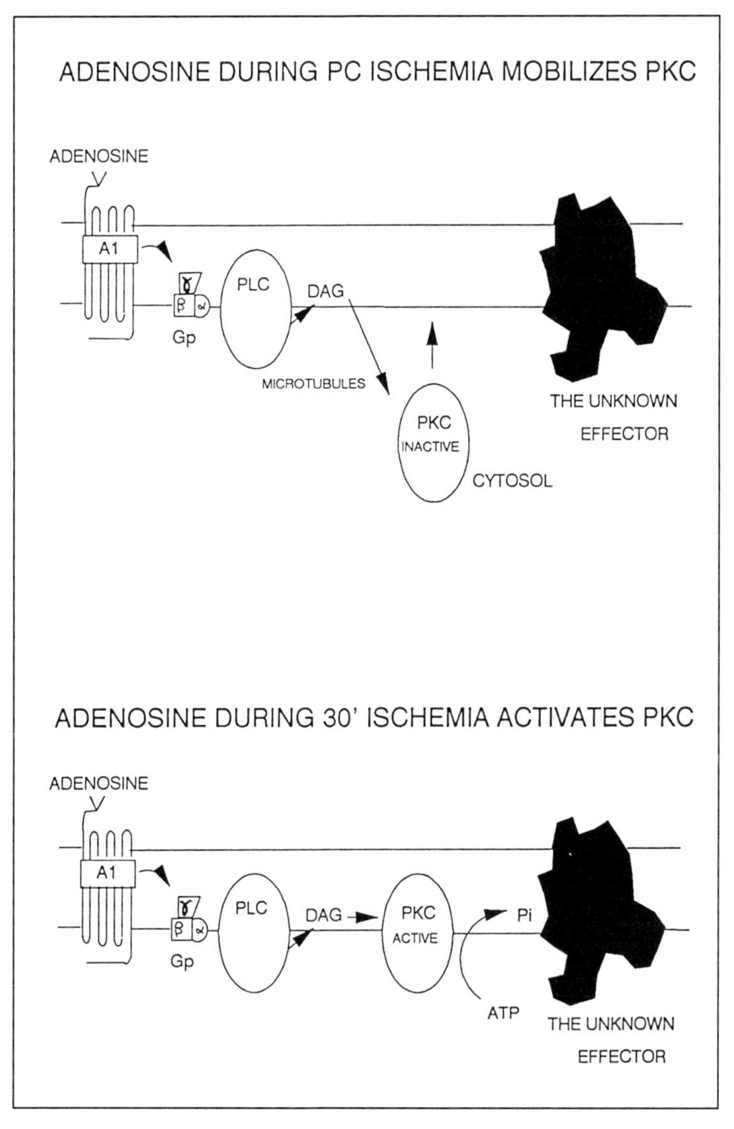

Fig. 11.4. Protein kinase C (PKC) hypothesis for protection of ischemic preconditioning. During the initial 5 minute ischemia adenosine produced by ischemic myocardium occupies its receptor which in turn results in cleavage of its G protein and subsequent activation of phospholipase C (PLC). The latter results in the metabolism of membrane phospholipids with production of diacylglycerol (DAG) which causes contraction of microtubules and translocation of inactive cytosolic PKC to the membrane where it is in its activated state. It is believed that the presence of the active enzyme in the membrane constitutes the memory function of ischemic preconditioning and differentiates preconditioned from nonpreconditioned cells. With renewed ischemia repopulation of adenosine receptors leads in turn to G protein cleavage, PLC activation, and DAG production. PKC which is already in its active state in the membrane can now quickly phosphorylate proteins and other structures, one of which is presumably the end-effector of the protection.

11.9. THE 5'-NUCLEOTIDASE THEORY OF PRECONDITIONING

Kitakaze et al[32,49,50] have proposed an alternative hypothesis for the "memory" of preconditioning. In their canine model brief ischemia appears to upregulate ectosolic and cytosolic 5'-nucleotidase which they suggest will increase adenosine production during a subsequent ischemia. Indeed, they have observed greatly increased amounts of adenosine washing out of reperfused tissue following ischemic preconditioning. Because α-adrenoceptor antagonists block both the increase in 5'-nucleotidase activity as well as the protection of ischemic preconditioning, they have proposed that ischemia releases norepinephrine which activates PKC with resulting upregulation of the adenosine-generating 5'-nucleotidase. This enzyme would then be poised to produce greater quantities of adenosine during the next ischemia. However, the validity of this scheme is questionable. Firstly, adenosine production by ischemic tissue has been found to actually decrease following ischemic preconditioning.[51] Van Wylen did his experiments in dog, but we have confirmed his observations in rabbits.[52] More importantly, if we pharmacologically precondition the rabbit heart with the α_1-adrenoceptor agonist phenylephrine, which also activates PKC (see below), and reinfuse the agonist just before the long ischemia, protection persists even when adenosine receptors are blocked,[53] thus implying that adenosine need not even be a part of the preconditioning mechanism.

11.10. OTHER RECEPTORS CAN PRECONDITION THE HEART

The PKC theory of preconditioning would predict that any receptor capable of stimulating PKC should be able to protect the heart (Fig. 11.5). α_1-Adrenoceptor and angiotensin agonists are known to stimulate receptors coupled to pertussis toxin-insensitive G_q,[10] while M_2-muscarinic receptor agonists mimic A_1-adenosine receptor agonists and activate pertussis toxin-sensitive G_i.[6,10] All of these receptor agonists reportedly can activate PKC and, when infused in lieu of the brief ischemia, protect ischemic myocardium equally well as ischemic preconditioning.[20,53-58] More recently it has been observed that brief infusions of either bradykinin[59,61,62] or endothelin also can protect ischemic myocardium. Polymyxin B blocks the protection induced by α_1-adrenoceptor,[53] angiotensin[55] and bradykinin[61] agonists, while chelerythrine also aborts bradykinin's[60] and endothelin's[62] protection. PKC blockade has not yet been tested in animals receiving M_2-muscarinic agents.

Endogenous adenosine still plays an important role in pharmacological preconditioning. For example, if the heart is preconditioned by transient exposure to the α_1-adrenoceptor agonist phenylephrine, adenosine is needed during the subsequent ischemia to reactivate PKC and protect the heart. As a result, blocking adenosine receptors during the

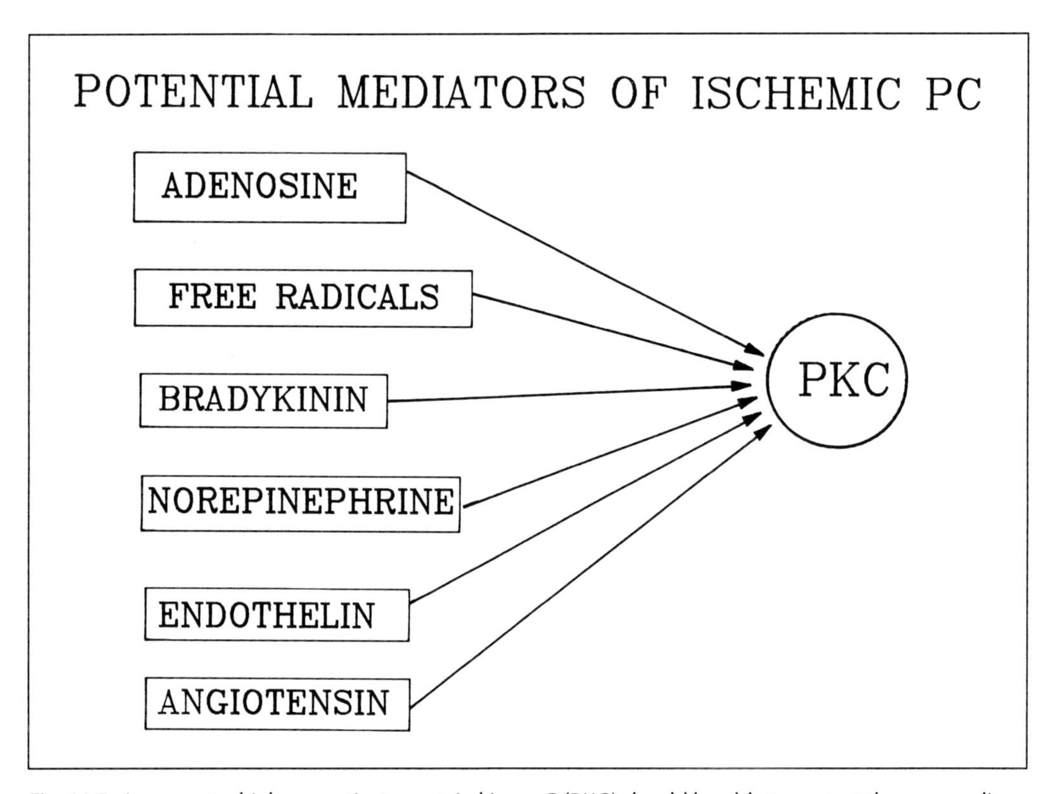

Fig. 11.5. Any agent which can activate protein kinase C (PKC) should be able to protect the myocardium and mimic ischemic preconditioning. To date the agonists known to upregulate PKC (adenosine, bradykinin, norepinephrine, endothelin, and angiotensin) have all been demonstrated in this laboratory to protect the heart, while others have shown the protective effect of ischemia-generated free radicals which probably activate phospholipases.

ischemic phase aborts phenylephrine's protection.[53] However, if stimulation of PKC during ischemia is maintained, despite adenosine receptor blockade, by including phenylephrine in the perfusate at the onset of ischemia, protection is restored.[53] Therefore, stimulation of any of these PKC-coupled receptors seems to be able to substitute for adenosine receptor activity during either the early or the late ischemic phase. These studies reveal that there is nothing unique about the adenosine receptor.

11.11. MULTIPLE RECEPTORS CONTRIBUTE TO ISCHEMIC PRECONDITIONING

During ischemia many of the agonists noted above are produced and released by the myocardium.[63-66] Both adenosine and bradykinin blockade can each effectively abort the protection of ischemic preconditioning in the rabbit.[4,59,61] Therefore, it is assumed that during a single

5 minute preconditioning ischemia (which is just above the ischemic threshold for protection) a combination of both adenosine and bradykinin receptor stimulation is required to reach a protective threshold of PKC stimulation, and interference with either would abort protection. Although a bradykinin antagonist can abort the protection of a single cycle of ischemic preconditioning,[59,61] this blockade can be overcome if four preconditioning occlusion/reperfusion cycles are used.[61] Presumably the increased adenosine release is able to activate sufficient PKC to reach a threshold level in the absence of bradykinin receptor activation. While norepinephrine is also known to be released during ischemia,[64] apparently not enough accumulates in the rabbit to be important as a trigger of protection since α-adrenoceptor blockade does not abort protection from a single 5 minute cycle of ischemia.[53,54] Free radicals produced during reperfusion can also stimulate PKC and, hence, may contribute to protection. Accordingly, free radical scavengers have been noted to abort protection in rabbits following a single 5 minute preconditioning cycle[67] but not after 4 cycles.[68] Figure 11.6 reveals how free radicals, adenosine and bradykinin can converge on PKC and contribute to protection.

11.12. BRADYKININ'S ANTI-INFARCT EFFECT DEPENDS UPON PKC

Ischemic preconditioning also protects against ischemia-related arrhythmias[69-72] and Parratt[73] has found that bradykinin is also involved in that protection. He and his colleagues have shown that the protection can be aborted by nitric oxide synthase blockers, suggesting that bradykinin causes production of nitric oxide which would likely act to increase cyclic GMP levels to provide the protection.[73] That does not seem to be the case for the bradykinin-induced anti-infarct effect, however. In isolated rabbit hearts we found that N_{ω}-nitro-L-arginine methyl ester (L-NAME) could not block bradykinin-induced protection against infarction, but that either the bradykinin B_2 receptor antagonist HOE 140 (icatibant) or the PKC antagonist polymyxin B could.[61] Thus, the anti-infarct effect of bradykinin seems to depend upon PKC as described above.

11.13. THE SIGNALING PATHWAY FOR ISCHEMIC PRECONDITIONING IS HIGHLY REDUNDANT

It is noteworthy that preconditioning of rabbit hearts with 10 minutes of global hypoxia causes protection that is mediated by both adenosine and catecholamines which apparently are released simultaneously. Protection in this model can be blocked only by concomitant infusion of an adenosine antagonist and α_1-adrenoceptor blocker.[74] The amounts of agonists released during ischemia may be dependent upon the specific stimulus being applied, and also may be different in other species.

Furthermore, necessary thresholds may vary among the animal models. Differing receptor densities and compositions have already been identified in the myocardium of several species.[75,76] Perhaps this explains why adenosine blockade cannot abort the protection of ischemic preconditioning in the rat.[71,77] While α_1-adrenoceptors appear to play a minor role in ischemic preconditioning in the rabbit, at least one report claims they play an important role in the rat heart.[78] However, these latter results are still somewhat controversial.[79] The presence of multiple receptor pathways for preconditioning reveals that the signaling system is highly redundant and will ensure that the heart will quickly adapt to an ischemic stress.

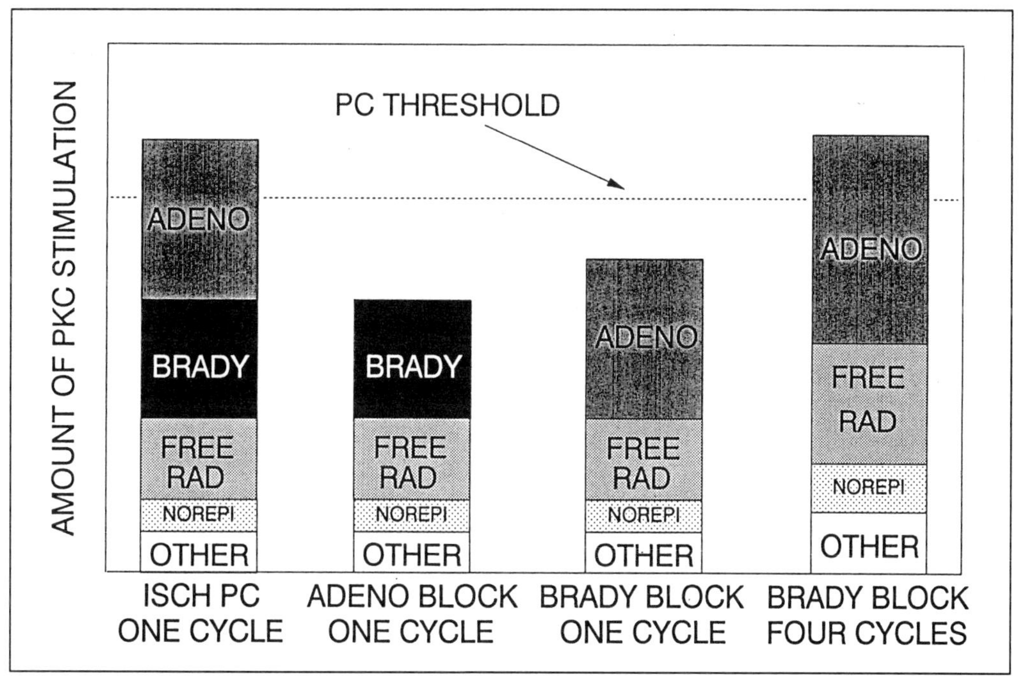

Fig. 11.6. Hypothesized scheme for activation of protein kinase C (PKC) during ischemic preconditioning (PC). One cycle of 5 minute ischemia/10 minute reperfusion leads to release of free radicals and several agonists of which two, adenosine and bradykinin, are major components. Each can activate PKC. Whereas no single agonist is released in sufficient quantity to itself activate enough PKC to exceed the threshold level necessary to protect ischemic myocardium, the combination of released agonists can. Because bradykinin and adenosine are the principal components released, blockade of either adenosine A_1- or bradykinin B_2-receptors results in insufficient activation of PKC by the remaining agonists to reach the threshold level, thus aborting protection. On the other hand, α_1-adrenergic blockade will not prevent protection because threshold can still be reached when the minor norepinephrine component is removed. Despite successful bradykinin receptor blockade, threshold can also be reached with four PC cycles by increasing the released quantities of the remaining components, especially adenosine.

11.14. TOLERANCE MAY BE AN OBSTACLE TO THERAPEUTIC APPLICATION

If A_1-adenosine agonists can trigger protection by initiating this cascade leading to PKC activation and phosphorylation of an as yet unknown effector protein, then one might expect continuous stimulation to maintain protection indefinitely. However, chronic exposure of rabbits to either of the selective A_1-adenosine agonists chlorocyclopentyl adenosine (CCPA)[80] or phenylisopropyl adenosine (PIA)[81] for several days results in the loss of the bradycardic action of the drug as well as the protection against infarction. Lest one mistakenly believe that this loss of effectiveness is a function only of exogenous adenosine agonists, multiple brief 5 minute coronary occlusions occurring every 30 minutes for 3-4 days and thus causing repeated exposure to endogenous adenosine also result in the inability of ischemic preconditioning to protect.[72] This latter study suggests that patients with recurrent ischemic episodes may be resistant to adenosine-based agents given as a cardioprotective intervention.

The site at which tolerance develops was the subject of a subsequent study. To determine whether continuous stimulation of the cascade had any effect on PKC, rabbits were infused with PIA for three days to make them tolerant to adenosine.[81] Their hearts were then removed and challenged with ischemia in vitro with phenylephrine included in the perfusate. Phenylephrine was chosen because it couples to G_q rather than G_i as do the adenosine receptors.[10] While a 5 minute period of ischemia could not protect the heart, the adrenoceptor agonist restored protection, implying PKC could still be activated. The effect of carbachol, an M_2-muscarinic agonist which interacts with G_i, was also investigated.[81] This agonist could likewise protect the ischemic heart attesting to the adequate functioning of the G protein in PIA-tolerant hearts. These data, therefore, indicated that downregulation must have occurred at the level of either the adenosine receptor itself or the interaction between the receptor and its G protein. Thus successful intervention in adenosine-tolerant hearts can still be possible by simply entering the cascade beyond the receptor level.

We do not yet know how to maintain the protection of preconditioning indefinitely. Each step in the signaling sequence that has been revealed to date requires an agonist for initiation of the interaction, and hence is vulnerable to development of tolerance. Perhaps a phosphatase inhibitor, which would interfere with dephosphorylation of the protective protein, could prolong the window of protection. Okadaic acid, a prototype phosphatase inhibitor, does indeed prolong the survival of isolated myocytes during ischemia.[82] Of course, the known phosphatase inhibitors are all tumor promoters and, therefore, would be unsuitable for clinical application. Nevertheless, because of the requirement for pretreatment in many clinical situations, our best hope is to try to identify an intervention whereby patients can be maintained in a preconditioned state continuously. That will probably require an antagonist rather than an agonist.

11.15. THE ELUSIVE END-EFFECTOR

The nature of the end-effector depicted in Figure 11.4 is not yet known. It has been suggested that the ATP-dependent K^+ channel may be this effector. Indeed, agents such as glibenclamide, which close the channel, can abort protection[25-27] while channel openers, such as pinacidil, can protect the heart.[27] However, we have found that the adenosine antagonist SPT can block pinacidil's protective effect.[27] Obviously if K^+ channels were the end effectors, one would not have expected adenosine blockade to affect the ability of a direct opener of K^+_{ATP} channels to protect. Several investigators have identified a link between adenosine A_1-receptors and K^+ channels,[83,84] while others have demonstrated that PKC activation can open K^+_{ATP} channels.[85,86] Therefore, the PKC and K^+_{ATP} channel hypotheses are not mutually exclusive. Positive identification of the end-effector may provide opportunities for devising a long-term treatment.

11.16. CONCLUSIONS

In summary, activation of PKC preconditions the heart while PKC antagonists block preconditioning. This evidence strongly suggests that PKC plays a major role in preconditioning. Both ischemic and pharmacological preconditioning appear to be equally dependent on this enzyme. Furthermore the "memory" function of preconditioning appears to coincide with the time course of translocation and activation of PKC. Much about this protection is explained by the involvement of PKC, although questions still exist. Perhaps the most pressing issue at present is identification of the end-effector. Discovery of this effector should not only help to harness preconditioning for clinical use, but may also provide important insights into why and how myocardial cells become irreversibly damaged.

REFERENCES

1. Murry CE, Jennings RB, Reimer KA. Preconditioning with ischemia: a delay of lethal cell injury in ischemic myocardium. Circulation 1986; 74:1124-1136.
2. Li GC, Vasquez JA, Gallagher KP et al. Myocardial protection with preconditioning. Circulation 1990; 82:609-619.
3. Schott RJ, Rohmann S, Braun ER et al. Ischemic preconditioning reduces infarct size in swine myocardium. Circ Res 1990; 66:1133-1142.
4. Liu GS, Thornton J, Van Winkle DM et al. Protection against infarction afforded by preconditioning is mediated by A_1 adenosine receptors in rabbit heart. Circulation 1991; 84:350-356.
5. Cohen MV, Downey JM. Ischemic preconditioning: can the protection be bottled? Lancet 1993; 342:6.
6. Fleming JW, Wisler PL, Watanabe AM. Signal transduction by G proteins in cardiac tissues. Circulation 1992; 85:420-433.
7. Tohse N, Kameyama M, Sekiguchi K et al. Protein kinase C activation enhances the delayed rectifier potassium current in guinea-pig heart cells. J Mol Cell Cardiol 1990; 22:725-734.

8. Venema RC, Raynor RL, Noland TA Jr et al. Role of protein kinase C in the phosphorylation of cardiac myosin light chain 2. Biochem J 1993; 294:401-406.
9. Qu Y, Torchia J, Sen AK. Protein kinase C mediated activation and phosphorylation of Ca^{2+} pump in cardiac sarcolemma. Can J Physiol Pharmacol 1992; 70:1230-1235.
10. Exton JH. Phosphoinositide phospholipases and G proteins in hormone action. Ann Rev Physiol 1994; 56:349-369.
11. Minneman KP. α_1-Adrenergic receptor subtypes, inositol phosphates, and sources of cell Ca^{2+}. Pharmacol Rev 1988; 40:87-119.
12. Terzic A, Pucéat M, Vassort G et al. Cardiac α_1-adrenoceptors: an overview. Pharmacol Rev 1993; 45:147-175.
13. Fedida D, Braun AP, Giles WR. α_1-Adrenoceptors in myocardium: functional aspects and transmembrane signaling mechanisms. Physiol Rev 1993; 73:469-487.
14. Baker KM, Singer HA. Identification and characterization of guinea pig angiotensin II ventricular and atrial receptors: coupling to inositol phosphate production. Circ Res 1988; 62:896-904.
15. Timmermans PBMWM, Wong PC, Chiu AT et al. Angiotensin II receptors and angiotensin II receptor antagonists. Pharmacol Rev 1993; 45:205-251.
16. Henrich CJ, Simpson PC. Differential acute and chronic response of protein kinase C in cultured neonatal rat heart myocytes to α_1-adrenergic and phorbol ester stimulation. J Mol Cell Cardiol 1988; 20:1081-1085.
17. Bogoyevitch MA, Parker PJ, Sugden PH. Characterization of protein kinase C isotype expression in adult rat heart: protein kinase C-ε is a major isotype present, and it is activated by phorbol esters, epinephrine, and endothelin. Circ Res 1993; 72:757-767.
18. Kohl C, Linck B, Schmitz W et al. Effects of carbachol and (-)-N^6-phenylisopropyladenosine on myocardial inositol phosphate content and force of contraction. Br J Pharmacol 1990; 101:829-834.
19. Liu GS, Richards SC, Olsson RA et al. Evidence that the adenosine A_3 receptor may mediate the protection afforded by preconditioning in the isolated rabbit heart. Cardiovasc Res 1994; 28:1057-1061.
20. Thornton JD, Liu GS, Downey JM. Pretreatment with pertussis toxin blocks the protective effects of preconditioning: evidence for a G-protein mechanism. J Mol Cell Cardiol 1993; 25:311-320.
21. Thornton JD, Thornton CS, Downey JM. Effect of adenosine receptor blockade: preventing protective preconditioning depends on time of initiation. Am J Physiol 1993; 265:H504-H508.
22. Ytrehus K, Liu Y, Downey JM. Preconditioning protects ischemic rabbit heart by protein kinase C activation. Am J Physiol 1994; 266:H1145-H1152.
23. Casnellie JE. Protein kinase inhibitors: probes for the functions of protein phosphorylation. Adv Pharmacol 1991; 22:167-205.
24. Harding EA, Jaggar JH, Squires PE et al. Polymyxin has multiple blocking actions on the ATP-sensitive potassium channel in insulin-secreting

cells. Pflügers Arch 1994; 426:31-39.
25. Gross GJ, Auchampach JA. Blockade of ATP-sensitive potassium channels prevents myocardial preconditioning in dogs. Circ Res 1992; 70:223-233.
26. Toombs CF, Moore TL, Shebuski RJ. Limitation of infarct size in the rabbit by ischemic preconditioning is reversible with glibenclamide. Cardiovasc Res 1993; 27:617-622.
27. Walsh RS, Tsuchida A, Daly JJF et al. Ketamine-xylazine anesthesia permits a K_{ATP} channel antagonist to attenuate preconditioning in rabbit myocardium. Cardiovasc Res 1994; 28:1337-1341.
28. Liu Y, Cohen MV, Downey JM. Chelerythrine, a highly selective protein kinase C inhibitor, blocks the antiinfarct effect of ischemic preconditioning in rabbit hearts. Cardiovasc Drugs Ther 1994; 8:881-882.
29. Goto M, Miura T, Sakamoto J et al. Infarct size limitation by adenosine A_1 receptor agonist was blocked by staurosporine. J Mol Cell Cardiol 1994; 26:CLII (Abstract).
30. Speechly-Dick ME, Mocanu MM, Yellon DM. Protein kinase C: its role in ischemic preconditioning in the rat. Circ Res 1994; 75:586-590.
31. Hu K, Nattel S. Signal transduction systems underlying ischemic preconditioning in rat hearts.Circulation 1994; 90(Suppl I):I-108 (Abstract).
32. Kitakaze M, Minamino T, Shinozaki Y et al. Activation of protein kinase C and subsequent activation of ectosolic 5'-nucleotidase as a major cause for the infarct size-limiting effect of ischemic preconditioning. Circulation 1994; 90(Suppl I):I-207 (Abstract).
33. Cave AC, Apstein CS. Inhibition of protein kinase C abolishes preconditioning against contractile dysfunction in the isolated blood perfused rat heart. Circulation 1994; 90(Suppl I):I-208 (Abstract).
34. Mitchell MB, Meng X, Ao L et al. Preconditioning of isolated rat heart is mediated by protein kinase C. Circ Res 1995; 76:73-81.
35. Li Y, Kloner RA. Does protein kinase C play a role in ischemic preconditioning in rat hearts? Am J Physiol 1995; 268:H426-H431.
36. Armstrong S, Downey JM, Ganote CE. Preconditioning of isolated rabbit cardiomyocytes: induction by metabolic stress and blockade by the adenosine antagonist SPT and calphostin C, a protein kinase C inhibitor. Cardiovasc Res 1994; 28:72-77.
37. Kolocassides KG, Galiñanes M. The specific protein kinase C inhibitor chelerythrine fails to inhibit ischemic preconditioning in the rat heart. Circulation 1994; 90(Suppl I):I-208 (Abstract).
38. Tsuchida A, Miura T, Miki T et al. Role of α1-adrenergic receptor and protein kinase C in infarct size limitation by ischemic preconditioning in rat heart. Circulation 1994; 90(Suppl I):I-647 (Abstract).
39. Vogt A, Barancik M, Weihrauch D et al. Protein kinase C inhibitors reduce infarct size in pig hearts in vivo. Circulation 1994; 90(Suppl I):I-647 (Abstract).
40. Przyklenk K, Sussman MA, Simkhovich BZ et al. Does ischemic preconditioning trigger translocation of protein kinase C in the canine model? Circulation 1995; 92:1546-1557.

41. Mitchell MB, Brew EC, Harken AH et al. Does protein kinase C mediate functional preconditioning? J Mol Cell Cardiol 1993; 25(Suppl III):S.64 (Abstract).
42. Vogt A, Barancik M, Weihrauch D et al. Activation of protein kinase C fails to protect ischemic porcine myocardium from infarction in vivo. J Mol Cell Cardiol 1994; 26:CXVIII (Abstract).
43. Van Winkle DM, Thornton JD, Downey DM et al. The natural history of preconditioning: cardioprotection depends on duration of transient ischemia and time to subsequent ischemia. Coron Artery Dis 1991; 2:613-619.
44. Miura T, Adachi T, Ogawa T et al. Myocardial infarct size-limiting effect of ischemic preconditioning: its natural decay and the effect of repetitive preconditioning. Cardiovasc Pathol 1992; 1:147-154.
45. Liu Y, Ytrehus K, Downey JM. Evidence that translocation of protein kinase C is a key event during ischemic preconditioning of rabbit myocardium. J Mol Cell Cardiol 1994; 26:661-668.
46. Strasser RH, Braun-Dullaeus R, Walendzik H et al. α_1-Receptor-independent activation of protein kinase C in acute myocardial ischemia: mechanisms for sensitization of the adenylyl cyclase system. Circ Res 1992; 70:1304-1312.
47. Simkhovich BZ, Kloner RA, Przyklenk K. Brief preconditioning ischemia does not trigger translocation of protein kinase C in canine myocardium. Circulation 1994; 90(Suppl I):I-208 (Abstract).
48. Przyklenk K, Sussman MA, Kloner RA. Fluorescence microscopy reveals no evidence of protein kinase C activation in preconditioned canine myocardium. Circulation 1994; 90(Suppl I):I-647 (Abstract).
49. Kitakaze M, Hori M, Takashima S et al. Ischemic preconditioning increases adenosine release and 5'-nucleotidase activity during myocardial ischemia and reperfusion in dogs: implications for myocardial salvage. Circulation 1993; 87:208-215.
50. Kitakaze M, Hori M, Kamada T. Role of adenosine and its interaction with α adrenoceptor activity in ischemic and reperfusion injury of the myocardium. Cardiovasc Res 1993; 27:18-27.
51. Van Wylen DGL. Effect of ischemic preconditioning on interstitial purine metabolite and lactate accumulation during myocardial ischemia. Circulation 1994; 89:2283-2289.
52. Goto M, Cohen MV, Van Wylen DGL et al. Attenuated purine production during subsequent ischemia in preconditioned rabbit myocardium is unrelated to the mechanism of protection. J Mol Cell Cardiol 1996; in press.
53. Tsuchida A, Liu Y, Liu GS et al. α_1-Adrenergic agonists precondition rabbit ischemic myocardium independent of adenosine by direct activation of protein kinase C. Circ Res 1994; 75:576-585.
54. Thornton JD, Daly JF, Cohen MV et al. Catecholamines can induce adenosine receptor-mediated protection of the myocardium but do not participate in ischemic preconditioning in the rabbit. Circ Res 1993; 73:649-655.

55. Liu Y, Tsuchida A, Cohen MV et al. Pretreatment with angiotensin II activates protein kinase C and limits myocardial infarction in isolated rabbit hearts. J Mol Cell Cardiol 1995; 27:883-892.
56. Hendrikx M, Toshima Y, Mubagwa K et al. Muscarinic receptor stimulation by carbachol improves functional recovery in isolated, blood perfused rabbit heart. Cardiovasc Res 1993; 27:980-989.
57. Yao Z, Gross GJ. Role of nitric oxide, muscarinic receptors, and the ATP-sensitive K^+ channel in mediating the effects of acetylcholine to mimic preconditioning in dogs. Circ Res 1993; 73:1193-1201.
58. Przyklenk K, Kloner RA. Acetylcholine acts as a "preconditioning-mimetic" in the canine model. J Am Coll Cardiol 1994; 23:396A (Abstract).
59. Wall TM, Sheehy R, Hartman JC. Role of bradykinin in myocardial preconditioning. J Pharmacol Exp Ther 1994; 270:681-689.
60. Brew EC, Mitchell MB, Rehring TF et al. Role of bradykinin in cardiac functional protection after global ischemia-reperfusion in rat heart. Am J Physiol 1995; 269:H1370-H1378.
61. Goto M, Liu Y, Yang X-M et al. Role of bradykinin in protection of ischemic preconditioning in rabbit hearts. Circ Res 1995; 77:611-621.
62. Wang P, Gallagher K, Downey JM et al. Pretreatment with enothelin-1 mimics ischemic precoditioning against infarction in isolated rabbit heart. J Mol Cell Cardiol 1996; in press.
63. Dorheim TA, Wang T, Mentzer RM Jr et al. Interstitial purine metabolites during regional myocardial ischemia. J Surg Res 1990; 48:491-497.
64. Schömig A. Catecholamines in myocardial ischemia: systemic and cardiac release. Circulation 1990; 82(Suppl II):II-13-II-22.
65. Noda K, Sasaguri M, Ideishi M et al. Role of locally formed angiotensin II and bradykinin in the reduction of myocardial infarct size in dogs. Cardiovasc Res 1993; 27:334-340.
66. Wang Q-D, Hemsén A, Li X-S et al. Local overflow and enhanced tissue content of endothelin following myocardial ischemia and reperfusion in the pig: modulation by L-arginine. Cardiovasc Res 1995; 29:44-49.
67. Tanaka M, Fujiwara H, Yamasaki K et al. Superoxide dismutase and N-2-mercaptopropionyl glycine attenuate infarct size limitation effect of ischemic preconditioning in the rabbit. Cardiovasc Res 1994; 28:980-986.
68. Iwamoto T, Miura T, Adachi T et al. Myocardial infarct size-limiting effect of ischemic preconditioning was not attenuated by oxygen free-radical scavengers in the rabbit. Circulation 1991; 83:1015-1022.
69. Shiki K, Hearse DJ. Preconditioning of ischemic myocardium: reperfusion-induced arrhythmias. Am J Physiol 1987; 253:H1470-H1476.
70. Hagar JM, Hale SL, Kloner RA. Effect of preconditioning ischemia on reperfusion arrhythmias after coronary artery occlusion and reperfusion in the rat. Circ Res 1991; 68:61-68.
71. Liu Y, Downey JM. Ischemic preconditioning protects against infarction in rat heart. Am J Physiol 1992; 263:H1107-H1112.
72. Cohen MV, Yang X-M, Downey JM. Conscious rabbits become tolerant to multiple episodes of ischemic preconditioning. Circ Res 1994; 74:998-1004.

73. Parratt JR. Protection of the heart by ischemic preconditioning: mechanisms and possibilities for pharmacological exploitation. Trends Pharmacol Sci 1994; 15:19-25.
74. Cohen MV, Walsh RS, Goto M et al. Hypoxia preconditions rabbit myocardium via adenosine and catecholamine release. J Mol Cell Cardiol 1995;27:1527-1534.
75. Shen Y-T, Vatner DE, Gagnon HE et al. Species differences in regulation of α-adrenergic receptor function. Am J Physiol 1989; 257:R1110-R1116.
76. Endoh M, Hiramoto T, Ishihata A et al. Myocardial α_1-adrenoceptors mediate positive inotropic effect and changes in phosphatidylinositol metabolism: species differences in receptor distribution and the intracellular coupling process in mammalian ventricular myocardium. Circ Res 1991; 68:1179-1190.
77. Li Y, Kloner RA. The cardioprotective effects of ischemic "preconditioning" are not mediated by adenosine receptors in rat hearts. Circulation 1993; 87:1642-1648.
78. Banerjee A, Locke-Winter C, Rogers KB et al. Preconditioning against myocardial dysfunction after ischemia and reperfusion by an α_1-adrenergic mechanism. Circ Res 1993; 73:656-670.
79. Asimakis GK, Inners-McBride K, Conti VR et al. Transient β adrenergic stimulation can precondition the rat heart against postischemic contractile dysfunction. Cardiovasc Res 1994; 28:1726-1734.
80. Tsuchida A, Thompson R, Olsson RA et al. The anti-infarct effect of an adenosine A_1-selective agonist is diminished after prolonged infusion as is the cardioprotective effect of ischemic preconditioning in rabbit heart. J Mol Cell Cardiol 1994; 26:303-311.
81. Hashimi W, Thornton JD, Downey JM et al. Chronic adenosine A_1-agonist exposure results in adenosine receptor rather than G protein downregulation. Circulation 1994; 90(Suppl I):I-107 (Abstract).
82. Armstrong SC, Ganote CE. Effects of the protein phosphatase inhibitors okadaic acid and calyculin A on metabolically inhibited and ischemic isolated myocytes. J Mol Cell Cardiol 1992; 24:869-884.
83. Kirsch GE, Codina J, Birnbaumer L et al. Coupling of ATP-sensitive K^+ channels to A_1 receptors by G proteins in rat ventricular myocytes. Am J Physiol 1990; 259:H820-H826.
84. Grover GJ, Sleph PG, Dzwonczyk S. Role of myocardial ATP-sensitive potassium channels in mediating preconditioning in the dog heart and their possible interaction with adenosine A_1-receptors. Circulation 1992; 86:1310-1316.
85. de Weille JR, Schmid-Antomarchi H, Fosset M et al. Regulation of ATP-sensitive K^+ channels in insulinoma cells: activation by somatostatin and protein kinase C and the role of cAMP. Proc Natl Acad Sci 1989; 86:2971-2975.
86. Wang YG, Lipsius SL. Acetylcholine pre-conditioning stimulates acetylcholine to activate a glibenclamide-sensitive K current dependent on SR Ca^{2+} release and PKC. Biophysical J 1994; 66:A429 (Abstract).

CHAPTER 12

The Mechanism of Preconditioning—What Have We Learned from the Different Animal Species?

Cherry L. Wainwright and Wei Sun

12.1. INTRODUCTION

The past 5-10 years has seen an increasing amount of interest in determining the mechanism of ischemic preconditioning, with the result that, while still some way from establishing the critical mechanism(s) involved in this endogenous cardioprotective response, we now have a greater insight into the phenomenon. Evidence for this is provided by the wealth of up-to-date information contained within the preceding chapters of this book. However, the route to our current level of understanding has been dogged with conflicting evidence from studies employing different animal species, methods of inducing preconditioning and assessment of different end-points, with the result that, as yet, the exact mechanism remains elusive.

The one common thread that ties all the studies together is that preconditioning, induced by whatever means, exerts a marked protection against a range of deleterious consequences of ischemia in all species tested thus far. Studies in pigs, dogs, rabbits, rats and even humans[1] have all demonstrated the preconditioning phenomenon to be the most powerful cardioprotectant ever encountered. Preconditioning can also be demonstrated in isolated tissue preparations and in cultured cell lines. The main conflict begins to arise when the mechanisms

Myocardial Preconditioning, edited by Cherry L. Wainwright and James R. Parratt.

which have been proposed to underlie this protection are considered—where evidence points strongly to one mechanism in a particular species, this may not apply to another species. Thus we are left with the question as to whether the mechanism of preconditioning is likely to be the same from one species to another and, if not, how do we then know which is most likely to be important in humans.

The aim of this chapter is to overview the different proposed mechanisms of preconditioning and, within this, to identify where the results from different species are in conflict. To achieve this, three main lines of investigation will be discussed (labile mediators, transduction/signaling mechanisms and cellular metabolic changes) in the hope that a final common link can be identified. Considering that the three main facets of preconditioning, that is a decrease in cellular injury (infarct size) and electrophysiological disturbances (arrhythmias) and an improvement in postischemic cardiac function, may not necessarily share a common mechanism, these will also be discussed.

12.2. ENDOGENOUS LABILE MEDIATORS

For the purpose of this overview, endogenous labile mediators are defined as substances produced by the tissues constituting the heart mass and which exert physiological responses at the level of the cardiac myocyte and/or coronary vasculature. To be considered as a potential candidate as an endogenous mediator of preconditioning, Parratt[2] suggested that the proposed mediator should fulfill the following criteria. First, it should be demonstrated to be released in response to a brief period of ischemic preconditioning. Second, brief exogenous administration of the mediator (or a stable analog) should be able to mimic the cardioprotective effects of a similarly brief period of ischemia. Finally, receptor antagonists, or inhibitors of the synthesis of the mediator, should be able to reverse the protective effects of preconditioning. Inherent in this last criterion is the requirement that the antagonist/synthesis inhibitor should not itself exacerbate the consequences of ischemia in the absence of preconditioning, since this could implicate the mediator as a modulator of ischemia in its own right, thus complicating determination of its potential role in preconditioning.

12.2.1. Adenosine

Of all the proposed endogenous mediators, adenosine is the most widely studied and is therefore worthy of discussion in its own right. The cardioprotective effects of adenosine against myocardial injury and electrophysiological disturbances resulting from ischemia/reperfusion alone (i.e., independent of preconditioning) has long been established (reviewed in ref. 3). This is an effect which have been demonstrated in a wide range of species, including dogs,[4,5] pigs[6,7] and rats.[8,9] Despite this effectiveness across species and against both injury and arrhythmias, its proposed role as a mediator of preconditioning is not consistent across this same species range, nor does it appear to mediate the protection against the full range of sequelae of ischemia/reperfusion.

The most convincing evidence implicating adenosine as a critical mediator of preconditioning comes from studies in rabbits (summarized in Table 12.1a). Here, brief exposure to either adenosine itself[10,11] or to adenosine agonists acting at various adenosine receptor subtypes,[10-13] or by increasing endogenous adenosine levels by, for example, preventing adenosine uptake[14] have all been shown to decrease infarct size (in vivo and in vitro; reviewed in ref. 15), reduce myocardial stunning[16] and decrease cellular injury in isolated myocytes[17] to a similar degree as that seen with preconditioning. Furthermore, increasing local adenosine concentrations with acadesine has been shown to decrease the temporal threshold for induction of preconditioning[18] and to extend the time window of the protective effect of preconditioning.[19] Similarly, adenosine antagonists have consistently been shown to reverse preconditioning in this species.[15] Thus there is little doubt that, in the rabbit at least, adenosine plays an important role in mediating preconditioning against these parameters. However, since the rabbit is not a very appropriate model for the assessment of ventricular arrhythmias, the role of adenosine as the mediator of the antiarrhythmic effects of preconditioning in this species has not really been studied. Interestingly, when we start to look at the evidence supporting a role for adenosine in other species, we begin to detect differences, not only between species but also with respect to consequences of ischemia. For example, in dogs, adenosine has been implicated in the infarct-limiting effect of preconditioning,[20-23] whereas studies assessing different end-points such as arrhythmias[24] and cardiac function[25] suggest that adenosine may not be important (summarized in Table 12.1b). These findings suggest that perhaps the importance of adenosine is end-point specific, playing an important role in the anti-infarct, but not the antiarrhythmic, effects of preconditioning. This implies that separate mechanisms exist for these two types of protection.

Adenosine exerts physiological effects at a variety of receptor subtypes—A_1, A_2, A_3 (and possibly others). The combined rabbit studies suggest that A_1-adenosine receptors may be the most important receptors in mediating preconditioning, although not all studies agree with this.[26] There is evidence emerging that part of the effect of adenosine-mediated preconditioning in rabbits may be linked to A_3-adenosine receptors.[13,17,27] While these receptors have been cloned (reviewed in ref. 28), their functional responses in the heart have yet to be clarified, although they have been linked to PKC activation.

In stark contrast to the findings in rabbits, the data emanating from studies in rats fails to suggest a significant involvement of adenosine in mediating any of the cardioprotective effects of preconditioning. Attempts to mimic preconditioning in rats with adenosine and/or stable receptor agonists, or to block preconditioning with adenosine

Table 12.1. Examples of studies on adenosine in preconditioning

(A) Studies implicating adenosine

Mediator	Implicated (+/–)	Species	Preconditioning Stimulus	Inhibitory Intervention	End Point	References
Adenosine	+	Dog	Ischemia/reperfusion	5'-nucleotidase inhibitor; A_1/A_2 adenosine antagonists	Infarct size	20, 21, 22, 23
	+	Dog	Adenosine i.c.		ATP depletion	104
	+	Rabbit	Ischemia/reperfusion	A_1/A_2, A_3 adenosine antagonists	Infarct size	10, 13, 105, 106, 107
	+	Rabbit	Ischemia/reperfusion +NTI inhibitor/acadesine		Infarct size	14, 18
	+	Rabbit	A_1, A_1/A_2, A_3 adenosine agonists		Infarct size	10, 11, 12, 13
	+	Rabbit	Ischemia/reperfusion	A_1/A_2 adenosine antagonists	Myocardial stunning	106
	+	Rabbit	Adenosine/dipyridamole		Myocardial stunning	16
	+	Rabbit	Hypoxia	A_1/A_2 adenosine antagonists	Infarct size	108
	+	Rabbit (isolated myocytes)	A_1, A_1/A_3, A_3 adenosine agonists	A_1, A_1/A_3, A_3 adenosine antagonists	Cellular injury	17, 27
	+	Human	Adenosine i.c.		ST-segment elevation during angioplasty	109

(B) Studies eliminating adenosine

Mediator	Implicated (+/–)	Species	Preconditioning Stimulus	Inhibitory Intervention	End Point	References
	–	Dog	Ischemia/reperfusion	A_1/A_2 adenosine antagonists	Arrhythmias	24
	–	Dog	Adenosine i.c.		Myocardial function	25
	–	Rabbit	Global ischemia/reperfusion; A_1 adenosine agonist	A_1/A_2 adenosine antagonist	Myocardial function	26
	–	Rat	Adenosine deaminase inhibitor		Infarct size	33
	–	Rat	Ischemia/reperfusion	A_1/A_2 adenosine antagonist	Infarct size	31
	–	Rat	Ischemia/reperfusion; adenosine	A_1/A_2 adenosine antagonist	Arrhythmias	29, 31
	–	Rat	Ischemia/reperfusion;	A_1/A_2 adenosine antagonist	Metabolic changes	110
	–	Rat	Adenosine/ A_1 agonist	A_1/A_2 adenosine antagonist	Metabolic changes	111
	–	Rat	Global ischemia; adenosine		Contractile function	112
	–	Pig	Ischemia/reperfusion	A_1/A_2 adenosine antagonist	Myocardial stunning (late)	113

NTI inhibitor = Nucleoside transport inhibitor.

antagonists, have universally failed to implicate adenosine, irrespective of whether infarct size, arrhythmias or cardiac function are studied as the end-point (summarized in Table 12.1b). What, then, could explain the marked species differences between rabbits and rats?

a. One hypothesis is that A_1-receptors do not mediate cardioprotection in rat hearts. This is not the case. Firstly, we have shown from studies in our own laboratories that the A_1-receptor agonist R-PIA exerts a marked antiarrhythmic effect in rat isolated hearts,[9] yet it is unable to substitute for the antiarrhythmic effect of preconditioning[29] (Fig. 12.1) using a similar infusion protocol to that employed in studies which have demonstrated its ability to mimic the anti-infarct effects of preconditioning in rabbit hearts.[11] Furthermore, the A_1-receptor antagonist, DPCPX, is able to reverse the antiarrhythmic effects of R-PIA in rat isolated hearts[9] but not the antiarrhythmic effects of preconditioning[29] (Fig. 12.2).

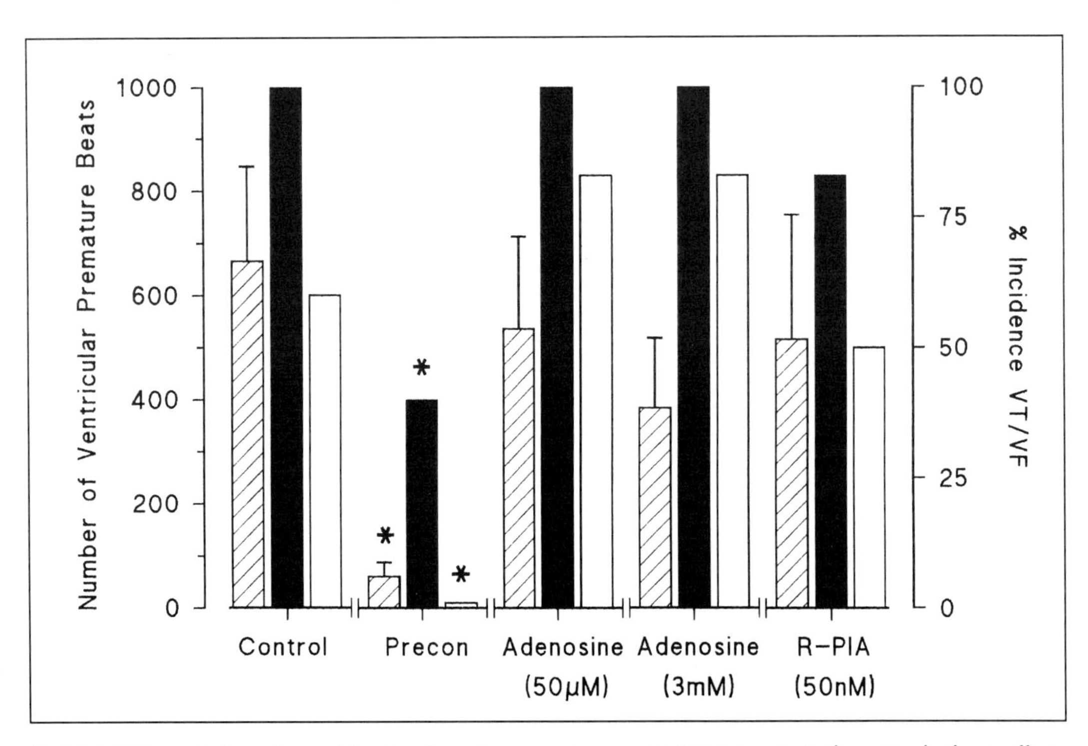

*Fig 12.1. Failure of adenosine and the A_1-adenosine receptor agonist, R-PIA, to mimic the antiarrhythmic effects of ischemic preconditioning in isolated rat hearts. Preconditioning was induced by 3 minute coronary artery occlusion followed by 10 minute reperfusion, prior to a long 30 minute coronary occlusion. Adenosine or R-PIA was infused for 3 minutes, followed by a 10 minute drug-free period prior to the 30 minute coronary artery occlusion. The figure illustrates the total number of ventricular premature beats (hatched bars) and the percent incidence of ventricular tachycardia (VT; solid bars) and ventricular fibrillation (VF: open bars) which occurred during a 30 minute period of ischemia. *P<0.05 compared with control (nonpreconditioned) group. Adapted from data in reference 29.*

Similarly, while adenosine has been shown to reduce infarct size in rat hearts in the absence of preconditioning,[30] adenosine receptor blockade is unable to attenuate the anti-infarct effects of preconditioning in this species.[31]

b. It is possible that endogenous adenosine is not released from rat hearts in sufficient quantities following a short period of ischemia to act as a mediator of preconditioning. However, significant adenosine release following a period of preconditioning ischemia from rat hearts has been demonstrated,[32] while raising levels to a similar degree with exogenous adenosine is not capable of mimicking preconditioning.[33]

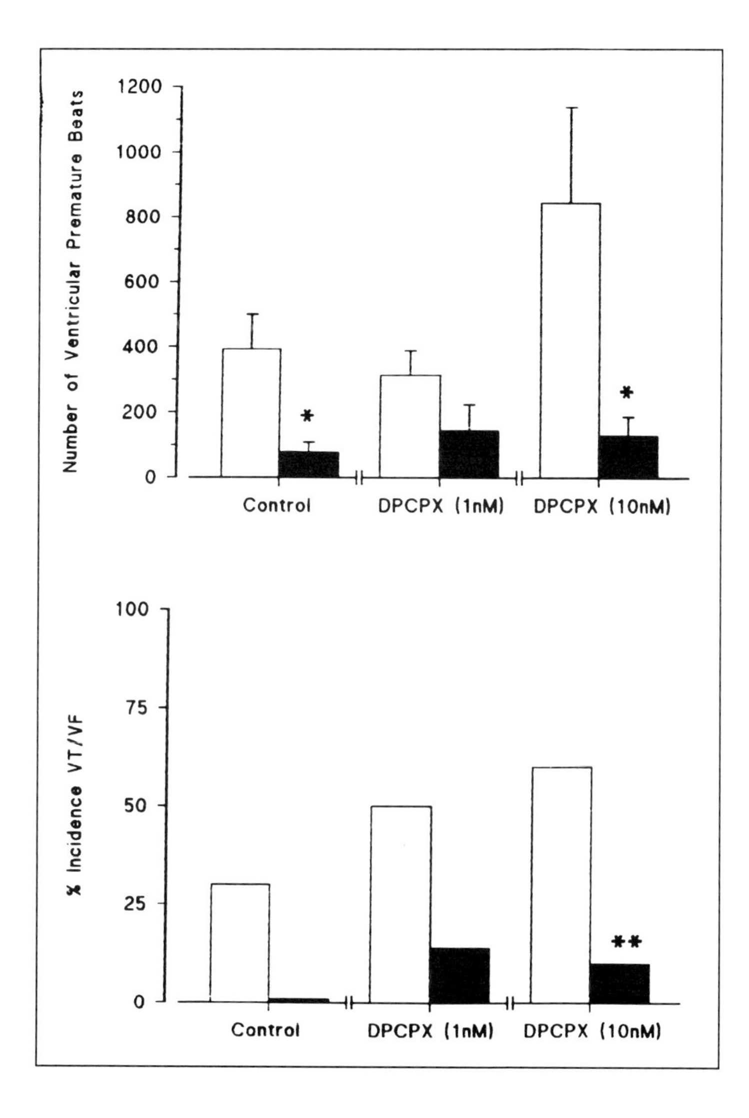

*Fig. 12.2. The effects of the A_1-adenosine receptor antagonist, DPCPX, on the total number of ventricular premature beats (upper panel) and the incidence of ventricular tachycardia and ventricular fibrillation (lower panel) during 30 minutes of coronary occlusion in rat isolated hearts with (solid bars) and without (open bars) antecedent preconditioning (3 minute occlusion and 10 minute reperfusion). *$P<0.05$; **$P<0.01$ compared with the appropriate nonpreconditioned group. Adapted from data in reference 29.*

c. Differences in collateral flow could not explain the species difference since both rats and rabbits have low collateral flow reserve.[34] Moreover, improvement in collateral flow may not be a major contributor to the cardioprotective effects of preconditioning.[35]

Despite the failure to implicate adenosine in preconditioning in rats, the very fact that rat hearts are equally susceptible to ischemic preconditioning as other species demonstrates a very clear species difference and highlights the importance of pursuing studies in a range of species. From the above, it appears that adenosine may well act as an initial preconditioning trigger to subsequent events leading to cardioprotection in rabbits and dogs, while the initial trigger is different in rats. Furthermore, the findings from studies with adenosine in dogs suggest that different triggers exist for the different protective facets of preconditioning.

12.2.2. Other Labile Mediators

In addition to adenosine, there are other labile mediators which could contribute to triggering the events leading to preconditioning. These include bradykinin, nitric oxide, and prostanoids (summarized in Table 12.2). Bradykinin has been implicated in the anti-infarct effect of preconditioning in a very few studies in rabbits,[36,37] but the majority of studies on bradykinin have been focused on arrhythmias. There is strong evidence from dog studies that bradykinin may be involved in this aspect of preconditioning, since intracoronary infusion of bradykinin has a profound antiarrhythmic effect, similar to that seen with preconditioning,[38] and preconditioning itself can be markedly attenuated by the B_2-receptor antagonist HOE 140[39] (icatibant). Once again, evidence from studies in rats is in marked contrast to this. HOE 140 does not prevent either the antiarrhythmic[40] or anti-infarct[41] effect of preconditioning in this species, despite the fact that, as with adenosine, a cardioprotective effect of bradykinin in the absence of preconditioning can be demonstrated.[41,42]

It has been postulated that the mechanism by which bradykinin preconditions the heart against arrhythmias in dogs is by release of nitric oxide, since both the antiarrhythmic effects of intracoronary bradykinin[43] and preconditioning[44,45] can be inhibited by the L-arginine analog L-NAME and by methylene blue. However, in rats there is no evidence that nitric oxide is involved in the protective effect of preconditioning against either ischemic[46] or reperfusion[47] arrhythmias. One possible explanation for this species difference in the importance of bradykinin and nitric oxide in preconditioning may lie in a difference between the response to B_2-receptor stimulation in rat and dog hearts, since in rat hearts there is evidence that, at least with respect to the vasodilator response to bradykinin, this involves cyclooxygenase products rather than NO release.[48] Interestingly, L-arginine analogs do

Table 12.2. Summary of studies on other labile mediators in preconditioning

Mediator	Implicated (+/–)	Species	Preconditioning Stimulus	Inhibitory Intervention	End Point	References
Bradykinin	+	Rabbit	Ischemia/reperfusion; BK infusion; captopril	HOE140	Infarct size	36, 37
	+	Dog	Ischemia/reperfusion; BK infusion	HOE140/ L-NAME	Arrhythmias	38, 39, 43
	+	Rat	Ischemia/reperfusion; BK infusion	Arg[Hyp-Thi-Phe]BK	Contractile function	114
	–	Rat	Ischemia/reperfusion; BK infusion	HOE140	Arrhythmias	46
	–	Rat	Ischemia/reperfusion; BK infusion	HOE140	Infarct size	41
Nitric Oxide	+	Dog	Ischemia/reperfusion	L-arginine analogs; methylene blue	Arrhythmias	44, 45
	–	Dog	Ischemia/reperfusion; ACh	L-arginine analogs	Infarct size	49
	–	Rabbit	Ischemia/reperfusion	L-arginine analogs	Infarct size	51
	–	Rat	Ischemia/reperfusion	L-arginine analogs	Infarct size	50
	–	Rat	Ischemia/reperfusion	L-arginine analogs	Arrhythmias	46, 47
Prostanoids	+	Dog	Ischemia/reperfusion	Cyclooxygenase inhibitor	Arrhythmias	115
	–	Rabbit	Ischemia/reperfusion	Cyclooxygenase inhibitor	Infarct size	117
	–	Rat	Ischemia/reperfusion	Cyclooxygenase inhibitor	Infarct size	116

HOE 140 = icatibant (B_2-receptor antagonist)

not prevent the anti-infarct effects of preconditioning in dogs[49] or rats,[50] suggesting once again that the protection afforded by preconditioning against the different consequences of ischemia probably do not share a common mechanism. Indeed, there is evidence to suggest that inhibition of nitric oxide production by the heart may reduce myocardial injury, rather than exacerbate it.[51]

12.3. TRANSDUCTION/SIGNALING MECHANISMS

The evidence described above clearly demonstrates that, while labile mediators such as adenosine, bradykinin and nitric oxide may be particularly important as initial triggers of some aspects of preconditioning in certain species, there is no simple explanation for the preconditioning phenomenon as a whole. As a consequence, research has turned to examine the secondary events following receptor activation by these mediators in an attempt to determine whether a common mechanism may exist. For the mediators described so far, the two most probable membrane-linked transducers are the pertussis-toxin sensitive G protein, G_i, and the ATP-dependent potassium (K_{ATP}) channel. Some of the mediators may be linked to both of these transducers, while others may act exclusively via only one. The intracellular events following either G protein or K_{ATP} channel activation are complex and, since they have been discussed in detail in other chapters within this book, are outwith the scope of this chapter. However, the one common event which may be important is activation of protein kinase C (PKC). Considering the aim of this chapter, the following section will concentrate on overviewing the variations in findings to support or eliminate a role for each of these systems between the different species and end-points studied.

12.3.1. K_{ATP} CHANNELS

The evidence implicating the K_{ATP} channel in the infarct-limiting effect of preconditioning has been dealt with in detail in the chapter by Grover (chapter 8). K_{ATP} channel activation may occur in response to adenosine A_1-receptor activation, since it has been demonstrated that this receptor is linked to K_{ATP},[52] at least in neonatal ventricular myocytes. The consequence of K_{ATP} channel opening would be to shorten the action potential duration, reduce calcium influx, depress contraction, reduce cellular energy consumption, preserve ATP and thus preserve myocardial function. Most of the evidence to support a role for the K_{ATP} channel in preconditioning emanates from studies using infarct size as the end point, although, in contrast to the studies on adenosine, most of these studies were performed in dogs (Table 12.3). In this species the K_{ATP} channel blocker glibenclamide can reverse the infarct limiting effect of preconditioning stimulated by either ischemia/reperfusion,[53-55] adenosine[22] or acetylcholine[49] (whose receptors are also linked to the K_{ATP} channel), while preactivation of the K_{ATP} channel with bimakalim can lower the temporal threshold for a period of ischemia to induce preconditioning.[56] Several studies in rabbits have also implicated the K_{ATP} channel in this facet of preconditioning[57-59] (Table 12.3). These findings are therefore consistent with the studies on the role of adenosine in the infarct-limiting effect of preconditioning in these species, suggesting that adenosine may be the trigger for preconditioning, acting via the K_{ATP} channel. However, one study in rabbits failed to support this hypothesis,[60] although follow-up studies suggest that this may be due to differences in anesthetic used.[57,59] There has been no systematic analysis of the role of the K_{ATP} channel in the infarct-limiting effect of preconditioning in rat hearts, perhaps due to the lack of evidence to implicate the possible trigger adenosine, although one study has demonstrated that glibenclamide cannot reverse this effect of preconditioning.[61]

There are few studies which have addressed the role of the K_{ATP} channel in the antiarrhythmic effects of preconditioning, but those which have been performed have failed to implicate this as a mechanism, irrespective of species studied (dogs[62] and rats[63]). One might expect that shortening of the action potential duration as a consequence of K_{ATP} channel opening would permit the development of re-entrant arrhythmias, which is the opposite effect to preconditioning. Indeed the proarrhythmic effects of K_{ATP} channel openers is well documented.[64,65] However, the effects of preconditioning on cardiac electrophysiology is complex and, at present, ill-defined[66] and requires further study before the role of the K_{ATP} channel in this aspect of preconditioning can be fully explained.

Thus, as with adenosine, there is a marked species difference between rats and rabbits (and dogs) with respect to evidence supporting a role for K_{ATP} channels as a mechanism of preconditioning. Although rat heart is known to be deficient in certain potassium channels, such as I_K,[67] it does possess the K_{ATP} channel[68] excluding this as a reason for the failure of glibenclamide to prevent preconditioning in this species. Furthermore, it has been demonstrated that K_{ATP} channels may indeed be important in mediating the protective effects of preconditioning on cardiac function in rat hearts,[69] but that this is dependent upon the method of induction of preconditioning.

12.3.2. G Proteins

A second receptor transducer which could transmit messages from the initial trigger to the cellular changes which result in preconditioning cardioprotection is activation of the inhibitory G protein, G_i, which is linked to both the A_1-adenosine receptor and the muscarinic M_2-receptor. A recent study has shown that responsiveness of pertussis toxin-sensitive G_i proteins towards receptor activation is reduced in ischemic tissue, but increased in tissue which has been preconditioned, resulting in an increase in the effectiveness of the receptor stimulants

Table 12.3. Summary of studies on transduction mechanisms in preconditioning

Transduction Mechanism	Implicated (+/–)	Species	Preconditioning Stimulus	Inhibitory Intervention	End Point	Reference
K_{ATP} Channels	+	Pig	Ischemia/reperfusion; K_{ATP} channel opener	Glibenclamide	Infarct size; Action potential duration	118, 119
	+	Dog	Ischemia/reperfusion	Glibenclamide; 5-HD	Infarct size	53, 54, 55
	+	Dog	KATP channel opener		Infarct size	55, 56
	+	Dog	Adenosine; acetylcholine	Glibenclamide; 5-HD	Infarct size	53, 54
	+	Dog	Ischemia/reperfusion ± K_{ATP} channel opener		Cardiac function; Infarct size	56, 120
	+	Rabbit	Ischemia/reperfusion	Glibenclamide	Infarct size	57, 58, 59
	+	Rabbit	Ischemia/reperfusion	Glibenclamide	Electrical uncoupling	121
	+	Human	Angioplasty	Glibenclamide	ST-segment	122
	+	Rat	Ventricular overdrive pacing	Glibenclamide	Cardiac function	69
	–	Dog	Ischemia/reperfusion	Glibenclamide	Arrhythmias	62
	–	Rabbit	Ischemia/reperfusion	Glibenclamide	Infarct size	60
	–	Rat	Ischemia/reperfusion	Glibenclamide	Infarct size	61
	–	Rat	Ischemia/reperfusion; K_{ATP} channel opener	Glibenclamide	Arrhythmias	63
	–	Rat	Ischemia/reperfusion	Glibenclamide	Cardiac function; ionic alterations	69, 123, 124

Transduction Mechanism	Implicated (+/–)	Species	Preconditioning Stimulus	Inhibitory Intervention	End Point	Reference
G Proteins	+	Dog	Ischemia/reperfusion		G-protein responsiveness	70
	+	Rabbit	Ischemia/reperfusion; carbachol	Pertussis toxin	Infarct size	71
	+	Rat (in vitro)	Ischemia/reperfusion	Pertussis toxin	Arrhythmias	74
	–	Pig	Ischemia/reperfusion		G-protein content	72
	–	Rat	Ischemia/reperfusion	Pertussis toxin	Infarct size	73
	–	Rat (in vitro blood perfused and in vivo)	Ischemia/reperfusion	Pertussis toxin	Arrhythmias	75, 76
PKC-Activation & Translocation	+	Rabbit	Ischemia/reperfusion; PKC activator	PKC inhibitors	Infarct size	81, 82, 83
	+	Rabbit myocyte	Hypoxia	PKC inhibitors	Ischemic contracture	125, 126
	+	Rat	Ischemia/reperfusion; PKC activator	PKC inhibitors	Infarct size	84, 85
	+	Rat	Ischemia/reperfusion	PKC inhibitors	Arrhythmias	84
	+	Rat	Ischemia/reperfusion; α-adrenoceptor stimulation	PKC inhibitors	Cardiac function; translocation of PKC-δ	86
	+	Rat	Hypoxia; PKC activator		Contractile recovery	87

(e.g., adenosine and acetylcholine) to induce the cellular responses which contribute to cardioprotection.[70] As with the K_{ATP} channel, studies which implicate G_i have been performed in the species (rabbit, and dog) where there is strong evidence for A_1-adenosine receptor activation as a mediator of the infarct-limiting effect of preconditioning,[70-72] whereas in rats inhibition of G_i function with *pertussis* toxin does not abrogate this effect of preconditioning.[73] The only studies of the role of G-protein activation in the antiarrhythmic effects of preconditioning have all been performed in rats. One study, performed in chrystalloid-perfused rat hearts, found that hearts from rats pretreated with *Bordetella pertussis* toxin could not be preconditioned by a standard ischemia/reperfusion preconditioning protocol,[74] whereas in experiments employing a similar preconditioning protocol using either blood-perfused[75] or in situ[76] hearts *Bordetella pertussis* toxin pretreatment did not affect the antiarrhythmic effect of preconditioning. One possible explanation for these conflicting findings in this species is that in the absence of blood there is a labile mediator (but probably not adenosine, since there is no positive evidence to implicate this as a mediator of preconditioning in this species) which is present in sufficient quantities to stimulate receptors linked to G_i, but that in the presence of blood this mediator is inactivated in some way. Taken together, however, the results of the studies on G-protein function in preconditioning again point to a possible role against infarct size in rabbits, and dogs (with limited evidence in pigs), but not in rats. A lack of functional G_i in rat heart cannot be put forward as an explanation for the species difference for two reasons. Firstly the results of the positive study mentioned above[74] demonstrate that under certain experimental conditions G_i may play a role in rats and, second, localization of G_i close to adenylyl cyclase in rat heart has been demonstrated,[77] arguing against a lack of the appropriate signaling system in this species.

12.3.3. Protein Kinase C Activation

In the absence of a common link between all species studied thus far to explain the mechanism responsible for preconditioning, the search has moved to events downstream of specific signaling events, such as A_1-adenosine receptor activation, and switching on of their respective transduction systems. It has been demonstrated that α_1-adrenoceptor stimulation may participate in at least some of the effects of preconditioning in rat hearts, such as a reduction in contractile dysfunction.[78] Activation of α_1-adrenoceptors stimulates phosphoinositide metabolism, resulting in increases in both IP_3 and diacylglycerol (DAG) levels, the latter being a primary activator of protein kinase C (PKC). PKC is a ubiquitous Ser-Thr kinase with multiple isoforms which are associated with a variety of receptors (including A_1-adenosine receptors and M_2 muscarinic receptors) and physiological effects, including regulatory effects on numerous intracellular functions such as contraction, metabolism and protein synthesis.[79,80] Thus PKC activation could occur downstream of all the initial triggers outlined so far and could conceivably provide the common link between the species. The studies on PKC in preconditioning have been reviewed in detail in the chapter by Cohen et al (chapter 11). In contrast to all the other potential mechanisms studied, the experiments to assess the role of PKC have yielded reasonably consistent results, rather than highlighting differences between species and end-points. For example in rabbit heart, the infarct-limiting effect of preconditioning can be blocked by inhibitors of protein kinase C[81,82] and mimicked by PKC activators;[83] the same results are found in rats[84,85] (Table 12.3). Furthermore, PKC inhibitors have been demonstrated to prevent the protective effects of preconditioning against ventricular arrhythmias[84] and contractile dysfunction[86] in rat heart and to prevent preconditioning against ischemic contracture in isolated rabbit myocytes.[87]

Thus, a potential mechanism has now been identified in the two species where the results of previous studies on other possible mechanisms have consistently been in conflict.

12.4. METABOLIC CHANGES

An alternative explanation for the initial preconditioning response is that, rather than release of a labile mediator which acts as a trigger for secondary responses ultimately resulting in cardioprotection, the metabolic consequences of ischemia (and reperfusion) itself may be a signal for this adaptation to stress. The metabolic changes which occur in the ischemic myocardium commence almost immediately when blood flow is reduced and include ATP depletion (which would activate K_{ATP} channels), acidosis and lactate accumulation, decreased buffering capacity, ionic instability and accumulation of toxic catabolites and cellular edema, all of which will ultimately contribute to electrophysiological disturbances and cell death. Furthermore, reperfusion (i.e., the reintroduction of blood and molecular oxygen) may contribute to injury by facilitating generation of oxygen-derived free radicals, which are extremely labile and active entities and which may play a role in both myocardial injury and arrhythmia development.

The role of free radical generation during the reperfusion period in the process of preconditioning appears to be as controversial as the evidence implicating them in some of the consequences of reperfusion[88,89] (Table 12.4). There is reasonable evidence to show that a brief period of ischemic preconditioning boosts the antioxidant reserve of the cardiac tissue, by increasing the amount of myocardial antioxidant enzymes such as superoxide dismutase, which is normally depleted by ischemia.[90-94] This effect has been implicated primarily in the delayed protection seen with preconditioning since oxidative stress has been shown to induce heat shock protein expression.[92] Conversely, at least one study has suggested that cardiac protection may be mediated in part by the free radicals themselves, since the protective effect of preconditioning against infarct size can be partially reversed by various free radical scavengers.[95] However, other studies have failed to show any reversal, by such scavengers, of the anti-infarct effects of preconditioning.[96-98]

Several studies have also assessed the effect of a period of preconditioning ischemia on the metabolic consequences of a prolonged ischemic period. Most of these have been performed in rats (Table 12.4). Together they suggest that preconditioning may increase the glycogen synthesizing capacity of the heart[99,100] and improve the buffering capacity during the prolonged ischemic episode,[101-103] thus reducing the metabolic damage and consequently improving cardiac function and increasing cell survival. Perhaps, then, the initial trigger for the preconditioning phenomenon in rats could be due to metabolic preservation.

12.5. CONCLUSIONS

From the above, but by no means comprehensive, review of studies into the mechanism of preconditioning two clear patterns seem to emerge. The first is that an obvious species difference exists with respect to the identity of the initial trigger which may be responsible for the infarct-limiting effect of preconditioning. Thus, in rabbits, dogs and, to some extent, pigs (although the number of studies in this species is limited) A_1-adenosine receptor activation appears to be a common and important trigger for later events, while in rats there appears to be no role for activation of these receptors in triggering any of the consequences of preconditioning. Despite the strong evidence in support of adenosine in rabbits and dogs, however, it would be naïve to assume that this is the only event which is capable of triggering the preconditioning process, since it can clearly be mimicked in these species by other stimuli, such as M_2-muscarinic receptor activation. The secondary event leading to preconditioning (i.e., the transduction mechanism) also appears to differ between rats and the other species, since there is reasonable evidence that K_{ATP} channels, in particular, and to some extent inhibitory G-protein (G_i) (both of which are coupled to A_1- and M_2-receptors) may act as transduction mechanisms, with rats again being the notable exception. Where commonality does appear, however, is with activation of PKC. This can be activated by a variety of receptor systems, as well as by nonreceptor-mediated events such as metabolic changes within the cells. Since there is some evidence to suggest that some metabolic changes play an important role in preconditioning in rat heart, this may hint at the initial trigger in this species.

The second message which is generated from this review is that the triggers for the different sequelae of preconditioning (e.g., infarct size limitation versus arrhythmia suppression) are not necessarily the same. This is made particularly clear from studies in dogs, where A_1-adenosine receptor and K_{ATP} channel activation may contribute to infarct size reduction, but there is no evidence to support its role in

Table 12.4. Studies on metabolic alterations in preconditioning

Metabolic Change	Implicated (+/–)	Species	Preconditioning Stimulus	Inhibitory Intervention	End Point	Reference
↑ Antioxidant Reserve	+	Rat	Hypoxia	Mn-SOD antisense	Cardiac function; antioxidant reserve	90, 91
	+	Rat	Oxidant stress		Cardiac function; HSP expression	92
	+	Dog	Ischemia/reperfusion		Neutrophil function; antioxidant activity	93, 94
	–	Rabbit	Ischemia/reperfusion		Antioxidant activity	127
	–	Pig	Ischemia/reperfusion		Antioxidant activity	72
Free Radical Generation	+	Rabbit	Ischemia/reperfusion	Oxyradical scavengers	Infarct size	95
	–	Rabbit	Ischemia/reperfusion	SOD	Infarct size	96
	–	Rabbit	Ischemia/reperfusion	SOD	Cardiac function; LDH release	97
	–	Rat	Ischemia/reperfusion	MPG	Infarct size	98
Phosphoprotein Phosphatase Activation	+	Rat	Ischemia/reperfusion		Myocardial glycogen metabolism	99, 100
↑ Buffering / ↓ Acidosis	+	Rat	Ischemia/reperfusion		Infarct size	101
	+	Rat	Ischemia/reperfusion		Contractile dysfunction	102, 103
Na^+/H^+ Exchange	–	Rat	Ischemia/reperfusion	Na^+/H^+ inhibitor	Infarct size	128
	–	Rat	Ischemia/reperfusion		Cardiac function; ATP content	129
Catabolite Accumulation	–	Pig	Ischemia/reperfusion; catabolite washout		Infarct size	130

arrhythmia reduction. Considering that the events in the ischemic myocardium which result in cellular injury and electrophysiological disturbances are likely to be different, then this may not be entirely surprising. However, since the role of PKC activation in the antiarrhythmic effects of preconditioning has yet to be elucidated, it is not possible to say at this time whether or not this may exist as a final common mechanism for these two facets of preconditioning.

References

1. Yellon DM, Alkhulaifi AM, Pugsley WB. Preconditioning the human myocardium. Lancet 1993; 342:276-277.
2. Parratt JR. Endogenous myocardial cardioprotective (antiarrhythmic) substances. Cardiovasc Res 1994; 27:693-702.
3. Forman MB, Velasco CE, Jackson EK. Adenosine attenuates reperfusion injury following regional myocardial ischemia. Cardiovasc Res 1993; 27:9-17.
4. Wainwright CL, Parratt JR. An antiarrhythmic effect of adenosine during myocardial ischemia and reperfusion. Eur J Pharmacol 1988; 146:183-194.
5. Olafsson B, Forman MB, Puet DW. Reduction of reperfusion injury in the canine preparation by intracoronary adenosine: importance of the endothelium and the 'no-reflow' phenomenon. Circulation 1987; 76: 1135-1145.
6. Wainwright CL, Parratt JR The effects of R-PIA, a selective A1-adenosine agonist, on hemodynamics and ischemic arrhythmias in pigs. Cardiovasc Res 1993; 27:84-89.
7. Wainwright CL, Parratt JR, Van Belle H. The antiarrhythmic effects of the nucleoside transporter inhibitor, R75231, in anesthetized pigs. Br J Pharmacol 1993; 109:592-599.
8. Fagbemi O, Parratt JR. Antiarrhythmic actions of adenosine in the early stages if experimental myocardial ischemia. Eur J Pharmacol 1984; 100:243-244.
9. Wainwright CL, Kang L. The effect of chronic pretreatment on the antiarrhythmic and hemodynamic effects of the A_1-adenosine agonist, R-PIA, in rat isolated hearts. Br J Pharmacol 1994; 112:504P (Abstract).
10. Liu GS, Thornton J, Van Winkle DM et al. Protection against infarction afforded by preconditioning is mediated by A_1 adenosine receptors. Circulation 1991; 84:350-356.
11. Thornton JD, Liu GS, Olsson RA et al. Intravenous pretreatment with A_1-selective adenosine analogs protects the heart against infarction. Circulation 1992; 85:659-665.
12. Hale SL, Bellows SD, Hammerman H et al. An adenosine A_1 receptor agonist R(-)-N-(2-phenylisopropyl)-adenosine (PIA), but not adenosine itself, acts as a therapeutic preconditioning mimetic agent in rabbits. Cardiovasc Res 1993; 27:2140-2145.
13. Liu GS, Richards SC, Olsson RA et al. Evidence that the adenosine A_3 receptor may mediate the protection afforded by preconditioning in the

isolated rabbit heart. Cardiovasc Res 1994; 28:1057-1061.
14. Itoya M, Miura T, Sakamoto J et al. Nucleoside transport inhibitors enhance the infarct size-limiting effect of ischemic preconditioning. J Cardiovasc Pharmacol 1994; 24:846-852.
15. Downey JM, Liu GS, Thornton JD. Adenosine and the antiinfarct effects of preconditioning. Cardiovasc Res 1993; 27:3-8.
16. Mosca SM, Gelpi RJ, Cingolani HE. Adenosine and dipyridamole mimic the effects of preconditioning. J Mol Cell Cardiol 1994; 26:1403-1409.
17. Armstrong S, Ganote CE. Adenosine receptor specificity in preconditioning of isolated rabbit cardiomyocytes: evidence of A_3 receptor involvement. Cardiovasc Res 1994; 28:1049-1056.
18. Tsuchida A, Liu GS, Mullane KM et al. Acadesine lowers temporal threshold for the myocardial infarct size limiting effect of preconditioning. Cardiovasc Res 1993; 27:116-120.
19. Tsuchida A, Yang X-M, Burckhartt B et al. Acadesine extends the window of protection afforded by ischemic preconditioning. Cardiovasc Res 1994; 28:379-383.
20. Kitakaze M, Hori M, Morioka T et al. Infarct size-limiting effect of ischemic preconditioning is blunted by inhibition of 5'-nucleotidase activity and attenuation of adenosine release. Circulation 1994; 89: 1237-1246.
21. Kitakaze M, Hori M, Takashima S et al. Ischemic preconditioning increases adenosine release and 5'-nucleotidase activity during myocardial ischemia and reperfusion in dogs: implications for myocardial salvage. Circulation 1993; 87:208-215.
22. Auchampach JA, Gross GJ. Adenosine A_1-receptors, K_{ATP} channels, and ischemic preconditioning in dogs. Am J Physiol 1993; 264:H1327-H1336.
23. Hoshida S, Kuzuya T, Nishida M et al. Adenosine blockade during reperfusion reverses the infarct limiting effect in preconditioned canine hearts. Cardiovasc Res 1994; 28:1083-1088.
24. Vegh A, Papp JG, Parratt JR. Pronounced antiarrhythmic effects of preconditioning in anesthetized dogs: is adenosine involved? J Mol Cell Cardiol 1995; 27:349-356.
25. Sekili S, Jeroudi MO, Tang XL et al. Effect of adenosine on myocardial 'stunning' in the dog. Circ Res 1995; 76:82-94.
26. Hendrikx M, Toshima Y, Mubagwa K et al. Improved functional recovery after ischemic preconditioning in the globally ischemic rabbit heart is not mediated by adenosine A_1 receptor activation. Bas Res Cardiol 1993; 88:576-593.
27. Armstrong S, Ganote C. Preconditioning of myocytes: Dose response evidence for two (A_1 and A_3) adenosine receptors. J Mol Cell Cardiol 1995; 27 (Suppl):A152 (Abstract).
28. Linden J. Cloned adenosine A_3 receptors: Pharmacological properties, species differences and receptor functions. Trends Pharm Sci 1994; 15:298-306.
29. Piacentini L, Wainwright CL, Parratt JR. The antiarrhythmic effect of

preconditioning, in rat isolated hearts, does not involve A_1 adenosine receptors. Br J Pharmacol 1992; 107:137P (Abstract).
30. Singh J, Garg KN, Garg D et al. Effect of adenosine and inosine on experimental myocardial infarction in rats. Ind J Exp Biol 1988; 26:771-774.
31. Li Y, Kloner RA. The cardioprotective effects of ischemic 'preconditioning' are not mediated by adenosine receptors in rat hearts. Circulation 1993; 87:1642-1648.
32. Urbanski NK, Beresewicz A. Ischemic preconditioning (PC) and adenosine release in isolated rat heart. J Mol Cell Cardiol 1995; 27 (Suppl): A152 (Abstract).
33. Li Y, Kloner RA. Adenosine deaminase inhibition is not cardioprotective in the rat. Amer Heart J 1993; 126:1293-1298.
34. Winkler B, Sass S, Binz K et al. Myocardial blood flow and myocardial infarction in rats, guinea pigs and rabbits. J Mol Cell Cardiol 1984; 16(Suppl 2):22 (Abstract).
35. Wainwright CL. Myocardial preconditioning as the hearts self-protecting response against the consequences of ischemia. Trends Pharm Sci 1992; 13:90-93.
36. Wall TM, Sheehy R, Hartman JC. Role of bradykinin in myocardial preconditioning. J Pharmacol Exp Ther 1994; 270:681-689.
37. Miki T, Kiura T, Ura N et al. Captopril potentiates infarct size-limiting effect of preconditioning via bradykinin-2 receptor activation. J Mol Cell Cardiol 1995; 27 (Suppl):A160 (Abstract).
38. Vegh A, Szekeres L, Parratt JR. Local intracoronary infusions of bradykinin profoundly reduce the severity of ischemia-induced arrhythmias in anesthetized dogs. Br J Pharmacol 1991; 104:294-295.
39. Vegh A, Papp JG, Parratt JR Attenuation of the antiarrhythmic effects of ischemic preconditioning by blockade of bradykinin B_2 receptors. Br J Pharmacol 1994; 113:1167-1172.
40. Sun W, Wainwright CL. The potential antiarrhythmic effects of exogenous and endogenous bradykinin in the ischemic rat heart in vivo. Coronary Art Dis 1994; 5:541-550.
41. Bugge E, Ytrehus K. Bradykinin can protect against infarction but is not the mediator of ischemic preconditioning in the rat heart. J Mol Cell Cardiol 1995; 27 (Suppl):A160 (Abstract).
42. Linz W, Martorana PA, Grotsch H et al. Antagonizing bradykinin obliterates the cardioprotective effects of bradykinin and angiotensin-converting enzyme (ACE) inhibition in ischemic hearts. Drug Dev Res 1990; 19:393-408.
43. Vegh A, Papp JG, Szekeres L et al. Prevention by an inhibitor of the L-arginine-nitric oxide pathway of the antiarrhythmic effects of bradykinin in anesthetized dogs. Br J Pharmacol 1993; 110:18-19.
44. Vegh A, Szekeres L, Parratt JR. Preconditioning of the ischemic myocardium; Involvement of the L-arginine nitric oxide pathway. Br J Pharmacol 1992; 107:648-652.

45. Vegh A, Papp JG, Szekeres L et al. The local intracoronary administration of methylene blue prevents the pronounced antiarrhythmic effect of ischemic preconditioning. Br J Pharmacol 1992; 107:910-911.
46. Sun W, Wainwright CL, Parratt JR. The antiarrhythmic effect of ischemic preconditioning is not prevented by inhibition of endothelium-derived nitric oxide in anesthetized rats. Br J Pharmacol 1994; 112:380P (Abstract).
47. Lu HR, Remeysen P, De Clerk F. Does the antiarrhythmic effects of ischemic preconditioning in rats involve the L-arginine nitric oxide pathway? J Cardiovasc Pharmacol 1995; 25:524-530.
48. Fulton D, Mahboubi K, McGiff JC et al. Cytochrome P450-dependent effects of bradykinin in the rat heart. Br J Pharmacol 1995; 114:99-102.
49. Yao Z, Gross GJ. Role of nitric oxide, muscarinic receptors, and the ATP-sensitive K^+ channel in mediating the effects of acetylcholine to mimic preconditioning in dogs. Circ Res 1993; 73:1193-1201.
50. Weselcouch EO, Baird AJ, Sleph P et al. Inhibition of nitric oxide synthesis does not affect ischemic preconditioning in isolated perfused rat hearts. Am J Physiol 1995; 268:H242-H249.
51. Woolfson RG, Patel VC, Neild GH et al. Inhibition of nitric oxide synthesis reduces infarct size by an adenosine-dependent mechanism. Circulation 1995; 91:1545-1551.
52. Kirsch GE, Condina J, Birnbaumer L et al. Coupling of ATP-sensitive K^+ channels to A_1 receptors by G proteins in rat ventricular myocytes. Am J Physiol 1990; 259:H820-H826.
53. Auchampach JA, Grover GJ, Gross GJ. Blockade of ischemic preconditioning in dogs by the novel ATP dependent potassium channel antagonist sodium 5-hydroxydecanoate. Cardiovasc Res 1992; 26:1054-1062.
54. Yao Z, Gross GJ. The ATP-dependent potassium channel: An endogenous cardioprotective mechanism. J Cardiovasc Pharmacol 1994; 24(Suppl 4):S28-S34.
55. Gross GJ, Auchampach JA. Blockade of ATP-sensitive potassium channels prevents myocardial preconditioning in dogs. Circ Res 1992; 70:223-233.
56. Yao Z, Gross GJ. The activation of ATP-sensitive potassium channel lower threshold for ischemic preconditioning in dogs. Am J Physiol 1994; 267:H1888-H1894.
57. Walsh RS, Tsuchida A, Daly JJF et al. Ketamine-xylazine anesthesia permits K(ATP) channel antagonist to attenuate preconditioning in rabbit myocardium. Cardiovasc Res 1994; 28:1337-1341.
58. Toombs CF, Moore TL, Shebuski RJ. Limitation of infarct size in the rabbit by ischemic preconditioning is reversible with glibenclamide. Cardiovasc Res 1993; 27:617-622.
59. Miura T, Goto M, Miki T et al. Glibenclamide, a blocker of ATP-sensitive potassium channels, abolishes infarct size limitation by preconditioning in rabbits anesthetized with xylazine/pentobarbital but not with pentobarbital alone. J Cardiovasc Pharmacol 1995; 25:531-538.

60. Thornton JD, Thornton CS, Sterling DL et al. Blockade of ATP-sensitive potassium channels increases infarct size but does not prevent preconditioning in rabbit hearts. Circ Res 1993; 72:44-49.
61. Liu GS, Downey JM. Ischemic preconditioning protects against infarction in rat heart. Am J Physiol 1992; 263:H1107-H1112.
62. Vegh A, Papp JG, Szekeres L et al. Are ATP sensitive potassium channels involved in the pronounced antiarrhythmic effects of preconditioning? Cardiovasc Res 1993; 27:638-643.
63. Lu H, Remeysen P, De Clerck F. The protection by ischemic preconditioning against myocardial ischemia- and reperfusion-induced arrhythmias is not mediated by ATP-sensitive potassium channels in rats. Coronary Art Dis 1993; 4:649-657.
64. Eilde AAM, Janse MJ. Electrophysiological effects of ATP-sensitive potassium channel modulation: implications for arrhythmogenesis. Cardiovasc Res 1994; 28:16-24.
65. de La Coussaye JE, Eledjam J-J, Bruelle P et al. Electrophysiologic and arrhythmogenic effects of the potassium channel agonist BRL 38227 in anesthetized dogs. J Cardiovasc Pharmacol 1993; 22:722-730.
66. Parratt JR, Kane KA. K_{ATP} channels in ischemic preconditioning. Cardiovasc Res 1994; 28:783-787.
67. Josephson IR, Brown AM. Inwardly rectifying single-channel and whole cell K^+ currents in rat ventricular myocytes. J Membrane Biol 1986; 94:19-35.
68. Wolleben CD, Sanguinetti MC, Siegl PKS. Influence of ATP-sensitive potassium channel modulators on ischemia-induced fibrillation in isolated rat hearts. J Mol Cell Cardiol 1989; 21:783-788.
69. Ferdinandy P, Szilvassy Z, Koltai M et al. Ventricular overdrive pacing-induced preconditioning and no-flow ischemia-induced preconditioning in isolated working rat hearts. J Cardiovasc Pharmacol 1995; 25:97-104.
70. Niroomand F, Weinbrenner C, Weis A. Impaired function of inhibitory G-proteins during acute myocardial ischemia of canine hearts and its reversal during reperfusion and a second period of ischemia. Possible mechanisms for the protective mechanism of ischemic preconditioning. Circ Res 1995; 76:861-870.
71. Thornton JD, Liu GS, Downey JM. Pretreatment with pertussis toxin blocks the protective effects of preconditioning: Evidence for a G-protein mechanism. J Mol Cell Cardiol 1993; 25:311-320.
72. Fu LX, Kirkeboen KA, Liang QM et al. Free radical scavenging enzymes and G-protein mediated receptor signalling systems in ischemically preconditioned porcine myocardium. Cardiovasc Res 1993; 27:612-616.
73. Liu Y, Downey JM. Preconditioning against infarction in the rat heart does not involve a pertussis toxin sensitive G protein. Cardiovasc Res 1993; 27:608-611.
74. Piacentini L, Wainwright CL, Parratt JR. The antiarrhythmic effect of ischemic preconditioning in isolated rat heart involves a *pertussis* toxin sensitive mechanism. Cardiovasc Res 1993; 27:674-680.

75. Lawson CS, Coltart DJ, Hearse DJ. The antiarrhythmic action of ischemic preconditioning in rat hearts does not involve functional G_i proteins. Cardiovasc Res 1993; 27:681-687.
76. Piacentini L, Wainwright CL, Parratt JR Effects of *Bordetella pertussis* toxin pretreatment on the antiarrhythmic action of ischemic preconditioning in anesthetized rats. Br J Pharmacol 1995; 114:755-760.
77. Schulze W, Kössler A, Hinsch KD et al. Immunocytochemical localization of G-proteins (α subunits) in rat heart tissue. Eur Heart J 1991; 12(Suppl F):132-134.
78. Banerjee A, Locke-Winter C, Rogers KB et al. Preconditioning against myocardial dysfunction after ischemia and reperfusion by an α_1-adrenergic mechanism. Circ Res 1993; 73:656-670.
79. Fedida D, Braun AP, Giles WR. α_1 Adrenoceptors in myocardium: functional aspects and transmembrane signalling mechanisms. Pharmacol Rev 1993; 73:469-487.
80. Terzic A, Puceat M, Vassort G et al. Cardiac α_1-adrenoceptors: an overview. Pharmacol Rev 1993; 45:147-175.
81. Liu Y, Ytrehus K, Downey JM. Evidence that translocation of protein kinase C is a key event during ischemic preconditioning of rabbit myocardium. J Mol Cell Cardiol 1994; 26:661-668.
82. Liu Y, Cohen MV, Downey JM. Chelerythrene, a highly selective protein kinase C inhibitor, blocks the antiinfarct effect of ischemic preconditioning in rabbit hearts. Cardiovasc Drugs Ther 1994; 8:881-882.
83. Ytrehus K, Liu Y, Downey JM. Preconditioning protects ischemic rabbit heart by protein kinase C activation. Am J Physiol 1994; 266: H1145-H1152.
84. Li Y, Kloner RA. Does protein kinase C play a role in ischemic preconditioning in rat hearts? Am J Physiol 1995; 268:H426-H431.
85. Speechly-Dick ME, Mocanu MM, Yellon DM. Protein kinase C: Its role in ischemic preconditioning in the rat. Circ Res 1994; 75:586-590.
86. Mitchell MB, Meng X, Ao L et al. Preconditioning of isolated rat heart is mediated by protein kinase C. Circ Res 1995; 76:73-81.
87. Webster KA, Discher DJ, Bishopric NH. Cardioprotection in an in vitro model of hypoxic preconditioning. J Mol Cell Cardiol 1995; 27:453-458.
88. Tosaki A, Das DK. Reperfusion-induced arrhythmias are caused by generation of free radicals. Cardiovasc Res 1994:424-432.
89. Euler DE. Reperfusion-induced arrhythmias are not caused by generation of free radicals. Cardiovasc Res 1994:424-432.
90. Engelman DT, Watanabe M, Engelman RM et al. Hypoxic preconditioning preserves antioxidant reserve in the working rat heart. Cardiovasc Res 1995; 29:133-140.
91. Yamashita N, Nishida M, Hoshida S et al. Induction of manganese superoxide dismutase in rat cardiac myocytes increases tolerance to hypoxia 24 hours after preconditioning. J Clin Invest 1995; 2193-2199.
92. Kukreja RC, Kontos MC, Loesser KE et al. Oxidant stress increases heat shock protein 70 mRNA in isolated perfused rat heart. Amer J Physiol

1995; 267:H2213-H2219.
93. Hoshida S, Kuzuya T, Yamashita N et al. Brief myocardial ischemia affects free radical generating and scavenging systems in dogs. Heart Vessels 1993; 8:115-120.
94. Hoshida S, Kuzuya T, Fuji H et al. Sublethal ischemia alters antioxidant activity in canine heart. Amer J Physiol 1993; 264:H33-H39.
95. Tanaka M, Fujiwara H, Yamasaki K et al. Superoxide dismutase and NI 2-mercaptopropionyl glycine attenuate infarct size limitation effect of ischemic preconditioning in the rabbit. Cardiovasc Res 1994; 28:980-986.
96. Iwamoto T, Miura T, Adachi T et al. Myocardial infarct size limiting effect of ischemic preconditioning was not attenuated by oxygen free radical scavengers in the rabbit. Circulation 1991; 83:1015-1022.
97. Omar BA, Hanson AK, Bose SK et al. Ischemic preconditioning is not mediated by free radicals in the isolated rabbits heart. Free Rad Biol Med 1991; 11:517-520.
98. Richard V, Tron C, Thuillez C. Ischemic preconditioning is not mediated by oxygen derived free radicals in rats. Cardiovasc Res 1993; 27:2016-2021.
99. McNulty PH, Luba MC. Transient ischemia induces regional myocardial glycogen synthase activation and glycogen synthesis in vivo. Am J Physiol 1995; 268:H364-H370.
100. Janier MF, Vanoverschelde JLJ, Bergmann SR. Ischemic preconditioning stimulates anaerobic glycolysis in the isolated rabbit heart. Am J Physiol 1994; 267:H1353-H1360.
101. Wolfe CL, Sievers RE, Visseren FLJ et al. Loss of myocardial protection after preconditioning correlates with time course of glycogen recovery within the preconditioned segment. Circulation 1993; 87:881-892.
102. de Albuquerque CP, Gerstenblith G, Weiss RG. Importance of metabolic inhibition and cellular pH in mediating preconditioning contractile and metabolic effects in rat hearts. Circ Res 1994; 74:139-150.
103. de Albuquerque CP, Gerstenblith G, Weiss RG. Myocardial buffering capacity in ischemia preconditioned rat hearts. J Mol Cell Cardiol 1995; 27:777-781.
104. Vander-Heide RS, Reimer KA, Jennings RB. Adenosine slows ischemic metabolism in canine myocardium in vitro: relationship to ischemic preconditioning. Cardiovasc Res 1993; 27:669-673.
105. Thornton JD, Thornton CS, Downey JM. Effect of adenosine receptor blockade: Preventing protective preconditioning depends on time of initiation. Am J Physiol 1993; 265:H504-H508.
106. Urabe K, Miura T, Iwamoto T et al. Preconditioning enhances myocardial resistance to postischemic myocardial stunning via adenosine receptor activation. Cardiovasc Res 1993; 27:657-662.
107. Miura T, Ogawa T, Iwamoto T et al. Dipyridamole potentiates the myocardial infarct size limiting effect of ischemic preconditioning. Circulation 1992; 86:9797-985.
108. Walsh RS, Borges M, Thornton JD et al. Hypoxia preconditions rabbit

myocardium by an adenosine receptor-mediated mechanism. Can J Cardiol 1995; 11:141-146.
109. Kerensky RA, Kutcher MA, Braden GA et al. The effects of intracoronary adenosine on preconditioning during angioplasty. Clin Cardiol 1995; 18:91-96.
110. Murphy E, Fralix, TA, London RE et al. Effects of adenosine antagonists on hexose uptake and preconditioning in perfused rat heart. Am J Physiol 1993; 265:C1146-C1155.
111. Asimakis GK, Inners-McBride K, Conti VR. Attenuation of postischemic dysfunction by ischemic preconditioning is not mediated by adenosine in the isolated rat heart. Cardiovasc Res 1993; 27:1522-1530.
112. Cave AC, Collis CS, Downey JM et al. Improved functional recovery by ischemic preconditioning is not mediated by adenosine in the globally ischemic isolated rat heart. Cardiovasc Res 1993; 27:663-668.
113. Sun JZ, Tang XL, Knowlton AA et al. Late preconditioning against myocardial stunning. An endogenous protective mechanism that confers resistance to postischemic dysfunction 24h after brief ischemia in conscious pigs. J Clin Invest 1995; 95:388-403.
114. Starkopf J, Ytrehus K. Bradykinin and ischemic preconditioning in functional protection of the isolated rat heart. J Mol Cell Cardiol 1995; 27 (Suppl):A160 (Abstract).
115. Vegh A, Szekeres L, Parratt, JR. Protective effects of preconditioning of the ischemic myocardium involve cyclooxygenase products. Cardiovasc Res 1990; 24:1020-1023.
116. Li Y, Kloner RA. Cardioprotective effects of preconditioning are not mediated by prostanoids. Cardiovasc Res 1992; 26:226-231.
117. Liu GS, Downey J, Stanley AWH. Cyclooxygenase products are not involved in the protection against myocardial infarction afforded by preconditioning in rabbits. Am J Cardiovasc Pathol 1992; 4:157-164.
118. Rohmann S, Weygandt H, Schelling P et al. Involvement of ATP-sensitive potassium channels in preconditioning protection. Bas Res Cardiol 1994; 89:563-576.
119. Schulz R, Rose J, Heusch G. Involvement of activation of ATP-dependent potassium channels in ischemic preconditioning in swine. Am J Physiol 1994; 267:H1341-H1352.
120. Geshi E, Ishioka H, Watanabe T et al. Effect of nicorandil on ischemic preconditioning. Ther Res 1994; 15:135-143.
121. Tan HL, Mazon P, Verberne HJ et al. Ischemic preconditioning delays ischemia induced cellular electrical uncoupling in rabbit myocardium by activation of ATP sensitive potassium channels. Cardiovasc Res 1993; 27:644-651.
122. Tomai F, Crea F, Gaspardone A et al. Ischemic preconditioning during coronary angioplasty is prevented by glibenclamide, a selective ATP-sensitive K^+ channel blocker. Circulation 1994; 90:700-705.
123. Grover GJ, Dzwonczyk S, Sleph PG et al. The ATP-sensitive potassium channel blocker glibenclamide (glyburide) does not abolish precondition-

ing in isolated ischemic rat hearts. J Pharmacol Exp Ther 1993; 265:559-564.
124. Fralix TA, Steenbergen C, London RE et al. Glibenclamide does not abolish the protective effect of preconditioning on stunning in the isolated perfused rat heart. Cardiovasc Res 1993; 27:630-637.
125. Armstrong S, Downey JM, Ganote CE. Preconditioning of isolated rabbit cardiomyocytes: induction by metabolic stress and blockade by the adenosine antagonist SPT and calphostin C, a protein kinase C inhibitor. Cardiovasc Res 1994; 28:72-77.
126. Armstrong S, Ganote CE. Preconditioning of isolated rabbit cardiomyocytes: effects of glycolytic blockade, phorbol esters, and ischemia. Cardiovasc Res 1994; 28:1700-1706.
127. Turrens JF, Thornton J, Barnard ML et al. Protection from reperfusion injury by preconditioning hearts does not involve increased antioxidant defenses. Am J Physiol 1992; 262:H585-H589.
128. Bugge E, Ytrehus K. Inhibition of sodium hydrogen exchange reduces infarct size in the isolated rat heart—a protective additive to ischemic preconditioning. Cardiovasc Res 1995; 29:269-274.
129. Mitani A, Yasui H, Tokunaga K. Effect of ischemic preconditioning on ischemia-induced contractile failure and accumulation of extracellular H^+ and K^+. Jap Circ J 1994; 58:894-902.
130. Sanz E, Dorado DG, Oliveras J et al. Dissociation between anti-infarct effct and anti-edema effect of ischemic preconditioning. Am J Physiol 1995; 268:H233-H241.

CHAPTER 13

Myocardial Stress Response, Cytoprotective Proteins and the Second Window of Protection Against Infarction

Gary F. Baxter, Michael S. Marber and Derek M. Yellon

13.1. INTRODUCTION

The cellular stress response is a multifactorial process, the nature of which is dependent on the organism, the metabolic status of the cell involved and the nature of the stress imposed. It is known that many cells and tissues, ranging from prokaryotes to highly organized and complex tissues such as myocardium are able to respond to a variety of metabolic stresses so that they become better able to withstand a subsequent period of metabolic stress several hours later.[1,2] The ischemic preconditioning phenomenon in myocardium is well recognized and could be regarded as a very particular form of acute stress response in myocytes (and possibly in other cellular components of heart tissue).[3,4] However, the stress response to hyperthermia and hypoxia, originally described in lower organisms such as yeasts and bacteria, differs fundamentally from ischemic preconditioning of myocardium in timecourse and quite probably in the mechanism of cellular

Myocardial Preconditioning, edited by Cherry L. Wainwright and James R. Parratt.

preservation. The presence of a rapid preconditioning mechanism in myocardium does not preclude the occurrence of other stress response mechanisms and there is now reason to believe that myocardium has the potential to respond to ischemic stress with at least two different adaptive routes to cytoprotection. The first of these is the preconditioning response that has been considered comprehensively in the preceding chapters of this volume, and which for the sake of clarity we will refer to as classic preconditioning. The other is analogous to the stress (or 'heat-shock') response, originally recognized in lower organisms. It is this latter form of adaptation that is thought to underlie the delayed phase of protection that develops following transient ischemic stress, many hours after classic preconditioning protection has disappeared.[5,6] We have coined the term "second window of protection" for this delayed protection following transient ischemia.[7] In this brief review, we will describe the background relevant to the discovery of the second window, review the evidence for its occurrence in myocardium and other tissues, discuss some of the cellular mechanisms that may be involved and speculate on its pathophysiological relevance.

13.2. THE STRESS RESPONSE, PROTECTION AND CROSS-TOLERANCE

A variety of noxious stimuli and environmental perturbations, including hyperthermia, ischemia, reperfusion, hydrogen peroxide, acidosis, ethanol, viral infection, exposure to heavy metals and arsenite, induce the synthesis of various proteins, including a group of proteins called heat shock or stress proteins (hsps; see refs. 1, 2, 8 and 9 for more comprehensive reviews of the nature and function of hsps). Constitutive hsps are expressed in the unstressed state but will be upregulated during stress. These form the majority of the hsp content of the 'normal' cell where they play key roles in a number of intracellular regulatory processes as molecular chaperones.[1,2,9] Alternatively, hsps may be exclusively inducible, i.e., they are only synthesized in response to stress. There are several families of hsps, which appear to serve different functions within the cell and conventionally classified according to their molecular mass as determined by SDS polyacrylamide gel electrophoresis.

The major hsps exhibit marked evolutionary conservation, the high degree of cross-species homology suggesting that they play crucial roles in the cellular response to injury. This suggestion has been confirmed by a number of studies in which manipulation of these proteins can markedly alter cellular vulnerability to stresses. Cells and tissues that accumulate hsps develop transient resilience to subsequent episodes of thermal stress, a phenomenon termed "acquired thermotolerance."[1,2,8,10,11] Furthermore, hsp induction by one stress can confer protection against a further different stress. For example, fibroblasts exposed to ethanol or arsenite accumulate hsps and develop "cross tolerance" to subsequent thermal stress.[12]

In addition to the induction of hsps, the regulation of many other proteins is affected by environmental and intracellular stresses, with variations in gene regulation depending on the nature of the stress.[13] These include the immediate early genes (proto-oncogenes) c-*myc*, c-*fos* and c-*jun*; genes for anti-oxidant proteins, especially superoxide dismutase and catalase; and genes for glucose regulated proteins which are closely related to hsps and are induced by stresses involving energy substrate deprivation.[13-15]

13.3. THE THERMAL STRESS RESPONSE AND MYOCARDIAL PROTECTION

In 1988 Currie et al reported that whole body hyperthermia in rats (42°C for 15 minutes) resulted in an elevation of myocardial hsp70i and catalase content 24 hours later.[16] At this timepoint the hearts from heat stressed animals showed enhanced tolerance to ischemia (better post ischemic contractile function and reduced creatine kinase efflux). This initial study was subsequently extended by us to the isolated rabbit heart.[17,18] Further experiments demonstrated that prior heat stress resulted in a 30-40% reduction in infarct size in anesthetized rats[19] and rabbits[20,21] and attenuated the incidence and severity of reperfusion-induced arrhythmias in the rat heart in vitro following both global ischemia[22] and regional ischemia.[23]

The mechanism of myocardial protection following the heat shock response is not understood but associations with hsp70i and catalase have been investigated.[21,23-26] Currie's group showed that inhibiting catalase activity with 3-aminotriazole resulted in loss of the protective effect associated with heat stress on infarct size.[25] However, Steare and Yellon[26] showed that 3-aminotriazole augmented the anti-arrhythmic effects of heat stress. Intriguingly, catalase mRNA levels in the heart are not increased following heat stress,[24] suggesting that the increase in catalase activity may result through post-translational mechanisms. Kukreja and Hess[27] have suggested an alternative explanation, namely that stress proteins modulate the activity of catalase by direct interaction with the enzyme. Some studies have suggested that the degree of myocardial protection observed after hyperthermia correlates with the level of hsp70i. Hutter et al showed that infarct size reduction in rats was related to the amount of stress protein induced by a graded hyperthermic pretreatment.[28] We have shown that the contractile recovery, following simulated ischemia, in right ventricular papillary muscles harvested from heat stressed rabbits was related to the amount of hsp70i in the neighboring papillary muscle.[29] Temporal associations of hsp70i and catalase content with myocardial protection have been investigated. Following hyperthermia, the expression of hsp70i is maximal at 24-48 hours and then declines although the protein is still detectable seven days after heat stress.[24] However, no protection was seen in an infarction model 40 hours after heat stress even though hsp70i was significantly elevated.[71] It is also clear that heat stress may have the potential for only modest protection. For example, Donnelly et al observed a reduction in infarct size following a 35 minute coronary occlusion in the rat, but not after a 45 minute occlusion.[30] Similarly in the rabbit, we have found that prior heat stress reduces infarction following a 30 minute coronary occlusion in situ[20] but not a 45 minute occlusion.[31]

13.4. EVIDENCE FOR THE "SECOND WINDOW OF PROTECTION" AFTER PRECONDITIONING

It has been known for some time that myocardial ischemia[32,33] and ischemia-reperfusion[34] can increase the expression in the myocardium of hsp70i, as well as other putative cytoprotective proteins. If the expression of these proteins is associated with enhanced resistance to ischemia, as the hyperthermia studies outlined above would suggest, then it could be inferred that ischemic preconditioning might be followed by protective protein induction. This hypothesis for adaptive cytoprotection in the myocardium is represented in Figure 13.1.

The induction of hsps by short periods of ischemia-reperfusion may in fact be a consequence of the free radical stress imposed by reperfusion. It has been shown that hsp70i mRNA is induced in isolated rat hearts perfused with a xanthine/xanthine oxidase mixture, a generator of the superoxide anion, and that the level of hsp induction is quantitatively similar to that induced by ischemia and reperfusion.[35] However, it is clear that classic preconditioning protection which appears and disappears within a very short time frame still occurs in the presence of an inhibitor of the synthesis of new protein.[36] It should be noted that the efficacy of the inhibitor was measured only with regard to total protein synthesis and therefore the study does not preclude the possibility that one or more hsps could be augmented under these conditions. Work in our own laboratory suggests that although hsp60 mRNA levels increase rapidly in response to ischemia and reperfusion, no change in either hsp60 or hsp70i protein was detected by Western blot analysis over the time course of classic ischemic preconditioning.[37] Although hsp involvement in classic preconditioning appears unlikely, until other hsp families, hsp translocation processes and post-translational structure and activity modifications (as opposed to hsp synthesis) are examined in greater detail, the involvement of hsps in the acute preconditioning response can not be definitively ruled out.

The known time-course of hsp appearance in myocardium following brief myocardial ischemia[34] would predict a later period of myocardial resistance to ischemia. We speculated that an ischemic preconditioning protocol would result in two phases of protection: the early phase of protection unrelated to protein synthesis; and a late phase of protection which would be established when the tissue hsp content/activity exceeded a critical threshold. We sought to test this hypothesis using a rabbit model of preconditioning with four 5 minute coronary occlusions followed by 24 hours recovery.[20] The animals were then subjected to an acute myocardial infarction protocol under anesthesia (30 minute regional ischemia, 120 minute reperfusion) after which infarct size was determined with triphenyltetrazolium staining. In the preconditioned animals, the amount of myocardium infarcting within

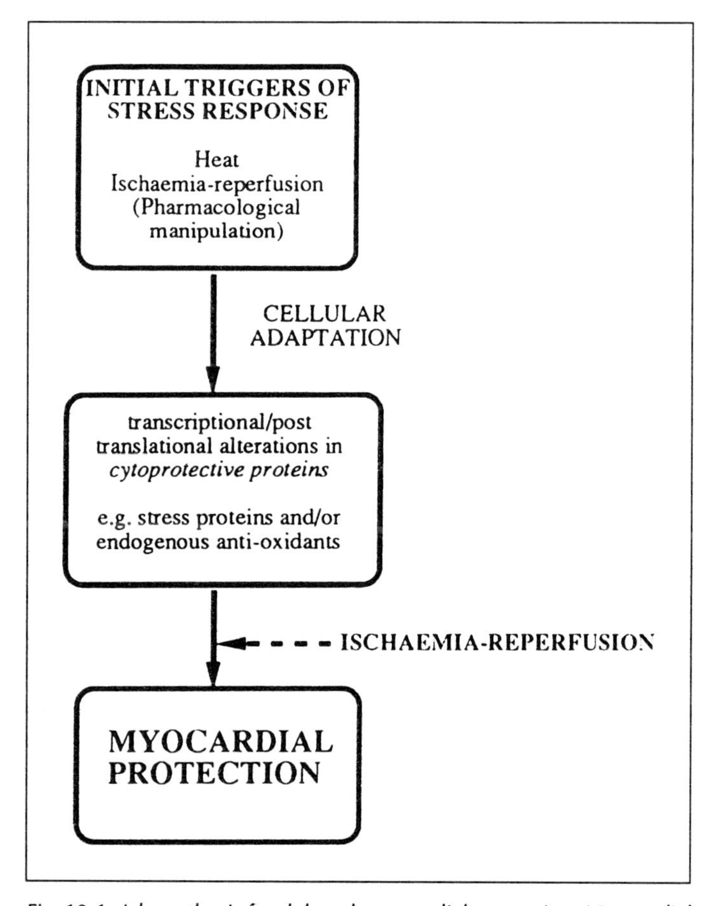

Fig. 13.1. A hypothesis for delayed myocardial protection. Myocardial response to transient stressful stimuli such as thermal stress and ischemia-reperfusion, that are not so severe as to cause irreversible injury, may include a series of intracellular adaptive processes. The more important of these may be transcriptional and post-translational modifications of cytoprotective proteins, particularly heat shock or stress proteins and intracellular anti-oxidant proteins. The gradual accumulation of these proteins during the hours following the initial trigger of the stress response would be associated with increased resilience of the cells to a subsequent stress, typically a severe ischemia-reperfusion event. This hypothetical scheme admits the possibility that the process of adaptive cytoprotection may be triggered by pharmacological stimuli.

the risk zone was 29% compared with 52% in control animals. This reduction in infarct size was very similar to the protection we observed following whole body hyperthermia. Both the ischemic pretreatment and hyperthermic pretreatment resulted in a two and a half-fold elevation of myocardial hsp70i. Additionally, ischemic pretreatment caused a two-fold elevation of the mitochondrial hsp60 which was not seen in animals pretreated with hyperthermia.

At the same time as this work was being conducted, workers in Osaka were examining the anti-oxidant responses of myocardium to sublethal ischemia. Hoshida et al[38] showed that four 5 minute coronary occlusions in the dog caused an increase 24-72 hours later in the subendocardial and subepicardial content and activity of the mitochondrial manganese-dependent superoxide dismutase (Mn-SOD). This study was followed by a subsequent investigation of the timecourse of myocardial protection against infarction following preconditioning.[39] Immediately after preconditioning there was a 63% reduction in infarct to risk volume ratio.(I/R) consistent with the classical preconditoning effect. Ninety minutes after preconditioning the protection against infarction was much reduced and there was only a 17% reduction in I/R. Twelve hours after preconditioning, I/R was 27% lower in preconditioned dogs, a reduction that was not significant. Twenty-four hours later, however, I/R was 46% lower in preconditioned dogs. This difference was statistically significant and not due to differences in regional myocardial blood flow in the preconditioned and control animals.

Together, these two reports constitute the first detailed descriptions of the second window of protection against infarction. Subsequently, we have confirmed our initial report in two separate studies undertaken to explore the signaling pathways involved in the second window,[40,41] described in more detail below. In conjunction with Cohen and Downey's group, we have also undertaken a study of delayed protection in a novel conscious animal model of coronary occlusion.[42] Here, we observed significant reduction of infarction following 30 minute regional ischemia in conscious rabbits, from 38% I/R in controls to 24% I/R ($p<0.01$) in rabbits pretreated 24 hours earlier with four 5 minute cycles of ischemia. Interestingly, the incidence of ventricular fibrillation (VF) during 30 minute coronary occlusion was also reduced in preconditioned rabbits compared with controls (0% v 43%, $p < 0.1$), a reduction that we had not previously noted in our anesthetized rabbit model. The attenuation of lethal arrhythmias during the second window of protection is consistent with the anti-arrhythmic effects reported by Vegh et al[43] in a canine model of coronary occlusion in which preconditioning was effected by rapid ventricular pacing 24 hours earlier (see chapter 14). Recently, Bolli's group have described a further aspect of the second window of protection, namely the attenuation of postischemic contractile dysfunction ("myocardial stunning") in an in vivo porcine model.[44]

Despite the growing body of experimental evidence in the literature to support the notion of a second window of protection against infarction, there have been two studies designed to investigate the possibility of a second window that have reported negative findings. Tanaka et al[45] undertook a study with a rabbit model of myocardial infarction in which early preconditioning protection was clearly present following four 5 minute coronary occlusions. However, when animals were allowed to recover for either 24 hours or 48 hours after preconditioning no anti-infarct effect was observed. This negative result was surprising to us in that the same species was used as in the other rabbit studies and an identical preconditioning protocol employed.[20,40,41] Nevertheless, on closer examination there were technical differences between these two studies that may explain these opposing results. In Tanaka's study, air was used to ventilate the animals and a left parasternal approach to the heart was used whereas we employed 100% oxygen as the ventilating gas and approached the heart through a mid-sternal incision. It may well be the case that after sternotomy, rabbits suffer less postoperative stress and have better respiratory capacity than after parasternal thoracotomy. Another negative report has been made recently that deserves comment. Schaper's group reported that four 5 minute preconditioning cycles were necessary to induce classical preconditioning in the pig but this protocol failed to elicit an infarct reducing effect 24 hours later.[46] Interestingly, a single 10 minute coronary occlusion did not elicit classic preconditioning reduction in infarction whereas other workers have found this to be adequate to precondition porcine myocardium against infarction.[47] This study draws attention to the possibility that the ischemic preconditioning stimulus required to elicit the second window of protection is somewhat greater than the minimum stimulus required to elicit early protection. It is also important to bear in mind here that the potency of the delayed protection may be limited by the duration of the long ischemic period, i.e., there may be protection against a 30 minute ischemic insult but not against a 60 minute insult.

13.5. TIMECOURSE OF THE SECOND WINDOW OF PROTECTION

Sublethal ischemia appears to evoke protection against subsequent (lethal) ischemia in a truly biphasic manner. The timecourse of the second window of protection has not so far been fully characterized. In exploring the temporal characteristics of the second window, we need to know something of its time of onset and, perhaps more especially, its duration. We have begun to address this question of duration in our in situ rabbit model of infarction by extending the interval between preconditioning and the infarction protocol beyond 24 hours. Observations in our laboratory, summarized in Figure 13.2, suggest that the second window of protection against infarction, in the rabbit

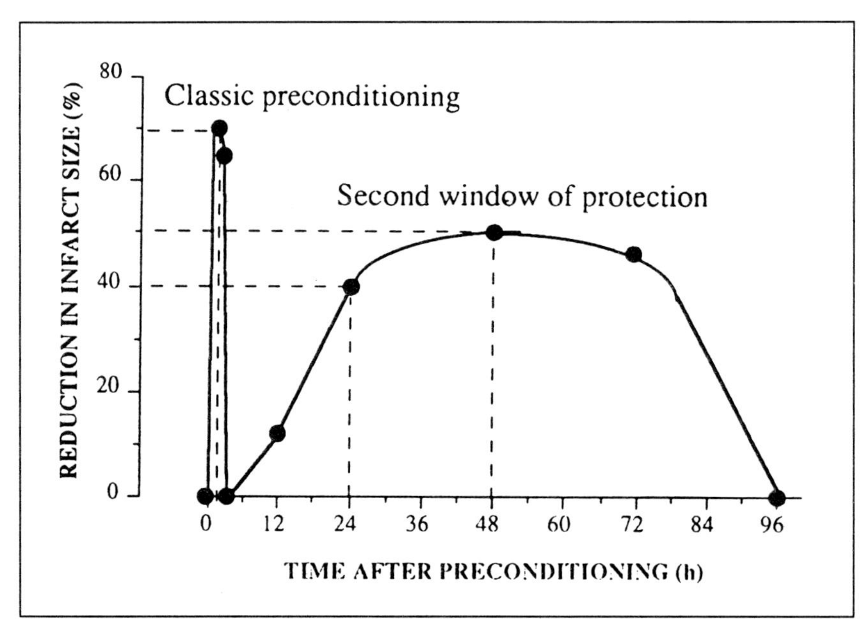

Fig. 13.2. Timecourse of myocardial resilience to infarction following ischemic preconditioning. This composite timecourse of protection against infarction has been determined from work in rabbit and canine myocardium. Following preconditioning with four 5 minute coronary artery occlusions, early preconditioning protection is observed (at time 0 hour) as a 70% reduction from control value in infarct size. This classic preconditioning effect wanes rapidly so that by 3 hours no protection is seen. The gradual evolution of a second window of protection then follows and is established in the dog[39] and rabbit[20,40-42] at 24 hours as a 35-45% reduction in infarct size. Preliminary evidence from our laboratory suggests that in the rabbit there is a 50% reduction in infarct size at 48 hours and at 72 hours but by 96 hours no protection is observed.[48]

at least, is of quite considerable duration. Indeed, the degree of protection conferred by preconditioning seems to increase beyond 24 hours, is maximal at 48-72 hours but is lost at 96 hours.[48] At first sight, this prolonged timecourse would appear to approximate quite well to the timecourse of stress protein presence following hyperthermia described by Karmazyn et al in the rat[25] although, apart from the elevation of hsp70i and hsp650 24 hours after preconditioning, at present we have little information concerning the timecourse of protein expression following ischemic pretreatment.

13.6. SIGNALING AND MEDIATION OF DELAYED PROTECTION

13.6.1. Triggers and Signaling Pathways

There is a wealth of evidence to suggest that in rabbits and pigs, one of the triggers of classic ischemic preconditioning is A_1-adenosine receptor activation, since adenosine receptor blockade (or augmented adenosine deamination) abolishes protection.[49,50] Our initial investigation of the signaling pathway(s) that might be involved in the development of the delayed protection following ischemic preconditioning examined the possibility that adenosine receptor activation was also a trigger for the second window in the rabbit.[40] We showed that the delayed protection was abolished if 8-(p-sulphophenyl)theophylline (8-SPT), an adenosine A_1- and A_2-receptor antagonist, was given during the preconditioning protocol. We were intrigued and surprised by this finding. We extended our investigation by substituting a selective A_1-adenosine agonist, 2-chloro-N^6-cyclopentyladenosine (CCPA), for preconditioning and found that there was enhanced resistance of the myocardium 24 hours later with a significant reduction in infarct size. We conducted further studies to confirm this delayed anti-ischemic effect of A_1-adenosine receptor activation and showed that following systemic treatment with CCPA the isolated rabbit heart was more resistant to zero-flow global ischemia, showing better contractile recovery and less lactate dehydrogenase efflux.[51] When the interval between CCPA pretreatment and coronary occlusion was extended to 72 hours, protection was observed against infarction in situ.[52] This anti-infarct effect was lost by 96 hours, a timecourse of protection identical to that seen following ischemic pretreatment. It is not clear how A_1-adenosine receptor activation evokes a second window of protection in the rabbit, and it is by no means certain that this is the only G protein-coupled receptor associated with the delayed protection. The hemodynamic sequelae of A_1 receptor activation may be profound. Following CCPA 0.1 mg/kg i.v. to anesthetized rabbits, we observed a marked bradycardia extending over 90 minutes and a significant, albeit briefer, reduction in mean arterial pressure. Although it is unlikely that these hemodynamic changes following A_1 receptor activation reduce myocardial perfusion sufficient to cause ischemic preconditioning, reflex and A_1-evoked catecholamine release might well occur and thus sympathetic activation of the myocardium could be involved in this pharmacologically-triggered second window. Meng et al have reported recently that a delayed myocardial protection is induced in the rat following noradrenaline pretreatment through an α_1-adrenoreceptor mechanism.[53]

Both the A_1-adenosine receptor and the α_1-adrenoreceptor are known to couple to protein kinase C (PKC). We have shown that PKC activation might play a pivotal role in the second window of protection in our rabbit model by administering chelerythrine, a protein kinase inhibitor selective for PKC, during the preconditoning stimulus.[41] Chelerythrine treatment had no effect on infarct size in control animals but completely abolished the delayed protection 24 hours later in preconditioned animals, providing indirect evidence that PKC activation plays a key role in the development of the delayed protection in this species. Preliminary evidence, therefore, suggests that both the early and late phases of protection following ischemic preconditioning in the rabbit could share similar proximal signal transduction pathways (see Fig. 13.3). It is not certain what isoforms of PKC are involved in the mediation of these different forms of protection. It is becoming clearer that different isoforms of PKC are activated in response to different pathophysiological and pharmacological stimuli. Strasser et al[54] showed that following brief periods of ischemia in rat hearts, PKC isoforms translocated to the sarcolemma. Furthermore, Banerjee's group has shown that other isoforms may undergo nuclear translocation in response to ischemia.[55] Such a nuclear translocation of PKC may be a fundamentally important step in the modulation of gene regulatory processes which are proposed to underlie the second window of protection (Fig. 13.3). Many oncogenes and nuclear transcription factors may be activated by PKC.[56] Meng et al reported that late myocardial protection following noradrenaline pretreatment was accompanied by altered levels of c-*fos*, c-*jun* and hsp70i mRNAs several hours after noradrenaline administration.[53] Further evidence to support PKC involvement in the second window of protection has come from the elegant studies of Yamashita and colleagues.[57] These workers used a rat cultured cardiomyocyte model of hypoxic preconditioning in which Mn-SOD content and activity were associated with enhanced tolerance to hypoxia 24 hours later. Treatment of myocytes during preconditioning with staurosporine attenuated the induction of Mn-SOD and abolished the protection 24 hours later.

13.6.2. Mediators

At the outset we began by stating that the stress response of any cell is a multifactorial event. It is indisputable that myocardial ischemia influences the expression of a host of gene products. The question of which gene products ultimately mediate the cytoprotection that characterizes the second window of protection is both exciting and pressing. The main candidate proteins that have so far been proposed by us and others are the stress proteins and endogenous antioxidants, of which there are many. Experimental associations between the appearance and disappearance of protection and the appearance and disappearance of protein activity can be persuasive, but ultimately do not provide the

Fig. 13.3 (opposite). Protein kinase C (PKC) signaling in delayed myocardial protection. Downey has proposed that classic preconditioning protection involves the stimulation of G_i protein-transduced receptors. The activation of phospholipase C (PLC) results in the generation of diacylglycerol (DAG) which activates PKC. The sarcolemmal translocation of PKC subtypes would permit the phosphorylation of some currently unknown membrane-associated protein which confers protection. The second window of protection, in the rabbit at least, may also involve PKC. Adenosine A_1 receptor stimulation via a G_i protein mechanism, as well as other stimuli, notably free radicals, may activate PKC subtypes to translocate to the nucleus. The phosphorylation of transcription factors by PKC could modify the transcription of genes coding for effectors of delayed protection.

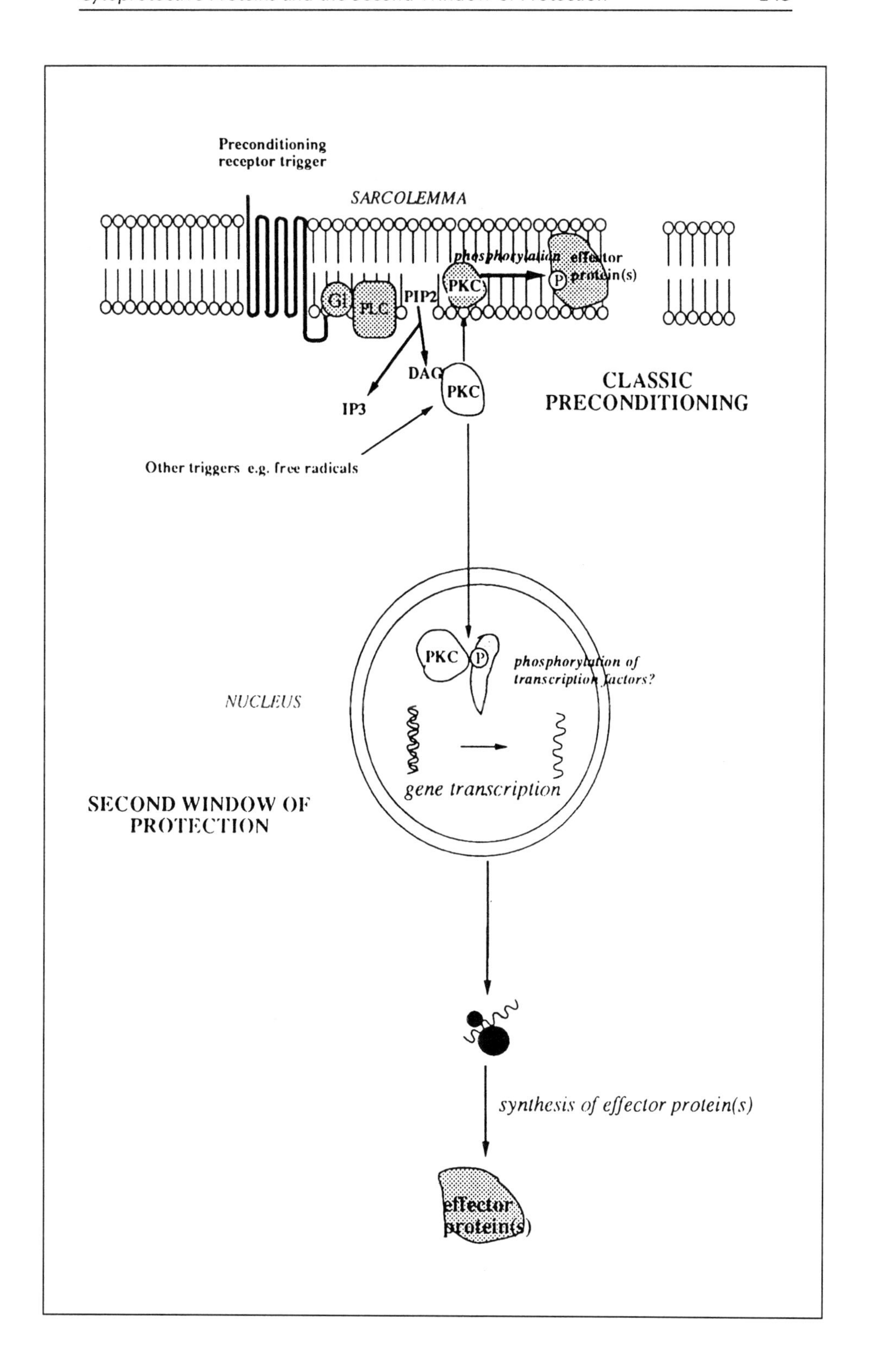
Preconditioning
receptor trigger
SARCOLEMMA
Gi
PLC
PIP2
PKC
phosphorylation
effector
protein(s)
P
DAG
PKC
IP3
CLASSIC
PRECONDITIONING
Other triggers e.g. free radicals
PKC
P
phosphorylation of
transcription factors?
NUCLEUS
gene transcription
SECOND WINDOW OF
PROTECTION
synthesis of effector protein(s)
effector
protein(s)

key to tackling the questions of causality and primacy. Thus, although there is accumulating evidence from transgenic animal[58-60] and cell transfection studies[61,62] that hsp70i expression in its own right confers protection against ischemia-reperfusion injury, the final mediators of protection in the second window of protection triggered by ischemia are not known. *It is intuitively likely that a complex trigger like preconditioning ischemia induces several protective mediators which act together to preserve cell viability during the prolonged ischemic period.* Other triggers of delayed protection (heat stress, endotoxin pretreatment, G protein-coupled receptor agonists), while retaining certain central pathophysiological routes in common, may induce different patterns of response that are manifested as a quantitatively similar degree of resistance to infarction. For example, we have shown that the degree of infarct reduction 24, 48 and 72 hours after A_1-adenosine receptor agonist pretreatment is similar to that seen following both ischemic preconditioning with four 5 minute coronary artery occlusions and heat stress. However, the pattern of stress protein induction following these stimuli appears to differ.[63] Although myocardial content of hsp70i is increased 24 hours after ischemic and thermal pretreatments, we were not able to detect changes in hsp70i 24 hours after CCPA treatment at a dose which conferred similar delayed protection (0.1 mg/kg). Interestingly, Mn-SOD content and activity is increased following all pretreatments although to a greater degree following CCPA. Furthermore, although heat stress increases catalase activity, neither ischemia nor CCPA pretreatments appear to augment this antioxidant. This Byzantine pattern of biochemical responses within the general framework of adaptive cytoprotection is undoubtedly a feature of: (1) complexity of the stress response in the heart; and (2) complexity and divergence of signal transduction which varies according to the nature of the priming stimulus. Nevertheless, it is perhaps not surprising that such a powerful form of endogenous protection is not unique to the heart. Delayed preconditioning protection has been defined in brain,[64] kidney[65] and intestinal mucosa[66] and in these tissues, as in the myocardium, common themes of time-course and anti-oxidant responses have been identified.

13.7. CLINICAL IMPLICATIONS OF SWOP AND FUTURE DIRECTIONS

It is perhaps not surprising that the intense research activity surrounding experimental preconditioning has kindled the interest of clinicians. Studies designed to subject the human heart to controlled sublethal ischemia suggest that short term adaptation exists.[3] Unfortunately a major limitation of these prospective intervention studies is the need to use a surrogate endpoint that does not involve irreversible myocardial injury.

More recently a number of investigators have begun to examine the influence that spontaneous pre-infarction angina may have on spontaneous subsequent infarction[67,68] (see chapter 7). The rationale for this approach is that in the human the sensation of angina is thought to be secondary to the accumulation of sufficient interstitial adenosine to activate A_1 receptors on cardiac nerve endings.[69] Therefore, angina is also likely to be an indication of myocyte A_1 receptor activation. Hence angina may be followed both by classic preconditioning protection and the second window of protection. Is there any evidence that this is the case?

Studies in the prethrombolysis era did not show consistently that pre-infarction angina had a favorable influence upon postinfarction mortality and left ventricular function.[70,71] There were numerous difficulties with these studies including the absence of thrombolytic reperfusion and adequate control for the differences that exist between patients with and without pre-infarction angina. In two recent studies these difficulties were overcome with a detailed clinical history and coronary angiography in the acute phase of infarction.[67,68] In both these studies pre-infarction angina was associated with a limitation of infarct size and an improvement of in-hospital outcome. Patients recruited for these studies had episodes of angina within 48[67] or 24[68] hours of the index acute myocardial infarct. In patients with angina within 24 hours, the mean time interval between the last episode of angina and the onset of infarction was 11.2 hours.[68] It is likely that this time interval was even greater in the study recruiting patients with angina within 48 hours of infarction. Since the benefits of classic preconditioning in animals, and probably humans, lasts only for 60 minutes, it is likely that the advantages seen in these studies[68,69] correspond not to the classic preconditioning phenomenon, but rather to the second window of protection.

13.8. CONCLUSION

The discovery of a second window of protection has added a new and exciting direction to our study of myocardial adaptation to ischemia, already a very active area of research. We believe that this delayed protection is the result of a complex stress response in the myocardium. An appreciation of the stress response and the process of adaptive cytoprotection may contribute to a clearer understanding of the pathophysiology of myocardial ischemic syndromes and, ultimately, may offer the potential for therapeutic exploitation.

REFERENCES

1. Lindquist SC. The heat shock response. Ann Rev Biochem 1986; 55:1151-1191.
2. Minowada G, Welch WJ. Clinical implications of the stress response. J Clin Invest 1995; 95:3-12.
3. Baxter GF, Yellon DM. Ischemic preconditioning of myocardium: a new paradigm for clinical cardioprotection? Br J Clin Pharmacol 1994; 38:381-387.

4. Parratt JR. Protection of the heart by ischemic preconditioning: mechanisms and possibilities for pharmacological exploitation. Trends Pharmacol Sci 1994; 15:19-25.
5. Marber MS, Yellon DM. Hsp70 in myocardial ischemia. Experientia 1994; 50:1075-1084.
6. Mestril R, Dillmann WH. Heat shock proteins and protection against myocardial ischemia. J Mol Cell Cardiol 1995; 27:45-52.
7. Yellon DM, Baxter GF. A 'second window of protection' or delayed preconditioning phenomenon: future horizons for myocardial protection? J Mol Cell Cardiol 1995; 27:1023-1034.
8. Schlesinger MJ. Heat shock proteins. J Biol Chem 1990; 265: 12111-12114.
9. Morimoto RI, Tissieres A, Georgopoulos C, eds. The Biology of Heat Shock Proteins and Molecular Chaperones. Cold Spring Harbor: Cold Spring Harbor Laboratory Press 1994.
10. Angelidis CE, Lazaridis I, Papoulatos GN. Constitutive expression of heat shock protein 70 in mammalian cells confers thermoresistance. Eur J Biochem 1991; 199:35-39.
11. Heads RJ, Latchman DS, Yellon DM. Stable high level expression of a transfected human HSP70 gene protects a heart-derived muscle cell line against thermal stress. J Mol Cell Cardiol 1994; 26:695-699.
12. Li GC. Induction of thermotolerance and enhanced heat shock protein synthesis in Chinese hamster fibroblasts by sodium arsenite and by ethanol. J Cell Physiol 1983; 115:116-122.
13. Das DK, Maulik N, Moraru II. Gene expression in acute myocardial stress. Induction by hypoxia, ischemia, reperfusion, hyperthermia and oxidative stress. J Mol Cell Cardiol 1995 27:181-193.
14. Donati YRA, Slosman DO, Polla BS. Oxidative injury and the heat shock response. Biochem Pharmacol 1993; 40:2571-2577.
15. Das DK, Engelman RM, Kimura Y. Molecular adaptation of cellular defences following preconditioning of the heart by repeated ischemia. Cardiovasc Res 1993; 27:578-584.
16. Currie RW, Karmazyn M, Kloc M et al. Heat shock response is associated with enhanced post-ischemic ventricular recovery. Circ Res 1988; 63:543-549.
17. Yellon DM, Pasini E, Cargnoni A et al. The protective role of heat stress in the ischemic and reperfused rabbit myocardium. J Mol Cell Cardiol 1992; 24:895-907.
18. Walker DM, Pasini E, Kuckukoglu S et al. Heat stress limits infarct size in the isolated perfused rabbit heart. Cardiovasc Res 1993; 27:962-967.
19. Donnelly TJ, Sievers RE, Vissern FLJ et al. Heat shock protein induction in rat hearts: a role for improved salvage after ischemia and reperfusion? Circulation 1992; 85:769-778.
20. Marber MS, Walker JM, Latchman DS et al. Cardiac stress protein elevation 24 hours following brief ischemia or heat stress is associated with resistance to myocardial infarction. Circulation 1993; 88:1264-1272.

21. Currie RW, Tanguay RM, Kingma JG. Heat shock response and limitation of tissue necrosis during occlusion-reperfusion in rabbit hearts. Circulation 1993; 87:963-971.
22. Mocanu MM, Steare SE, Evans M et al. Heat stress attenuates free radical release in the isolated perfused rat heart. Free Rad Biol Med 1993; 15:459-463.
23. Steare SE, Yellon DM. The protective effect of heat stress against reperfusion arrhythmias in the rat. J Mol Cell Cardiol 1993; 25:1471-1481.
24. Currie RW, Tanguay RM. Analysis of RNA for transcripts for catalase and SP71 in rat hearts after in vivo hyperthermia. Biochem Cell Biol 1991; 69:375-382.
25. Karmazyn M, Mailer K, Currie RW. Acquisition and decay of heat shock enhanced post-ischemic ventricular recovery. Am J Physiol 1990; 259:H424-H431.
26. Steare SE, Yellon DM. Increased endogenous catalase activity caused by heat stress does not protect the isolated rat heart against exogenous hydrogen peroxide. Cardiovasc Res 1994; 28:1096-1101.
27. Kukreja RC, Hess ML. The oxygen free radical system: from equations through membrane-protein interactions to cardiovascular injury and protection. Cardiovasc Res 1992; 26:641-655.
28. Hutter MA, Sievers RE, Barbosa V et al. Heat shock protein induction in rat hearts: a direct correlation between the amount of heat shock protein induced and the degree of myocardial protection. Circulation 1994; 89:355-360.
29. Marber MS, Walker JM, Latchman DS et al. Myocardial protection after whole body heat stress in the rabbit is dependent on metabolic substrate and is related to the amount of the inducible 70-kD heat stress protein. J Clin Invest 1994; 93:1087-1094.
30. Donnelly TJ, Sievers RE, Vissern FLJ et al. Heat shock protein induction in rat hearts. A role for improved myocardial salvage after ischemia and reperfusion? Circulation 1992; 85:769-778.
31. Yellon DM, Iliodromitis E, Latchman DS et al. Whole body heat stress fails to limit infarct size in the reperfused rabbit heart. Cardiovasc Res 1992; 26:342-346.
32. Dillmann WH, Mehta HB, Barrieux A et al. Ischemia of the dog heart induces the appearance of a cardiac mRNA coding for a protein with migration characteristics similar to the heat shock/stress protein 71. Circ Res 1986; 59:110-114.
33. Mehta HB, Popovich BK, Dillman WH. Ischemia induces changes in the level of mRNAs coding for stress protein 71 and creatine kinase M. Circ Res 1988; 63:512-517.
34. Knowlton AA, Brecher P, Apstein CS. Rapid expression of heat shock protein in the rabbit after brief cardiac ischemia. J Clin Invest 1990; 87:139-147.
35. Kukreja RC, Kontos MC, Loesser KE et al. Oxidant stress increases heat shock protein 70 mRNA in isolated perfused rat heart. Am J Physiol 1994;

267:H2213-H2219.
36. Thornton J, Striplin S, Liu GS et al. Inhibition of protein synthesis does not block myocardial protection afforded by preconditioning. Am J Physiol 1990; 259:H1822-H1825.
37. Heads RJ, Latchman DS, Yellon DM. Differential stress protein mRNA expression during early ischemic preconditioning in the rabbit heart and its relationship to adenosine receptor function. J Mol Cell Cardiol 1995; 27:2133-2148.
38. Hoshida S, Kuzuya T, Fuji H et al. Sublethal ischemia alters myocardial antioxidant activity in canine heart. Am J Physiol 1993; 264:H33-H39.
39. Kuzuya T, Hoshida S, Yamashita N et al. Delayed effects of sublethal ischemia on the acquisition of tolerance to ischemia. Circ Res 1993; 72:1293-1299.
40. Baxter GF, Marber MS, Patel VC et al. Adenosine receptor involvement in a delayed phase of myocardial protection 24 hours after ischemic preconditioning. Circulation 1994; 90:2993-3000.
41. Baxter GF, Goma FM, Yellon DM. Involvement of protein kinase C in the delayed cytoprotection following sublethal ischemia in rabbit myocardium. Br J Pharmacol 1995; 115:222-224.
42. Yang X-M, Baxter GF, Yellon DM et al. Second window of protection in conscious rabbits. J Mol Cell Cardiol 1995; 27:A27 (Abstract).
43. Vegh A, Papp JG, Parratt JR. Prevention by dexamethasone of the marked antiarrhythmic effects of preconditioning induced 20 hours after rapid cardiac pacing. Br J Pharmacol 1994; 113:1081-1082.
44. Sun J-Z, Tang X-L, Knowlton AA et al. Late preconditioning against myocardial stunning. An endogenous protective mechanism that confers resistance to postischemic dysfunction 24 h after brief ischemia in conscious pigs. J Clin Invest 1995; 95:388-403.
45. Tanaka M, Fujiwara H, Yamasaki K et al. Ischemic preconditioning elevates cardiac stress protein but does not limit infarct size 24 or 48 h later in rabbits. Am J Physiol 1994; 267:H1476-H1482.
46. Strasser R, Arras M, Vogt A et al. Preconditioning of porcine myocardium: how much ischemia is required for induction? What is its duration? Is a renewal of effect possible? Circulation 1994; 90(suppl):I-109 (Abstract).
47. Schulz R, Rose J, Heusch G. Involvement of ATP-dependent potassium channels in ischemic preconditioning in swine. Am J Physiol 1994; 267:H1341-H1352.
48. Baxter GF, Goma FM, Yellon DM. Duration of the 'second window of protection' following ischemic preconditioning in the rabbit. J Mol Cell Cardiol 1995; 27:A162 (Abstract).
49. Downey JM, Liu GS, Thornton JD. Adenosine and the anti-infarct effects of preconditioning. Cardiovasc Res 1993; 27:3-8.
50. Schulz R, Rose J, Post H et al. Involvement of endogenous adenosine in ischemic preconditioning in swine. Pflugers Archiv—Eur J Physiol 1995; 430:273-282.

51. Baxter GF, Kerac M, Zaman MJ et al. Delayed myocardial protection: responses to global ischemia 24 hours after adenosine A_1 receptor activation. J Mol Cell Cardiol 1994; 26:CLII (Abstract).
52. Baxter GF, Yellon DM. Temporal characterization of the 'second window of protection': prolonged anti-infarct effect after adenosine A_1 receptor activation. Circulation 1994; 90(suppl):I-475 (Abstract).
53. Meng X, Brown JM, Ao L et al. Norepinephrine induces late cardiac protection preceded by oncogene and heat shock protein overexpression. Circulation 1993; 88(suppl):I-633 (Abstract).
54. Strasser RH, Braun-Dullaeus R, Walendzik H et al. α_1 receptor-independent activation of protein kinase C in acute myocardial ischemia: mechanisms for sensitization of the adenylyl cyclase system. Circ Res 1992; 70:1304-1312.
55. Mitchell MB, Meng X, Ao L et al. Preconditioning of isolated rat heart is mediated by protein kinase C. Circ Res 1995; 76:73-81.
56. Hug H, Sarre TF. Protein kinase C isoenzymes: divergence in signal transduction. Biochem J 1993; 291:329-343.
57. Yamashita N, Nishida M, Hoshida S et al. Induction of manganese superoxide dismutase in rat cardiac myocytes increases tolerance to hypoxia 24 hours after preconditioning. J Clin Invest 1994; 94:2193-2199.
58. Currie RW, Plumier J-C L, Ross BM et al. Transgenic mice expressing high levels of the human Hsp70 have improved post-ischemic myocardial recovery. Circulation 90:I-377 (Abstract).
59. Radford NB, Fina M, Benjamin IJ et al. Enhanced functional and metabolic recovery following ischemia in intact hearts from hsp70 tansgenic mice. Circulation 1994; 90:I-G (Abstract).
60. Marber MS, Mestril R, Chi S-H et al. Overexpression of the rat inducible 70 kiloDalton heat stress protein in a transgenic mouse increases the resisitance of the heart to ischemic injury. J Clin Invest 1995; 95:1446-1456.
61 Mestril R, Chi S-H, Sayen MR et al. Expression of inducible stress protein 70 in rat heart myogenic cells confers protection against simulated ischemia induced injury. J Clin Invest 1994; 93:759-767.
62. Heads RJ, Yellon DM, Latchman DS. Differential cytoprotection against heat stress or hypoxia following expression of specific stress protein genes in myogenic cells. J Mol Cell Cardiol 1995; 27:1669-1678.
63. Heads RJ, Baxter GF, Latchman DS et al. Delayed protection in rabbit heart following ischemic preconditioning is associated with modulation of hsp27 and superoxide dismutase at 24 hours. J Mol Cell Cardiol 1995; 27:A163 (Abstract).
64. Kitagawa K, Matsumoto M, Kuwabara K et al. 'Ischemic tolerance' phenomenon detected in various brain regions. Brain Res 1991; 561:203-211.
65. Yoshioka T, Bills T, Moore-Jarrett T et al. Role of intrinsic antioxidant enzymes in renal oxidant injury. Kidney Int 1990; 38:282-288.
66. Osborne DL, Aw T, Cephinskas G et al. Development of ischemia-reperfusion tolerance in the rat small intestine. An epithelium-independent

event. J Clin Invest 1994; 1910-1918.
67. Kloner RA, Shook T, Pryzklenk K et al. Previous angina alters in-hosptial outcome in TIMI-4. A clinical correlate to preconditioning? Circulation 1995; 91:37-47.
68. Ottani F, Galvani M, Ferrini D et al. Prodromal angina limits infarct size. A role for ischemic preconditoning. Circulation 1995; 91:291-297.
69. Crea F, Gaspardone A, Kaski JC et al. Relation between stimulation site of cardiac afferent nerves by adenosine and distribution of cardiac pain: results of a study in patients with stable angina. J Am Coll Cardiol 1992; 20:1498-1502.
70. Behar S, Reicher-Reiss H, Allmader E. The prognostic significance of angina pectoris preceeding the occurence of a first acute myocardial infarction in 4166 consecutive hospitalized patients. Am Heart J 1992; 123:1481-1486.
71. Muller DW, Topol EJ, Califf RM et al. Relationship between antecedent angina pectoris and short-term prognosis after thrombolytic therapy for acute myocardial infarction. Thrombolysis and angioplasty in myocardial infarction (TAMI) study group. Am Heart J 1990; 119:224-231.

CHAPTER 14

Delayed Ischemic Preconditioning Induced by Drugs and by Cardiac Pacing

Agnes Vegh and James R. Parratt

14.1. INTRODUCTION

Recent evidence already summarized in this volume (chapter 13) has revealed the exciting possibility that a heart can be protected several hours after a preconditioning stimulus induced, for example, by short periods of coronary artery occlusion. The first suggestion for delayed myocardial protection following a 'classical' preconditioning stimulus (four brief repeated episodes of 5 minute coronary artery occlusions) in dogs was the report[1] that this resulted in a reduction in infarct size both when the coronary artery was reoccluded immediately after the preconditioning stimulus and also 24 hours later. There was no protection observed when the artery was reoccluded 3 or 12 hours after the preconditioning stimulus. These experiments thus indicated that there were two phases to myocardial protection elicited by brief periods of ischemia. This has been referred to by Yellon and his colleagues as the 'second window of protection' or SWOP.[2,3]

14.2. DELAYED MYOCARDIAL PROTECTION BY DRUGS

The above studies, however, were not the earliest indications that the heart can be protected against ischemia, or other injurious stimuli, several hours or even days after a particular intervention. Historically,

Myocardial Preconditioning, edited by Cherry L. Wainwright and James R. Parratt.

studies resulting from catecholamine administration almost certainly represent the earliest examples of 'delayed cardioprotection.' Selye, Rona, Balazs and Poupa, working independently demonstrated that myocardial resistance developed against toxic doses of isoprenaline if rats were pretreated with smaller doses.[4-8] Of particular interest was the finding that coronary artery ligation, of either the left or right coronary artery, also protected against the toxic effects of isoprenaline.[4] Rona referred to this as 'delayed myocardial resistance or protection.'[5] This was not associated with β-adrenoceptor down-regulation and it lasted several days or even weeks. Somewhat later, Beckman and his colleagues observed that dogs that had developed long term tolerance to adrenaline were also resistant to coronary embolization with microspheres.[9] Particularly relevant to the present discussion was their finding that, although 11 out of 31 of their control dogs fibrillated following embolization, only one of 14 adrenaline-tolerant dogs fibrillated after embolization over the same time period. All these deaths occurred early, within 15 minutes of the commencement of embolization and were mainly due to ventricular fibrillation occurring during the early (phase 1) period of ischemia. Some of the later evidence suggesting delayed protection by catecholamines is reviewed elsewhere in this volume (Ravingerova, chapter 10).

A concept of delayed cardioprotection by drugs has been pioneered by Laszlo Szekeres in Szeged. In the early 1980s he described the effects of a stable derivative of prostacyclin, 7-oxo-prostacyclin. The Szeged group made the important discovery that the administration of this prostacyclin derivative in a variety of species (rats, guinea pigs, cats, rabbits) resulted in what they described as a 'late-appearing, long-lasting' cardioprotection which had a time course similar to that described several years later for the delayed effect of ischemic preconditioning and for cardiac pacing. The protection induced by 7-oxo-prostacyclin depended on protein synthesis and was manifested in a variety of ways, such as enhanced recovery of contractile function following a period of ischemia and reperfusion, protection against the calcium paradox, protection against the ultrastructural changes resulting from global ischemia and, of immediate relevance, protection against ischemia- and reperfusion-induced ventricular arrhythmias.[10-15] One example will suffice. In isolated hearts removed from rats pretreated with 7-oxo-prostacyclin, the incidence of sustained ventricular fibrillation (VF) on reperfusion, following a 0.5 hour period of regional ischemia, was reduced from 70% to 0%.[15]

Another, more recent, example of delayed cardioprotection by an acute intervention are studies with bacterial endotoxin. It has long been known that endotoxin, such as that derived from *Escherichia coli*, depresses the myocardial responses to exogenous catecholamines.[16] However, the stimulus for a possible cardioprotective effect of bacterial endotoxin against the consequences of myocardial ischemia arose from the finding that nitric oxide generation is involved in the pronounced antiarrhythmic effect of 'classical' ischemic preconditioning in dogs[17,18] (see also chapter 3) and the observation that nitric oxide can be induced by endotoxin in a variety of cells including vascular smooth muscle cells[19,20] and cardiac myocytes.[21] It was argued that if nitric oxide was involved in the protective effects of ischemic preconditioning then one might expect that hearts removed from animals given bacterial endotoxin would show a similar resistance to ischemia. This was indeed found to be so.[22] Hearts were removed from rats administered endotoxin several hours previously and were then perfused at constant flow and subjected to coronary artery occlusion. The severity of the resultant ventricular arrhythmias was reduced compared with those from hearts taken from vehicle (saline) treated hearts at the same time. This protection was particularly marked 8 and 24 hours after endotoxin administration (Fig. 14.1) but was not present if the rats had been pretreated with dexamethasone.[22] This 'cross tolerance' was

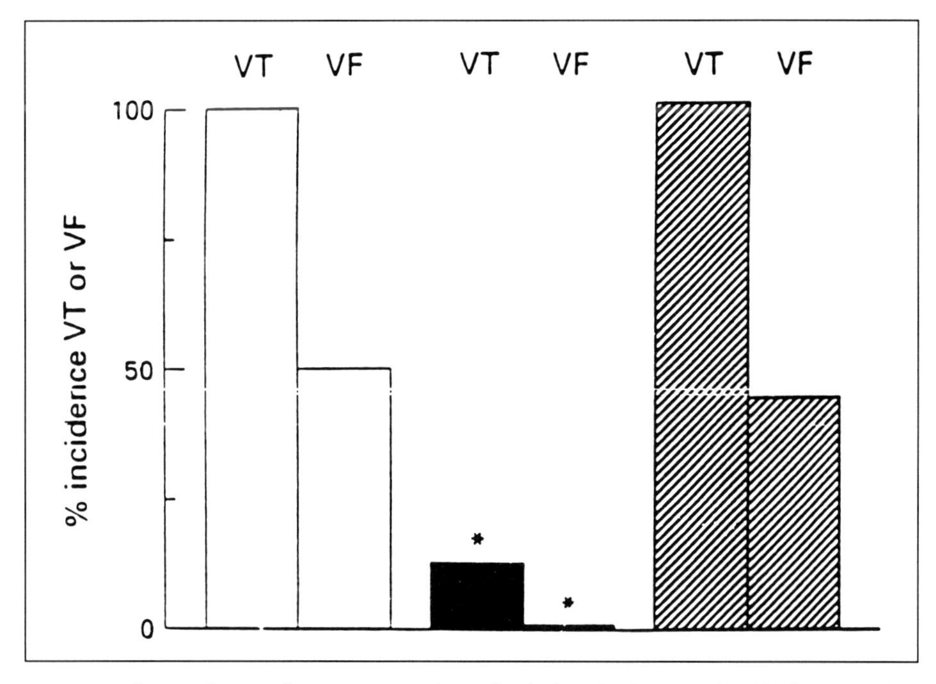

*Fig. 14.1. The incidence of ventricular tachycardia (VT) and of ventricular fibrillation (VF) in hearts removed from rats administered either saline (open columns), endotoxin (solid columns) or a combination of endotoxin and dexamethasone (hatched columns) 8 hours previously and subjected to coronary artery occlusion. There is a marked reduction in the incidence of these arrhythmias in hearts removed from rats given endotoxin. This protection is abolished by the prior administration of dexamethasone. n = 8-9 for all groups. *P<0.05 compared to saline controls. Reproduced with permission from Wu S et al, Br J Pharmacol 1994; 113:1083-1084.*

believed to be due to the induction of nitric oxide synthase or cyclooxygenase II. There have been other studies with endotoxin, or the nontoxic derivative of endotoxin, monophosphoryl lipid A which demonstrate protection against ischemia-reperfusion injury particularly as assessed by improved contractile recovery after reperfusion. More pertinent to the present study was the finding that monophosphoryl lipid A reduces infarct size in dogs when this was assessed 24 hours after administration.[23] The mechanisms of this particular protection are unclear; changes in myocardial catalase were not significant in this study but the contribution of inhibition of noradrenaline release (or depletion of noradrenaline stores), induction of cyclooxygenase II or nitric oxide synthase remains to be elucidated.

14.3. DELAYED MYOCARDIAL PROTECTION BY CARDIAC PACING

Rapid (overdrive) ventricular pacing both in anesthetized dogs[24] and in conscious rabbits[25,26] reduce the consequences of subsequent periods of regional ischemia including epicardial ST-segment elevation, changes in the degree of inhomogeneity of electrical activation within the ischemic area, coronary artery occlusion-induced ventricular arrhythmias, changes in left ventricular-end diastolic pressure and endocardial ST-segment changes induced by subsequent periods of overdrive pacing. This protection has been associated with an increase in cyclic GMP.[26]

In the conscious rabbit model ventricular pacing reduced the ischemic changes during subsequent periods of pacing 24 and 48 hours (but not 72 hours) after the initial preconditioning pacing period[25] (Fig. 14.2). The time course of this effect (early protection within the 1 hour time frame after the initial pacing stimulus, lost between 1 and 2 hours, regained around 24 hours and then lost again after 72 hours) is similar to that seen following preconditioning by short periods of regional ischemia resulting from coronary artery occlusion (see above). More conclusive evidence comes from our own studies in the canine model which examine the effects of rapid right ventricular pacing (4 x 5 minutes at 220 beats/min) on the consequences of coronary artery occlusion initiated at different times after the pacing 'preconditioning' stimulus.[26] We examined particularly the effects of such pacing on arrhythmia severity following coronary artery occlusion and reperfusion, on epicardial ST-segment changes and on the degree of inhomogeneity of electrical activation. The time course showed that the effects of pacing on these parameters was marked at 1 hour,[27] lost between 1 and 4 hours but regained again 20-24 hours later.[27,28] The time course is illustrated in Figure 14.3. There was some evidence for protection remaining both at 48 and 72 hours (for example, a reduction in the number of ventricular premature beats) and this is also clear from an examination of the changes in electrical inhomogeneity (Fig. 14.4).

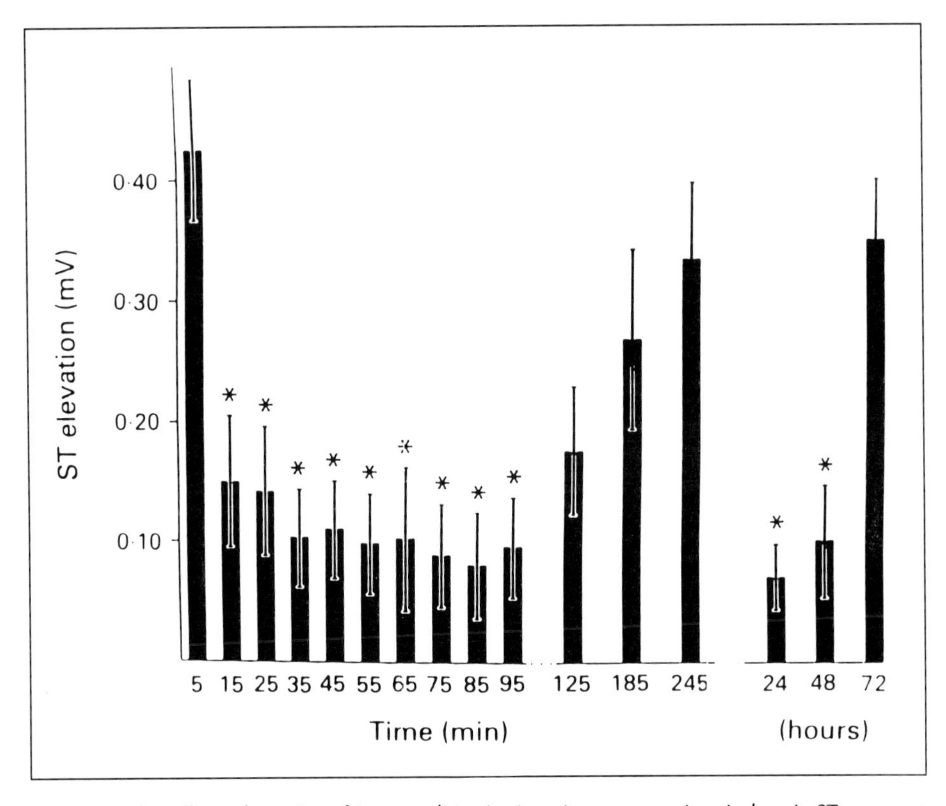

*Fig. 14.2. The effect of a series of 'preconditioning' pacings on transient ischemic ST-segment elevation recorded from an intracavital electrogram. Ten pacings were performed at a rate of 500 beats/min, with 5 minute intervals between the pacing periods. Note the marked decline in ST-segemnt elevation during serial pacing. This protection gradually disappeared (at 185 and 245 minutes) but reappeared after 24 hours and 48 hours. The protection was not seen after 72 hours. *P<0.05 compared to ST-segment elevation before preconditioning. Reproduced with permission from Szekeres L et al, Cardiovasc Res 1993; 27:593-596.*

However, the marked reduction in the incidence of ventricular fibrillation observed 24 hours after pacing (reduced from 50% in the control dogs, in which the pacing electrode was placed in the lumen of the right ventricle but the animals were not paced, to 0% in the paced animals) was lost at 48 and 72 hours (incidences of ventricular fibrillation at these times, 56% and 50% respectively).

The mechanisms of this delayed protection achieved by cardiac pacing remain unclear but there are two results which suggest that the mechanism may be similar to that involved in 'classical' protection against ventricular arrhythmias following coronary artery occlusion in the canine model. The first was that the protection at 24 hours was not observed if dexamethasone had been administered before the pacing stimulus.[28] The second, suggesting a role for bradykinin as both a trigger and

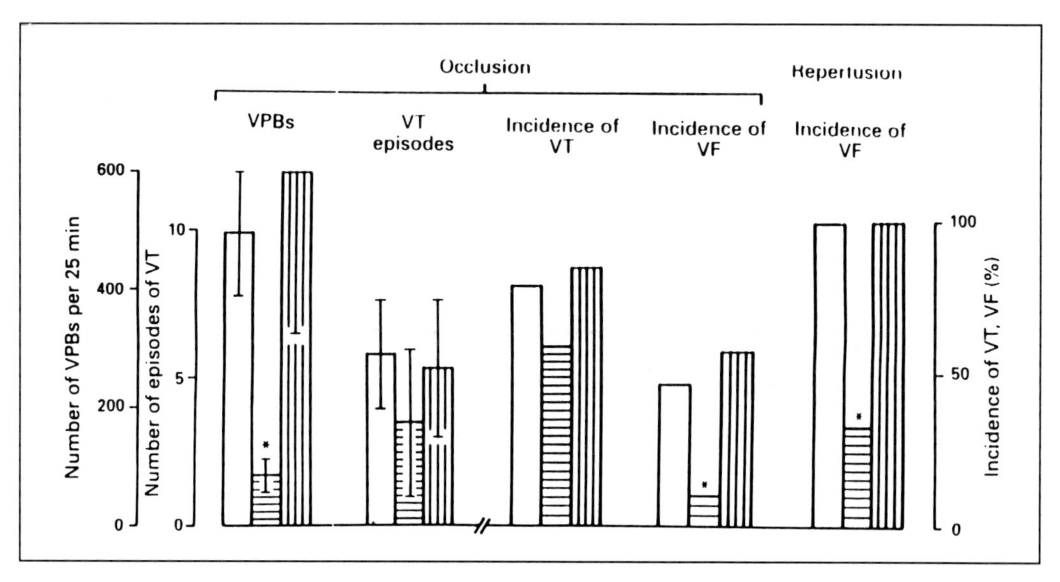

*Fig. 14.3. Ventricular arrhythmias during a 25 minute occlusion of the left anterior descending coronary artery in anesthetized dogs subjected, 20 hours previously, to four 5 minute periods of global ischemia induced by rapid cardiac pacing (columns with horizontal stripes; n = 10) and subjected to the same protocol but given dexamethasone (4 mg/kg) 45 minutes previously (columns with vertical stripes; n = 7). The control dogs (open columns; n = 20) had a pacing catheter in place in the lumen of the right ventricle, but were not subjected to pacing. Pacing markedly reduced the severity of ventricular arrhythmias, but this effect was abolished by prior administration of dexamethasone. VPB's, ventricular premature beats; VT, ventricular tachycardia; VF, ventricular fibrillation. Values shown are means ± s.e.mean. *P<0.05 compared to controls. Reproduced with permission from Vegh A et al, Br J Pharmacol 1994; 113:1081-1082.*

mediator of this protection is the finding[29] that icatibant, a drug which inhibits the effects of bradykinin on B_2 receptors, markedly attenuated the protection when given *either* before the preconditioning pacing stimulus or after the stimulus but immediately prior to the coronary artery occlusion. These results might suggest that bradykinin is involved (perhaps by stimulating the translocation of protein kinase C from the cytosol to the nucleus) and also that nitric oxide (or a prostanoid) may be involved, since dexamethasone prevents the induction of enzymes such as nitric oxide synthase and cyclooxygenase II. The effect of dexamethasone on this delayed protection is illustrated in Figure 14.3 (for arrhythmias) and Figure 14.4 (for ST-segment changes).

14.4. CONCLUSIONS

The possibility that the time window for protection against arrhythmias can be extended to at least 24 hours is an exciting development and warrants further intensive investigation.[30] Particularly apposite to any clinical application of these findings would be the possibility of extending the time window by re-enforcing the preconditioning

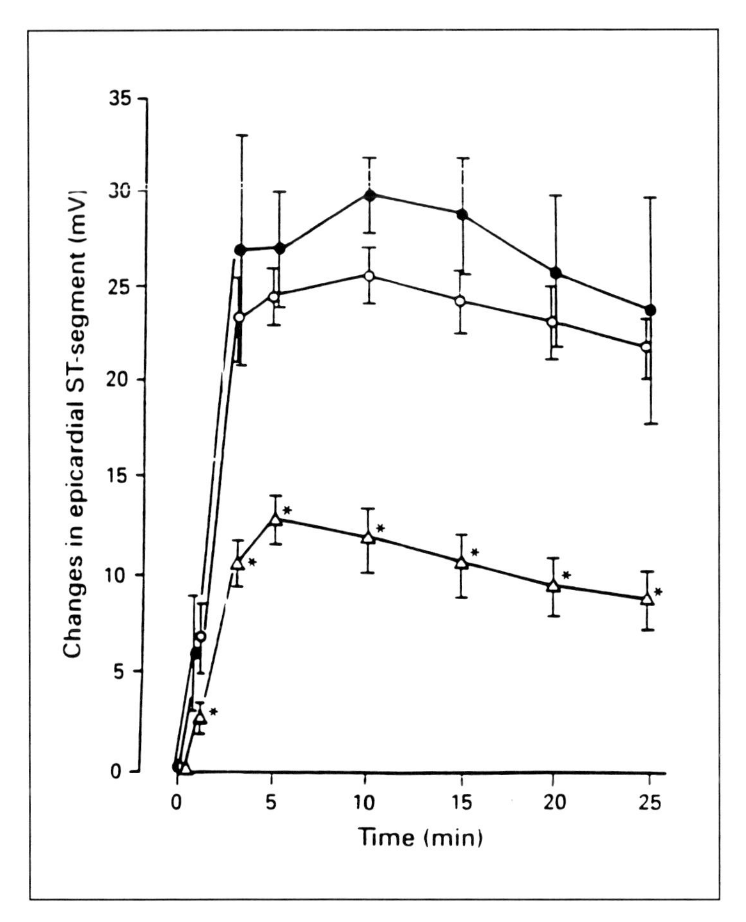

*Fig. 14.4. Changes in the epicardial ST-segment elevation during a 25 minute occlusion of the left anterior descending coronary artery in dogs subjected to ventricular pacing 20 hours previously (Δ; n = 10), of dogs also subjected to pacing but in the presence of dexamethasone (●; n = 7) and in control dogs on which the pacing electrode was positioned in the right ventricle but the dogs were not paced (O; n = 20). The marked attenuation of this index of myocardial ischemia which resulted from prior cardiac pacing was abolished if dogs were subjected to the same pacing protocol after dexamethasone administration. Values shown are means ± s.e. means. *P<0.05 compared to controls. Reproduced from with permission from J Mol Cell Cardiol 1995: 27:991-1000.*

stimulus. We believe that if we understood the mechanisms involved they could almost certainly be exploited.

Acknowledgments

Most of our own experiments described and reviewed in this chapter were carried out in the laboratories of the Department of Pharmacology at the Albert Szent-Györgyi Medical University of Szeged and we wish to thank the two chairmen involved over this period (1988 to the present), Professors Laszlo Szekeres and Julius Gy Papp, for their encouragement and support. We also wish to acknowledge the financial support of initially, the Royal Society, the Scottish Home and Health Department and the Wellcome Trust and currently the British Council, the European Economic Community (Scientific Network grant No. CIPA CT92 4009) the Hungarian National Committee for Technical Development and the Hungarian State Government (OTKA). In the Glasgow department we are particularly indebted to Mrs. Margaret Laird for producing the manuscript. Among our younger co-workers we particularly appreciate the dedication and enthusiasm of Dr 'Charlie' (K) Kaszala.

References

1. Tsunehiko K, Hoshida S, Yamashita N et al. Delayed effects of sublethal ischemia on the acquisition of tolerance to ischemia. Circ Res 1993; 72:1293-1299.
2. Marber MS, Latchman DS, Walker JM et al. Cardiac stress protein elevation 24 hours after brief ischemia or heat stress is associated with resistance to myocardial infarction. Circulation 1993; 88:1264-1272.
3. Baxter GF, Marber MS, Patel VC et al. Adenosine receptor involvement in a delayed phase of myocardial protection 24 hours after ischemic preconditioning. Circulation 1994; 90:2993-3000.
4. Selye H, Veilleux R, Grasso S. Protection, by coronary ligature, against isoproterenol-induced myocardial necroses. Proc Soc Exp Circ NY 1960; 104:343-345.
5. Rona G, Kahn DS, Chappel CL. Studies on infarct-like myocardial necrosis produced by isoproterenol: a review. Rev Can Biol 1963; 22:241-255.
6. Dusek J, Rona G, Kahn DS. Myocardial resistance: a study of its development against toxic doses of isoproterenol. Arch Path 1970; 89:79-83.
7. Dusek J, Rona G, Kahn DS. Myocardial resistance to isoprenaline in rats: Variation with time. J Pathol 1971; 105:279-282.
8. Balazs T. Cardiotoxicity of isoproterenol in experimental anaimals. Influence of stress, obesity, and repeated dosing. Myocardiology 1972; 1:770-778.
9. Beckman CB, Niazi Z, Dietzman RH et al. Protective effects of epinephrine tolerance in experimental cardiogenic shock. Circ Shock 1981; 8:137-149.
10. Szilvassy Z, Szekeres L, Udvary E et al. On the 7-oxo-PgI_2 induced lasting protection against ouabain arrhythmias in anesthetised guinea-pigs. Biomed Biochim Acta 1988; 47:10/11, S35-S38.
11. Szekeres L, Németh, Papp JGy et al. Neue entwicklungen der antiarrhythmischen therapie. In: Perspektiven der Arrhythmiebehandlung. Berlin, Heidelberg: Springer-Verlag 1988:23-34.

12. Szekeres L, Németh M, Papp JGy et al. Short incubation with 7-oxo-prostacyclin induces long lasting prolongation of repolarisation time and effective refractory period in rabbit papillary muscle preparation. Cardiovasc Res 1990; 1:37-41.
13. Szekeres L, Szilvássy, Udvary É et al. 7-oxo-PgI_2 induced late appearing and long-lasting electrophysiological changes in the heart in situ of the rabbit, guinea-pig, dog and cat. J Mol Cell Cardiol 1989; 21:545-554.
14. Dzurba A, Ziegelhoeffer A, Breier A et al. Increased activity of sarcolemmal (Na^+ K^+)-ATPase is involved in the late cardioprotective action of 7-oxo-prostacyclin. Cardioscience 1991; 2:105-108.
15. Ravingerová, Tribulová N, Ziegelhöffer A et al. Suppression of reperfusion induced arrhythmias in the isolated rat heart: pretreatment with 7-oxo-prostacyclin in vivo. Cardiovasc Res 1993; 27:1051-1055.
16. Parratt JR. Myocardial and circulatory effects of *E. coli* endotoxin: modification of responses to catecholamines. Br J Pharmacol 1973; 47:12-25.
17. Parratt JR, Vegh A. Pronounced antiarrhythmic effects of ischemic preconditioning. Cardioscience 1994; 5:9-18.
18. Parratt JR, Vegh A, Papp JGy. Pronounced antiarrhythmic effects of ischemic preconditioning—are there possibilities for pharmacological exploitation? Pharmacol Res 1995; 31:225-234.
19. Julou-Schaeffer G, Gray GA, Fleming I et al. Loss of vascular responsiveness induced by endotoxin involves the L-arginine pathway. Am J Physiol 1990; 259:H1038-H1043.
20. Parratt JR, Stoclet JC, Furman BL. Substances mainly derived from vascular endothelium (endothelium-derived relaxing factor, or nitric oxide, and endothelin) as chemical mediators in sepsis and endotoxemia. In: Naugebauer E, Holaday JW, eds. Handbook of Mediators of Septic Shock. Boca Raton: CRC Press, 1993:381-393.
21. Schulz R, Nava E, Moncada S. Induction and potential biological relevance of Ca^{2+}-independent nitric oxide synthase in the myocardium. Br J Pharmacol 1992; 103:575-580.
22. Wu S, Furman BL, Parratt JR. Attenuation by dexamethasone of endotoxin protection against ischemia-induced ventricular arrhythmias. Br J Pharmacol 1994; 113:1083-1084.
23. Yao Z, Auchampach JA, Pieper GM et al. Cardioprotective effects of monophosphoryl lipid A, a novel endotoxin analog, in the dog. Cardiovasc Res 1993; 27:832-838.
23. Xavier J, Bloom S, Pledger G et al. Determinants of resistance to the cardiotoxicity of isoproterenol in rats. Toxicol App Pharmacol 1983; 69:199-205.
24. Vegh A, Szekeres L, Parratt JR. Transient ischemia induced by rapid cardiac pacing results in myocardial preconditioning. Cardiovasc Res 1991; 25:1051-1053.
25. Szekeres L, Papp J Gy, Szilvássy Z et al. Moderate stress by cardiac pacing may induce both short term and long term cardioprotection. Cardiovasc Res 1993; 27:593-596.

26. Szilvassy Z, Ferdinandy P, Bor P et al. Ventricular overdrive pacing-induced anti-ischemic effect: a conscious rabbit model of preconditioning. Amer Physiol Soc 1994; H2033-H2041.
27. Kaszala K. Vegh A, Parratt JR et al. Time course of pacing-induced preconditioning in dogs. J Mol Cell Cardiol 1995; 27:A145 (Abstract).
28. Vegh A, Papp JGy, Parratt JR. Prevention by dexamethasone of the marked antiarrhythmic effects of preconditioning induced 20h after rapid cardiac pacing. Br J Pharmacol 1994; 113:1081-1082.
29. Vegh A, Kaszala K, Papp J Gy et al. Delayed myocardial protection by pacing-induced preconditioning: a possible role for bradykinin. Br J Pharmacol 1995; 116:288P.
30. Parratt JR, Szekeres L. Delayed protection of the heart against ischemia. Trends Pharmac Sci 1995; 16:351-355.

CHAPTER 15

Cardioprotective Effects of Chronic Hypoxia: Relation to Preconditioning

František Kolář

15.1. INTRODUCTION

Preconditioning with ischemia originally referred to the observation that brief sublethal coronary occlusions, each followed by reperfusion, limit the infarct size in dogs after a subsequent longer period of ischemia.[1] Since then, this remarkable phenomenon has been demonstrated in a variety of experimental models and the concept of preconditioning broadened to include the temporal protection of the heart against other adverse consequences of ischemia and reperfusion, such as arrhythmias or contractile dysfunction. This term is now used to describe also the increased tolerance of the heart to various types of stress mediated by brief stimuli other than ischemia (e.g., acute hypoxia, drugs).[2] Occasionally, preconditioning has been improperly used in a more general meaning that includes any intervention, even long-term, which increases the resistance of the heart against subsequent injury. As ischemic preconditioning represents the most efficient form of temporal protection, it has attracted a great deal of attention and considerable progress has been achieved in understanding this phenomenon over the past nine years. At the same time, however, other forms of protection have been largely ignored.

It has been known for many years that the resistance of the heart against injury can be increased by long-term adaptive processes. Among them, the most common and clinically important is the cardiac adaptation to chronic hypoxia. It seems appropriate, therefore, to include a

Myocardial Preconditioning, edited by Cherry L. Wainwright and James R. Parratt.

chapter in this book which briefly reviews the protective effects of chronic hypoxia on the heart and their possible mechanisms.

Chronic myocardial hypoxia, as a consequence of a disproportion between oxygen delivery and demand at the tissue level, may have several different causes: a decrease in the arterial PO_2 (hypoxic hypoxia), a reduction of coronary blood flow (ischemic hypoxia) or a decrease in the oxygen transport capacity of the blood at normal PO_2 (anemic hypoxia). Hypoxic hypoxia is the main pathophysiological feature of various cardiopulmonary diseases, such as chronic obstructive lung disease and cyanotic congenital heart defects. In two situations myocardial hypoxia can be considered as physiological: in the hearts of high altitude residents and in the fetal heart which is exposed to arterial PO_2 as low as one third of the normal level.[3] Though the heart obviously has the capability to adapt to various forms of hypoxia, this chapter relates only to the effects of hypoxic hypoxia induced by prolonged high altitude exposure.

Chronic high altitude hypoxia (HAH) is either naturally encountered in the mountain environment or can be simulated under laboratory conditions in a hypobaric or normobaric chamber; its character is either permanent or intermittent. Prolonged exposure to any type of HAH induces a variety of adaptive changes that help to maintain heart function and to cope with subsequent stress. On the other hand, it may also exert adverse effects, such as right ventricular hypertrophy due to pulmonary hypertension and myocardial fibrosis which may lead to heart failure. The situation can be principally different in low-landers acclimatized to HAH (phenotypical adaptation) and in species indigenous to high altitude (genotypical adaptation). As compared to simulated HAH, other factors such as cold or physical activity have to be taken into account at natural high altitude, though hypoxia is the main stimulus. A detailed discussion of these general aspects can be found elsewhere.[4-7]

15.2. PROTECTIVE EFFECTS OF CHRONIC HIGH ALTITUDE HYPOXIA (HAH)

It is interesting to note that the history of cardioprotection mediated by chronic hypoxia is quite different from that of preconditioning. Whereas the latter phenomenon was discovered in the laboratory, the experimental investigation of cardiac effects of HAH was stimulated by clinical-epidemiological observations in the late fifties that the incidence of myocardial infarction and mortality from coronary heart diseases is significantly reduced in people residing at high altitudes (for review see refs. 3, 4). Kopecky and Daum[8] in 1958 were the first to demonstrate the protective effect of simulated HAH on the heart under laboratory conditions. They observed that cardiac muscle isolated from rats exposed every other day for 6 weeks to an altitude of 7000 m recovered its contractile function during reoxygenation following acute anoxia better than that of normoxic animals. This finding was confirmed by others using different acclimatization procedures;[9-11] a similar protective effect can be induced by relatively brief daily exposures to hypoxia (4 hours/day, a total of 24 exposures up to 7000 m).[12,13] Recently, an increased tolerance to global ischemia, manifested as improved recovery of the contractile function following reperfusion, was shown in isolated perfused hearts of rats acclimatized to permanent normobaric hypoxia (3 weeks, PO_2 corresponding to 5500 m).[14] Meerson et al[15] however, did not find any improvement in recovery of heart function by intermittent hypoxia (5 hours/day, 40 days at 4000 m) in a similar ischemia/reperfusion model.

Intermittent exposure to a high altitude of 7000 m increases the tolerance of the rat hearts to the toxic effects of isoprenaline, as indicated by the reduced size of myocardial necrotic lesions.[9,16] It has also been reported that rats acclimatized to intermittent HAH (6000 m) developed a smaller myocardial necrosis following coronary occlusion than control animals kept at sea level.[17] In another study, however, this effect was observed only in rats permanently exposed to the same altitude, but not in those acclimatized intermittently.[18] The isolated perfused hearts of rats acclimatized to natural hypoxic conditions at 3500 m maintained higher myocardial levels of ATP, phosphocreatine and glycogen after coronary occlusion, but their function did not differ from controls.[19] Last, but not least, the incidence and severity of ischemic or reperfusion arrhythmias were also significantly reduced in rats after a prolonged exposure to HAH.[20,21]

An important feature of acclimatization to chronic hypoxia is that its protective effects persist long after the removal of the animals from the hypoxic environment. For example, a better recovery of cardiac contractile function following anoxia/reoxygenation can be demonstrated in acclimatized rats 4 months after the restoration of normoxia, while the regression of other hypoxia-induced adaptive changes, such as polycythemia, pulmonary hypertension, right ventricular hypertrophy etc., occurs within a few weeks.[22] Similarly, the increased resistance of the heart to isoprenaline-induced necrosis does not return to control values even 6 months after the last hypoxic exposure.[16]

It follows from the above outline that the protective effects of HAH have been studied largely in acclimatized rats and the data on larger species are missing. Protection against the adverse consequences of subsequent acute hypoxia, ischemia or administration of isoprenaline has been determined using different end points, such as lethal cell injury, contractile dysfunction or incidence of arrhythmias. Though the majority of studies clearly showed an increased resistance of chronically hypoxic hearts against injury, it is unclear to what extent the results depend on the degree, duration, intensity and character (permanent, intermittent, natural, simulated) of the hypoxic exposure. Moreover, the conditions needed to provide maximum and long-lasting myocardial protection are unknown and may differ with respect to the type of stress involved and the type of damage determined as the end-point.

15.3. PROPOSED MECHANISMS OF PROTECTION BY HAH

15.3.1. Oxygen Transport to Tissue

Chronic hypoxia may potentially exert its protective effect on the heart by activating mechanisms which improve myocardial oxygenation. It is not clear to what extent they contribute to increased tolerance of the heart in vivo against the adverse consequences of hypoxia and other stress stimuli. Obviously, these mechanisms may only be of limited importance, because the cardioprotective effects of acclimatization to HAH were demonstrated under in vitro conditions using nonperfused cardiac muscle preparations.

Myocardial vascularity and coronary blood flow

Although the effect of chronic hypoxia on proliferation of coronary vessels has been examined in a great number of studies, their results are inconsistent. A moderate increase in capillary density and capillary-to-fiber ratio was observed in the hearts of guinea pigs native to Andean mountains and in those acclimatized to simulated altitude during maturation.[23,24] In the latter study,[24] this effect was absent in right ventricles with a large degree of hypertrophy. Rakušan et al[25] did not find any change in the left ventricular capillarization in guinea pigs native to high altitude. An increased area occupied by capillaries has been reported in the hearts of dogs born at simulated high altitude.[26] In rats, higher capillarization was observed in the right, but not in the left, ventricular myocardium of animals born at simulated hypoxia or exposed to it only postnatally.[27,28] Others, however, have reported a decrease in the right ventricular capillary density in rats under similar experimental conditions.[29,30]

An indication of changes in myocardial vascularization can be also obtained by measuring coronary blood flow at maximal vasodilation. Turek et al[31] observed a greater specific coronary flow during ventilation of hypoxic gas mixture in rats acclimatized to simulated HAH than in normoxic controls. This effect, indicating improved vascular capacity of the heart, was present in both hypertrophic right ventricle and in nonhypertrophic left ventricle. Similar changes, measured as a vasodilatory response to adenosine, were demonstrated in rabbit and dog hearts during acclimatization to simulated hypoxia.[32,33] In contrast, no influence of HAH on the minimal coronary resistance was found in calves.[34]

The possibility that coronary collateral circulation is stimulated by chronic HAH remains insufficiently examined. An increased incidence of interarterial anastomotic connections was reported in piglets after prolonged exposure to simulated high altitude.[35] However, functional measurement in dogs did not reveal any effect of HAH on collateral flows.[33]

Thus, some experimental data indicate increased, while others point to an unchanged or even decreased, vascularization of the myocardium following HAH. Mutual comparison of these studies is difficult due to variations in experimental models and other factors. Among these, the interspecies differences in acclimatization to hypoxia may play an important role. Rodents, for example, have been shown to have a relatively higher capacity for myocardial vascular growth than other mammals.[36] Another important factor may be the age at which the animals were exposed to HAH and the duration of exposure. For instance, increased maximal coronary blood flow was demonstrated in rats acclimatized during postnatal life, but not in those born in simulated hypoxia.[31] Moreover, while coronary flow was increased in dogs after a one-month hypoxic exposure, it did not differ from controls two months later.[33] Coronary angiogenesis and increased coronary blood flow induced by hypoxia are probably the most important compensatory changes aiming at normalization of tissue PO_2 in newcomers during acclimatization, but their role decreases later on as other adaptive mechanisms develop.[31,33] This view is supported by essentially normal or even decreased values of coronary blood flow in high altitude residents.[3]

Hematocrit and hemoglobin

An increased oxygen-carrying capacity of the blood by elevated hematocrit and concentration of hemoglobin was traditionally considered as an effective adaptive mechanism to chronic hypoxia. It results mainly from the stimulation of formation and differentiation of erythroid precursor cells in the bone marrow by erythropoietin which is produced in hypoxic kidneys.[37] Most animal species, including the rat, develop marked polycythemia associated with a shift of the oxygen dissociation curve to the right due to increased concentrations of 2,3-diphosphoglycerate following exposure to HAH. Similar changes favoring the release of oxygen to the tissue can be observed in high altitude human subjects.[6,38]

Not all species, however, respond to chronic hypoxia in this way. Cattle, for example, which possess a low overall resistance to hypoxia, do not develop polycythemia.[34,39] South American camelids and guinea pigs native to high altitude exhibit only a minor erythropoietic response and a leftward shift in oxygen dissociation curve, indicating increased affinity of hemoglobin for oxygen.[6,40] This system avoids the potentially harmful influence of increased blood viscosity due to polycythemia and favors survival at extreme hypoxia. Interestingly, it has been shown using a theoretical model that a combination of high affinity for oxygen and lower hemoglobin concentrations in the guinea pig resident at high altitude has a similar beneficial effect on myocardial oxygenation at low PO_2, to a combination of a low affinity for oxygen and high hemoglobin concentration in rats acclimatized to HAH.[41] These divergent responses of the rat and guinea pig to chronic hypoxia are thought to reflect differences between phenotypical adaptation in low-landers exposed to HAH and genotypical adaptation in species indigenous to high altitude.[6,38]

Myoglobin

The role of myoglobin consists of storing oxygen and facilitating its transport to tissues. Numerous studies have demonstrated an increased concentration of myoglobin in myocardium as a result of chronic hypoxia (for review see ref. 3), but significant differences exist depending on species, age etc. For example, rats born at simulated high altitude had the same concentration of myoglobin in the right ventricle as controls, but in the left ventricle its concentration was lower than in the controls.[42] On the other hand, rats or guinea pigs born in normoxia and exposed to high altitude later on have a higher concentration of myoglobin in the right ventricle and unchanged concentration in the left ventricle as compared with normoxic animals.[42,43]

15.3.2. Energetic Metabolism

Chronic hypoxia is associated with a variety of adaptive changes at the level of myocardial energy production and utilization that enable the heart to cope with the lack of oxygen and work more economically. According to Moret,[3] these changes may involve an increased capacity for aerobic metabolism, as indicated by the larger number of mitochondria, higher concentrations or activities of certain mitochondrial enzymes or cytochromes, improved ability of mitochondria to generate energy in the presence of low ADP concentration, higher efficiency of oxidative phosphorylation and preferential use of carbohydrates as substrate etc. It is outside the scope of this chapter to review the wealth of studies dealing with the influence of chronic HAH on myocardial energetic metabolism; detailed information can be found elsewhere.[3,38] It should be mentioned, however, that the available information is not consistent; the situation is complicated, for example, by differences between experimental models and conditions studied. It is often impossible to distinguish the direct effects of hypoxia from those of hypertrophy and other factors associated with exposure to HAH.[44,45] Furthermore, the differences between the effects of acclimatization in low-landers and genotypical adaptation in high altitude residents remain unclear.[5,38]

It is conceivable that adaptive changes at the level of myocardial energy production can play an important role in increased tolerance of chronically hypoxic hearts against subsequent stress, but any serious evidence supporting the possible link between these phenomena is missing. The experiments of McGrath and Bullard[10] should perhaps be mentioned in this context. These studies demonstrated that the protective effect of acclimatization to HAH on postanoxic contractile dysfunction of isolated right ventricular strips was removed by the glyceraldehyde-3-phosphate dehydrogenase inhibitor iodoacetate. It was therefore suggested that the influence of acclimatization might be due to increased glycolytic capacity. Because ATP produced by glycolytic pathways is considered to maintain the integrity of the sarcolemma,[46] this mechanism may be beneficial for hypoxic myocyte survival.

15.3.3. Neurohumoral Factors

Chronic hypoxia alters the function of the autonomic nervous system and endocrine glands; some of these changes might have a beneficial influence on myocardial resistance against injury. For example, the exposure to HAH is initially associated with increased adrenergic activity and elevated plasma concentration of catecholamines[47,48] which result in continuous chronotropic and inotropic stimulation of the heart. Acclimatization after prolonged exposure is characterized by a reduced β-adrenoceptor responsiveness[47,49] due to downregulation of myocardial β-adrenoceptors,[50,51] while the intracellular signal transduction does not seem to be altered.[51] Chronic hypoxia may also lead to increased degradation of catecholamines, as indicated by the elevated activity of catechol-o-methyltransferase.[52] It is conceivable that these changes could be antiarrhythmic and protect the myocardium by decreasing its oxygen demand. Obviously, the downregulation phenomenon may be of principal importance, at least when considering the protective effect of HAH on myocardial injury due to toxic doses of isoprenaline.[16]

The involvement of the parasympathetic system in acclimatization to chronic hypoxia is poorly understood. While an increased parasympathetic activity and decreased atrial muscarinic receptor density were reported in chronically hypoxic guinea pigs,[53] an opposite change, i.e., upregulation of muscarinic receptors with an increase in the affinity for the agonist was observed in atria[54] and ventricles[55] of rats. Muscarinic stimulation induced by acclimatization to HAH might contribute to a reduced β-adrenoceptor responsiveness of the heart and decrease its susceptibility to arrhythmias.

It has been shown convincingly that the long-term exposure of animals to HAH decreases thyroid function, although the mechanism is not clear (for review see ref. 38). Induction of hypothyroidism may be one of the causes of cardioprotection provided by chronic HAH. It has been shown that hypothyroid myocardium is less susceptible than controls to postanoxic contractile dysfunction, as examined in isolated right ventricular muscle of the rat.[56] Conversely, excess thyroid hormone increases cardiac sensitivity to acute lack of oxygen.[57,58] The cardioprotective effect of hypothyroidism seems to be related to the improved efficiency of myocardial energy utilization,[59-61] but the mechanism is probably complex and is poorly understood.

15.3.4. Stress Proteins

The synthesis of heat shock or stress proteins (hsp), induced in the myocardium by a variety of stimuli, seems to be related to increased ability of the heart to tolerate subsequent stress. It has been shown that acute hypoxia is able to trigger the induction of hsp 70 both in isolated cardiomyocytes[62] and in the heart in vivo,[63] though it may be less potent than another type of stress, e.g., heat shock.[64] It is not clear whether stress proteins can play a role in myocardial protection mediated by acclimatization to chronic hypoxia. The exposure of rats to a simulated altitude of 5500 m for two weeks did not affect the expression of hsp 70 mRNA in the heart, but the level of mRNA coding for another hsp protein, heme oxygenase, was increased.[65] In another study[15] an increased synthesis of two hsp 70 isoforms was demonstrated in the hearts of rats acclimatized for 40 days to intermittent high altitude of 4000 m. In this setting, however, the tolerance of the isolated perfused heart to postischemic contractile dysfunction and creatine kinase release was not improved by acclimatization and the tolerance to heat stress-induced injury was only moderately increased.

15.3.5. Prostaglandins

Prostaglandins, and especially prostacyclin, have been shown to play a role in protecting the myocardium: their antiarrhythmic effect is well documented (for review see ref. 66). Production of prostaglandins is markedly increased during acute hypoxia.[67,68] Exposure of man to HAH initially leads to increased plasma concentrations of these substances which return to control values after eight days at an altitude of 4350 m.[69] In rats exposed intermittently to a similar altitude in a hypobaric chamber, myocardial levels of prostaglandins were elevated even after 40 days of acclimatization.[70] The latter study also demonstrated a permanently increased concentration of thromboxane which, however, is arrhythmogenic.[71]

The possibility that prostacyclin contributes to the cardioprotection provided by acclimatization to intermittent HAH has been recently examined.[11] A single dose of 7-oxo-prostacyclin increased the tolerance of the isolated heart muscle to postanoxic contractile dysfunction similarly in controls and chronically hypoxic rats. The protective effects of acclimatization and prostacyclin were additive, suggesting that different mechanisms are involved and that prostacyclin does not contribute to protection mediated by HAH.

15.3.6. Adenosine

Adenosine has attracted considerable attention in the past few years with regard to its cardioprotective properties. Convincing evidence has accumulated indicating a central role for this endogenous substance as a trigger and mediator of preconditioning in many species, with the exception of rats.[2] The protective effects of adenosine may involve many possible mechanisms which have been recently reviewed in detail in a special issue of *Cardiovascular Research* (1993; 27:No1).

Formation of adenosine is closely coupled with tissue oxygenation[72] and its release from the heart during hypoxia is well documented.[73-75] There is little information available as yet about the possible involvement of this substance in acclimatization to chronic hypoxia. It has been shown recently that prolonged exposure of rats to a simulated high altitude leads to a decrease in myocardial A_1-adenosine receptor density without any change in the affinity for the agonist.[54] Adenosine is known to stimulate glycolysis in the heart[76] and it has been suggested that increased glycolytic capacity may account for the protective effect of HAH.[10] Adenosine released during hypoxia may also act as a stimulus for the growth of new vessels[77,78] and some reports indicate that angiogenesis may occur in the chronically hypoxic heart.[24,27,31] In addition, this substance may be related to HAH induced polycythemia because it stimulates the production of erythropoietin.[37] Adenosine is also known to antagonize many effects of catecholamines[79,80] and thereby protect the heart against excessive β-adrenoceptor stimulation. It should be noted, however, that no direct evidence exists that adenosine really takes part in any of the cardioprotective effects of acclimatization to chronic hypoxia.

Some additional chemical mediators, e.g., bradykinin or nitric oxide, have also been suggested to act as endogenous myocardial protective substances under certain circumstances.[80] Their possible role in mediating the protective effect of chronic hypoxia has not been examined.

15.3.7. Other Possible Mechanisms

There are many additional changes induced by acclimatization to HAH that might potentially contribute to the protection of the heart against subsequent damage, but the available information is only fragmentary. Thus, it has been demonstrated that the simulated high altitude leads to a shift of the myosin isoenzyme composition in the rat myocardium towards the lower ATPase activity V_3 form.[81] As this change is known to be associated with improved economy of force development,[59] the hearts of acclimatized animals may be expected to consume less energy for contraction.

Acclimatization to HAH lowers the specific activities of several sarcolemmal ATPases and, at the same time, it increases their affinity for ATP.[82] The latter can be considered as an adaptive mechanism at the enzyme level that enables more efficient utilization of available ATP and thereby helps to preserve membrane transport functions under conditions of low energy production.

In the hearts of rats acclimatized to intermittent HAH, higher activities of antioxidant enzymes such as catalase and superoxide dismutase were observed together with decreased susceptibility to reperfusion arrhythmias.[21] In another study, however, no appreciable change in myocardial total antioxidant concentration was found in a similar model of chronic hypoxia.[83]

15.4. CHRONIC HYPOXIA VS. PRECONDITIONING—CONCLUSIONS

There seems to be sufficient evidence that the acclimatization to chronic HAH protects the heart against subsequent damage, but the mechanism of this phenomenon is far from clear. In fact, the majority of related studies are descriptive and only a few have attempted to determine how the protection may be accomplished. It is likely that the mechanism is complex and the contribution of various factors might differ depending on the type of stress stimulus and on the type of subsequent injury evaluated as the end-point. Whatever the principal mechanisms, they must be intrinsic to the cardiac tissue, since the protection can be demonstrated even in isolated nonperfused preparations. However, any current speculative explanation is weakened by the observation that cardioprotection persists several months after the end of hypoxic exposure when all known adaptive changes have already returned to normal. This finding certainly deserves more attention.

Unlike chronic hypoxia, preconditioning leads to protection which lasts only transiently, though its efficiency seems to be much higher. An important question arises whether these two phenomena can share, at least partly, the same cardioprotective mechanism. It is evident that the information available is insufficient to answer this question. It cannot be excluded that some of the factors proposed to play a central role in preconditioning may also contribute to protection mediated by chronic HAH. On the other hand, changes that require prolonged hypoxic exposure to develop obviously cannot take a part in the rapid action of preconditioning. Only a single recent study has addressed the problem whether preconditioning can be demonstrated in hearts following acclimatization to chronic hypoxia. Using the isolated perfused rat heart, Tajima et al[14] observed that the protective effect of ischemic preconditioning against subsequent postischemic contractile dysfunction is additive to that provided by acclimatization to a simulated high altitude of 5500 m. It suggests that these two phenomena are independent and utilize different mechanisms. It needs to be proven whether this is true also for other manifestations of injury, such as arrhythmias and myocardial necrosis. Thus, to avoid possible confusion, the protection mediated by chronic hypoxia should not be considered as a special kind of "long-term preconditioning."

References

1. Murry CE, Jennings RB, Reimer KA. Preconditioning with ischemia: a delay of lethal cell injury in ischemic myocardium. Circulation 1986; 74:1124-1136.

2. Lawson CS, Downey JM. Preconditioning: state of art myocardial protection. Cardiovasc Res 1993; 27:542-550.
3. Moret PR. Hypoxia and the heart. In: Bourne GH, ed. Hearts and Heart-like Organs. New York: Academic Press, 1980:333-387.
4. Heath D, Williams DR. Man at High Altitude. Edinburgh: Churchill Livingstone, 1981.
5. Oštádal B, Widimsky J. Intermittent Hypoxia and Cardiopulmonary System. Prague: Academia, 1985.
6. Monge C, Leon-Velarde F. Physiological adaptation to high altitude: oxygen transport in mammals and birds. Physiol Rev 1991; 71:1135-1172.
7. Oštádal B, Kolář F, Pelouch V et al. Intermittent high altitude and the cardiopulmonary system. In: Nagano M, Takeda N, Dhalla NS, eds. The Adapted Heart. New York: Raven Press, 1994:173-182.
8. Kopecky M, Daum S. Tissue adaptation to anoxia in rat myocardium (in Czech). Čs Fysiol 1958; 7:518-521.
9. Poupa O, Krofta K, Procházka J et al. Acclimatization to simulated high altitude and acute cardiac necrosis. Fed Proc 1966; 25:1243-1246.
10. McGrath JJ, Bullard RW. Altered myocardial performance in response to anoxia after high-altitude exposure. J Appl Physiol 1968; 25:761-764.
11. Ziegelhöffer A, Grünermel J, Dzurba A et al. Sarcolemmal cation transport systems in rat hearts acclimatized to high altitude hypoxia: influence of 7-oxo-prostacyclin. In: Oštádal B, Dhalla NS, eds. Heart Function in Health and Disease. Boston: Kluwer 1992:219-228.
12. McGrath JJ, Procházka J, Pelouch V et al. Physiological responses of rats to intermittent high altitude stress. Effect of age. J Appl Physiol 1973; 34:289-293.
13. Widimsky J, Urbanová D, Ressl J et al. Effect of intermittent altitude hypoxia on the myocardium and lesser circulation in the rat. Cardiovasc Res 1973; 7:798-808.
14. Tajima M, Katayose D, Bessho M et al. Acute ischemic preconditioning and chronic hypoxia independently increase myocardial tolerance to ischemia. Cardiovasc Res 1994; 28:312-319.
15. Meerson FZ, Malyshev IY, Zamotrinsky AV. Differences in adaptive stabilization of structures in response to stress and hypoxia relate with the accumulation of hsp70 isoforms. Mol Cell Biochem 1992; 111:87-95.
16. Faltová E, Mráz M, Pelouch V et al. Increase and regression of the protective effect of high altitude acclimatization on the isoprenaline-induced necrotic lesions in the rat myocardium. Physiol Bohemoslov 1987; 36:43-52.
17. Meerson FZ, Gomzakov OA, Shimkovich MV. Adaptation to high altitude hypoxia as a factor preventing development of myocardial ischemic necrosis. Am J Cardiol 1973; 31:30-34.
18. Turek Z, Kubát K, Ringnalda BEM et al. Experimental myocardial infarction in rats acclimated to simulated high altitude. Basic Res Cardiol 1980; 75:544-553.

19. Opie LH, Duchosal F, Moret PR. Effect of increased left ventricular work, hypoxia, or coronary artery ligation on hearts from rats at high altitude. Eur J Clin Invest 1978; 8:309-315.
20. Meerson FZ, Ustinova EE, Orlova EH. Prevention and elimination of heart arrhythmias by adaptation to intermittent high altitude hypoxia. Clin Cardiol 1987; 10:783-789.
21. Meerson FZ, Arkhipenko IV, Rozhitskaia II et al. Opposite effects on antioxidant enzymes of adaptation to continuous and intermittent hypoxia (in Russian). Byull Eksp Biol Med 1992; 114:14-15.
22. Oštádal B, Procházka J, Pelouch V et al. Pharmacological treatment and spontaneous reversibility of cardiopulmonary changes induced by intermittent high altitude hypoxia. Progr Resp Res 1985; 29:17-25.
23. Valdivia E. Total capillary bed of the myocardium in chronic hypoxia. Fed Proc 1962; 21:221 (Abstract).
24. Kayar SR, Banchero N. Myocardial capillarity in acclimation to hypoxia. Pflügers Arch 1985; 404:319-325.
25. Rakušan K, Turek Z, Kreuzer F. Myocardial capillaries in guinea pigs native to high altitude (Junin, Peru, 4,105 m). Pflügers Arch 1981; 391:22-24.
26. Becker EL, Cooper RG, Hataway GD. Capillary vascularization in puppies born at a simulated altitude of 20,000 feet. J Appl Physiol 1955; 8:166-168.
27. Turek Z, Grandtner M, Kreuzer F. Cardiac hypertrophy, capillary and muscle fiber density, muscle fiber diameter, capillary radius and diffusion distance in the myocardium of growing rats adapted to a simulated altitude of 3,500 m. Pflügers Arch 1972; 335:19-28.
28. Grandtner M, Turek Z, Kreuzer F. Cardiac hypertrophy in the first generation of rats native to simulated high altitude. Pflügers Arch 1974; 350:241-248.
29. Clark DR, Smith P. Capillary density and muscle fibre size in the hearts of rats subjected to simulated high altitude. Cardiovasc Res 1978; 12:578-584.
30. Smith P, Clark DR. Myocardial capillary density and muscle fibre size in rats born and raised at simulated high altitude. Br J Exp Path 1979; 60:225-230.
31. Turek Z, Turek-Maischeider M, Claessens RA et al. Coronary blood flow in rats native to simulated high altitude and in rats exposed to it later in life. Pflügers Arch 1975; 355:49-62.
32. Holmes G, Epstein ML. Effect of growth and maturation in a hypoxic environment on maximum coronary flow rates in isolated rabbit hearts. Pediat Res 1993; 33:527-532.
33. Scheel KW, Seavey E, Gaugl JF et al. Coronary and myocardial adaptations to high altitude in dogs. Am J Physiol 1990; 259:H1667-H1673.
34. Manohar M, Parks CM, Busch MA et al. Regional myocardial blood flow and coronary vascular reserve in unanesthetized young calves exposed to a simulated altitude of 3500 m for 8-10 weeks. Circ Res 1982; 50:714-726.

35. Reiner L, Freudenthal RR, Greene MA et al. Interarterial coronary anastomosis in piglets at simulated high altitude. Arch Pathol 1972; 93:198-208.
36. Scheurer J, Buttrick P. The cardiac hypertrophic responses to pathologic and physiologic loads. Circulation 1987; 75(Suppl I):63-68.
37. Scholz H, Schurek H-J, Eckardt K-U et al. Role of erythropoietin in adaptation to hypoxia. Experientia 1990; 46:1197-1201.
38. Harris P. Myocardial metabolism. In: Heath D, Williams DR, Man at High Altitude. Edinburgh: Churchill Livingstone 1981:196-208.
39. Bisgard GE. Pulmonary hypertension in cattle. Adv Vet Sci Comp Med 1977; 21:151-172.
40. Banchero N. Cardiovascular responses to chronic hypoxia. Ann Rev Physiol 1987; 49:465-476.
41. Turek Z, Rakusan K. Computer model analysis of myocardial tissue oxygenation: a comparison of high altitude guinea pig and rat. In: Leon-Velarde F, Arregui A, eds. Hipoxia: Investigationes Basicas y Clinicas. Lima: UPCH, 1993:141-154.
42. Turek Z, Ringnalda BEM, Grandtner M et al. Myoglobin distribution in the heart of growing rats exposed to a simulated altitude of 3,500 m in their youth or born in the low pressure chamber. Pflügers Arch 1973; 340:1-10.
43. Bui MV, Banchero N. Effects of chronic exposure to cold or hypoxia on ventricular weights and ventricular myoglobin concentration in guinea pigs during growth. Pflügers Arch 1980; 385:155-160.
44. Barrie SE, Harris P. Effects of chronic hypoxia and dietary restriction on myocardial enzyme activities. Am J Physiol 1976; 231:1308-1313.
45. Bass A, Oštádal B, Procházka J et al. Intermittent high altitude-induced changes in energy metabolism in the rat myocardium and their reversibility. Physiol Bohemoslov 1989; 38:155-161.
46. Weiss J, Hiltbrand B. Functional compartmentation of glycolytic versus oxidative metabolism in isolated rabbit heart. J Clin Invest 1985; 75:436-447.
47. Maher JT, Manchanda SC, Cymerman A et al. Cardiovascular responsiveness to beta-adrenergic stimulation and blockade in chronic hypoxia. Am J Physiol 1975; 228:477-481.
48. Richalet JP. The heart and the adrenergic system. In: Sutton JR, Coates G, Remmers JE, eds. Hypoxia, the Adaptations. Philadelphia: Dekker 1990:231-245.
49. Oštádal B, Ressl J, Urbanová et al. The effect of beta-adrenergic blockade on pulmonary hypertension, right ventricular hypertrophy and polycythemia, induced in rats by intermittent high altitude hypoxia. Basic Res Cardiol 1978; 73:422-432.
50. Voelkel NF, Hegstrand L, Reeves JT et al. Effects of hypoxia on density of β-adrenergic receptors. J Appl Physiol 1981; 50:363-366.
51. Kacimi R, Richalet J-P, Corsin A et al. Hypoxia-induced downregulation of β-adrenergic receptors in rat heart. J Appl Physiol 1992; 73:1377-1382.

52. Maher JT, Denniston JC, Wolfe DL. Mechanism of the attenuated cardiac response to beta-adrenergic stimulation in chronic hypoxia. J Appl Physiol 1978; 44:647-651.
53. Crockatt LH, Lund DD, Schmid PG et al. Hypoxia-induced changes in parasympathetic neurochemical markers in guinea pig heart. J Appl Physiol 1981; 50:1017-1021.
54. Kacimi R, Richalet J-P, Crozatier B. Hypoxia-induced differential modulation of adenosinergic and muscarinic receptors in rat heart. J Appl Physiol 1993; 75:1123-1128.
55. Wolfe BB, Voelkel NF. Effect of hypoxia on atrial muscarinic cholinergic receptors and cardiac parasympathetic response. Biochem Pharmacol 1983; 32:1999-2002.
56. Martin LG, Westenberger GE, Bullard RW. Thyroidal changes in the rat during acclimatization to simulated high altitude. Am J Physiol 1971; 221:1057-1063.
57. Palacios I, Sagar K, Powell WJ. Effect of hypoxia on mechanical properties of hyperthyroid cat papillary muscle. Am J Physiol 1979; 237: H293-H298.
58. Seppet EK, Eimre MA, Kallikorm AP et al. Effect of exogenous phosphocreatine on heart muscle contractility modulated by hyperthyroidism and extracellular calcium concentration. J Appl Cardiol 1988; 3:369-380.
59. Holubarsch CH, Goulette RP, Litten RZ et al. The economy of isometric force development, myosin isozyme pattern and myofibrillar ATPase activity in normal and hypothyroid rat myocardium. Circ Res 1985; 56:78-86.
60. McDonough KH, Chen V, Spitzer J. Effect of altered thyroid status on in vitro cardiac performance in rats. Am J Physiol 1987; 252:H788-H795.
61. Seppet EK, Kadaya LY, Hata T et al. Thyroid control over membrane processes in rat heart. Am J Physiol 1991; 26 (Suppl):66-71.
62. Iwaki K, Chi S-H, Dillmann WH et al. Induction of HSP70 in cultured rat neonatal cardiomyocytes by hypoxia and metabolic stress. Circulation 1993; 87:2023-2032.
63. Howard G, Geoghegan TE. Altered cardiac tissue gene expression during acute hypoxic exposure. Mol Cell Biochem 1986; 69:155-160.
64. Mestril R, Chi S-H, Sayen MR et al. Isolation of a novel inducible rat heat-shock protein (HSP70) gene and its expression during ischemia/hypoxia and heat shock. Biochem J 1994; 298:561-569.
65. Katayose D, Isoyama S, Fujita H et al. Separate regulation of heme oxygenase and heat shock protein 70 mRNA expression in the rat heart by hemodynamic stress. Biochem Biophys Res Commun 1993; 191: 587-594.
66. Parratt JR. Eicosanoids and arrhythmogenesis. In: Vaughan-Williams EM, ed. Antiarrhythmic Drugs. Berlin: Springer Verlag 1989:569-589.
67. De Deckere EAM, Nugteren DH, Ten Hoor F. Prostacyclin is the major prostaglandin released from the isolated perfused rabbit and rat heart. Nature 1977; 268:160-163.

68. Karmazyn M, Dhalla NS. Physiological and pathophysiological aspects of cardiac prostaglandins. Can J Physiol Pharmacol 1983; 61:1207-1225.
69. Richalet JP, Hornych A, Rathat C et al. Plasma prostaglandins, leukotrienes and thromboxane in acute high altitude hypoxia. Respir Physiol 1991; 85:205-215.
70. Pshennikova MG, Kuznetsova VA, Kopylov IN et al. The role of prostaglandin system in the cardioprotective effect of adaptation to hypoxia in stress (in Russian). Kardiologiya 1992; 32:61-64.
71. Curtis MJ, Pugsley MK, Walker MJA. Endogenous chemical mediators of ventricular arrhythmias in ischemic heart disease. Cardiovasc Res 1993; 27:703-719.
72. Schrader J, Deussen A, Smolenski RT. Adenosine is a sensitive oxygen sensor in the heart. Experientia 1990; 46:1172-1175.
73. Rubio R, Berne RM. Release of adenosine by the normal myocardium in dogs and its relationship to the regulation of coronary resistance. Circ Res 1969; 25:407-415.
74. Fenton RA, Dobson JG. Measurement by fluorescene of interstitial adenosine levels in normoxic, hypoxic, and ischemic perfused rat hearts. Circ Res 1987; 60:177-184.
75. Xu J, Tong H, Wang L et al. Endogenous adenosine, A_1 adenosine receptor, and *pertussis* toxin sensitive guanine nucleotide binding protein mediate hypoxia induced AV nodal conduction block in guinea pig heart in vivo. Cardiovasc Res 1993; 27:134-140.
76. Wyatt DA, Edmunds MC, Rubio R et al. Adenosine stimulated glycolytic flux in isolated perfused rat hearts by A_1-adenosine receptors. Am J Physiol 1989; 257:H1952-H1957.
77. Ziada AM, Hudlicka O, Tyler KR et al. The effect of long-term vasodilatation on capillary growth and performance in rabbit heart and skeletal muscle. Cardiovasc Res 1984; 18:724-732.
78. Meininger CJ, Schelling ME, Granger HJ. Adenosine and hypoxia stimulate proliferation and migration of endothelial cells. Am J Physiol 1988; 255:H554-H562.
79. Boachie-Ansah G, Kane KA, Parratt JR. Is adenosine an endogenous myocardial protective (antiarrhythmic) substance under conditions of ischemia? Cardiovasc Res 1993; 27:77-83.
80. Parratt JR. Endogenous myocardial protective (antiarrhythmic) substances. Cardiovasc Res 1993; 27:693-702.
81. Pelouch V, Ošťádal B, Procházka J et al. Effect of high altitude hypoxia on the protein composition of the right ventricular myocardium. Progr Resp Res 1985; 20:41-48.
82. Ziegelhöffer A, Procházka J, Pelouch V et al. Increased affinity to substrate in sarcolemmal ATP-ases from hearts acclimatized to high altitude hypoxia. Physiol Bohemoslov 1987; 36:403-415.
83. Lebedev AV, Sadredtinov SM, Pelouch V et al. Free radical membrane scavengers in myocardium of rats of different age exposed to chronic hypoxia. Biomed Biochim Acta 1989; 48:S122-S125.

INDEX

Page numbers in italics denote figures (f) or tables (t).

A

Acadesine, 6, 209
Acetylcholine, 71-72, 157
Aquired thermotolerance, 234
Adenosine, 186, 208-209, *210t-211t*, 212-214, *212f, 213f*
  antiarrhythmic effect of preconditioning, 44, 46
  chronic high altitude hypoxia, 268-269
  receptor, 4-7, *5f*, 70, 153-156, 157, 175, 187-188, 194, 198-199, *199f*, 209, 212-214, *212f, 213f*
    ATP sensitive potassium channels (K_{ATP}), 8-10, 70, 130, 138-139, 153, 155-156, 201, 216-217
    delayed myocardial protection, 240-242, *243f*
      ischemia/reperfusion injury, 27
    memory of preconditioning, 196-197
    phospholipase C-protein kinase C, 10-12, *11f*
    tolerance to A_1-adenosine agonists, 200
Adenylyl cyclase, 148, 149, 150, 151, 152, 157, 169, 179
Adrenergic receptors
  α–, 70-71, 138, 157, 170, 171, 173, 174, 175, 178, 179, 186, 198-199
  β–, 170, 171, 173, 179, 186, 267
β–Adrenoceptor kinase (βARK), 149, 152
3-Aminotriazole, 235
Anaerobic glycolysis, 66-67, *66f*
Angina, 118-120, 123-124, 245
Angioplasty. *See* Balloon angioplasty.
Angiotensin, 186, 196
Angiotensin II receptor, 157
Angiotensin-converting enzyme (ACE), 48
Antioxidants, 69, 221, 269-270
Arrhythmias. *See* Ventricular arrhythmias.
Asimakis GK, 171
ATP, 64, *65f*, 269
ATP sensitive potassium channels (K_{ATP}), 8-10, 70, 201. *See also* K^+ channels, ATP-dependent.
Autonomic nervous system, 267

B

Balazs T, 252
Balloon angioplasty, 120-121, 124
Banerjee A, 70, 71, 171, 173, 242
Bankwala Z, 174
Beckman CB, 252
Billman GE, 53
Bimakalim, 134, 135, 139, 217
Bogoyevitch MA. 186, 194
Bolli (initial not given), 238
Bradykinin, 46-50, *47f, 49f, 50f*, 71, 196, 197-199, *199f*, 214, *215t*, 255-256
Bullard RW, 266

C

c-fos, 242
c-jun, 242
Ca^{2+}, 186
  cathecholamines, 168
  channels, 148
  ischemia/reperfusion injury, 22, 26, 27, 28
Calphostin C, 189
cAMP, 53, 169
Carbachol, 150, 157
Cardiac pacing
  delayed myocardial protection, 254-257, *255f, 256f, 257f*
Cardiac surgery, 122-123, 124
Cardiomyocyte. *See also* Ventricular antiarrhythmics.
  mechanical reperfusion injury, 20, *24f, 25f*
    mechanical fragility, 20-21, 23-24, 26-27
    mechanical stress
      cell-to-cell mechanical interaction, 23
      hypercontracture, 21-22, 27-28, 133-134
      osmotic swelling, 22-23, 28, *29f*
  stretch, 85, 87-88, *87f*, 91
Catalase, 235, 254, 269
Catecholamines, 186
  chronic high altitude hypoxia, 267
  myocardial injury, 169-171
  myocardial ischemia, 168-169
  myocardial preconditioning, 180, 252
    G proteins, 178-179
    protection against arrhythmias, 175, *176f*, 177-179, *177f, 178f*
    protection against contractile dysfunction, 171, *172f*, 173
    reduction of necrosis, 173-175, *174f*
CCPA (2-chloro-N^6-cyclopentyl adenosine), 200, 241, 244
cGMP, 53, 54
Chelerythrine, 189, 241
Cholera toxin, 148
Chronic high altitude hypoxia (HAH), 262
  cardioprotection, 262-264
    mechanisms, 269-270
      energy metabolism, 266-267
      neurohumoral factors, 267
      oxygen transport to tissue, 264-266
  myocardial preconditioning, 270
Circumflex preconditioning/LAD occlusion protocol, 85
Colchicine, 194
Cole WC, 134

Contractile dysfunction, 62, *63f*, 64, 74
  anaerobic glycolysis, 66-67, *66f*
  catecholamines, 171, *172f*, 173
  intracellular pH, 67-68, *67f*
  ischemic contracture, 64, *65f*
  mechanisms of preconditioning-induced protection
    extracellular mediators and receptor stimulation, 69-70
      adenosine receptor stimulation, 70
      α_1–adrenergic receptors, 70-71
      bradykinin receptors, 71
      K^+ channels, ATP-dependent, 72
      muscarinic receptors, 71-72
    intracellular second messengers and protein kinase C (PKC), 72-73
  preconditioning protocol, 61-62
Coronary artery
  chronic high altitude hypoxia, 264-263
  collaterals, 119, 121, 122, 214
  endothelium. *See* Ventricular antiarrhythmics.
  partial occlusion without intervening reperfusion, 104-106, *106f*, *107f*
Creatine kinase, 68
Cribrier A, 121
Cromakalim, 72, 132, 133, 134, 137, 139
Currie RW, 235
Cyclooxygenase, 51-52, 132, 254, 256
Cytoskeleton
  reperfusion, 21

D
Daum S, 262
Desipramine, 169
Deutsch E, 120, 121
Diacylglycerol (DAG), 10, 169, 186
Dipyridamole (DIP), 6-7, *6f*
Dobutamine, 4
Dofetilide, 139
Donnelly TJ, 236
Downey JM, 70, 71, 72
DPCPX, 139, 2121, *213f*

E
Ecto-5'-nucleotidase, 7
Endothelin, 196
Endothelium. *See* Ventricular antiarrhythmics.
Endotoxin, 252, 253, *253*, 254
Erythropoietin, 265
Escande D, 135

F
Forskolin, 150, 171
Free radicals, 198, 221
  scavengers, 198, 222

G
G proteins, 130, 147-149
  myocardial ischemia, 169
  myocardial ischemia, 149-150
    G_i proteins, 150-151
    G_s proteins, 150
  preconditioning, 151-152, *153f*, *154t*, *155f*, *156f*, 158, 217, *219t*, 220
    catecholamine-induced, 178-179
    coupled receptors, 152-157, 187-188
    protein kinase C (PKC), 157
Gadolinium chloride, 87-88, 108
Glibenclamide, 8-9, *9t*, 72, 108, 156, 201, 216, 217
Glyburide, 132, 133, 134, 135, 136, 137, 138, 140
Glycogen, 26, 222
Glycolysis, 267
Goto M, 71
Gross GJ, 8, 72, 129, 132, 133, 134, 135, 136
Grover GJ, 8, 10, 72, 216
GTPase, 151
Guanylyl cyclase, 51

H
Haag M, 179
Harris AS, 104
Hearse DJ, 40
Heat shock proteins (hsp), 234
  myocardial protection, delayed, 236-239, *237f*
Hemoglobin
  chronic high altitude hypoxia, 265-266
Henrich CH, 186
Hess ML, 235
Hirai T, 118
HOE 140. *See* Icatibant.
Hoshida S, 238
Hsp 60, 236, 237
Hsp 70, 89, 235, 236, 237, 242, 244
  chronic high altitude hypoxia, 268
Hu K, 179
Hutter MA, 235
Hypoxia
  chronic, 261-262
  high altitude hypoxia. *See* Chronic high altitude hypoxia (HAH).

I
Icatibant, 48, *49f*, *50f*, 52, 214, 255
Inositol 1,4,5-trisphosphate (IP_3), 10, 169, 186
Iodoacetate, 267
Ionotropic agents, 4
Ischemia/reperfusion injury. *See also* Mechanical reperfusion injury.
  heat shock proteins, 236-239, *237f*
Ischemic contracture, 64, *65f*, 133-134
Isoprenaline, 150, 152, 171, *172f*, 252, 263
Iwase T, 150, 151

J
Jennings RB, 10

K

Karmazyn M, 240
K_{ATP} (ATP sensitive potassium channels), 8-10, 70, 72, 130, 139-140, 201, 216-217, *218t*
adenosine, 138-139, 153, 155-156
blockers, 135-138
openers, 130-135, *131f*
Ketamine, 8, 136
Kinins, 47-48
Kirsch CE, 8, 138
Kitakaze M, 7, 69, 139, 196
Kloner RA, 119
Knowlton AA, 92
Kohl C, 11
Kolocassides KG, 64, 67
Komori S, 38
Kopecky M, 262
Kremastinos DT, 91, 92
Kukreja RC, 235

L

L-NAME (N^G-nitro-L-arginine-methyl ester), 50-51, 198
Lactate, 26, 28, *29f*, 67-68
Lazdunski (initial not given), 138
Liu GS, 5, 71, 73
Locke-WinterCR, 171
Lucchesi (initial not given), 42

M

MARCKS, 27
Matsuda M, 4, 118
McGrath JJ, 266
Mechanical reperfusion injury, 20, *24f*, *25f*, 29-30, *30f*
mechanical fragility, 20-21, 23-24, 26-27
mechanical stress
cell-to-cell mechanical interaction, 23
hypercontracture, 21-22, 27-28
osmotic swelling, 22-23, 28, *29f*
Meclofenamate, 52, 132, 133, 137
Meerson FZ, 263
Memory of preconditioning, 2-3, 73, 135, 189
5'-nucleotidase theory, 196
protein kinase C (PKC), 192-193
Meng X, 241, 242
Methylene blue, 51
Mitochondria, 134, 140
Mn-SOD. *See* Superoxide dismutase (SOD).
Monoamine oxidase, 168
Moret PR, 266
Murry CE, 1, 62, 79, 80, 89, 97, 185, 187
Muscarinic receptors, 151, 157, 186, 196, 267
Myocardial infarct, 123. *See also* Preconditioning at a distance.
catecholamine-related reduction of necrosis, 173-175, *174f*
reperfusion therapy, 115-116
size reduction, 1, *3f*, 80, *81f*, 82, 83-85, *84f*
mechanisms, 82-83
pertussis toxin, 155-156, *155f*
Myocardial injury
catecholamines, 169-171
Myocardial ischemia
catecholamine release, 168-169
Myocardial preconditioning, 1-3, *3f*, 19-20, 97-99, *192f*, 198-199, *199f*, 222, 224
at a distance, 79-80, 88-92, *90f*, *91*, *93f*
mechanisms, 85, *86f*, 87-88, *87f*
catecholamines, 180
G proteins, 178-179
protection against arrhythmias, 175, *176f*, 177-179, *177f*, *178f*
protection against contractile dysfunction, 171, *172f*, 173
reduction of necrosis, 173-175, *174f*
chronic high altitude hypoxia (HAH), 270
humans, 116-117
angina, 118-120, 123-124
balloon angioplasty, 120-121
cardiac surgery, 122-123
infarction, 123
'warm-up' phenomenon, 121-122
metabolic changes, 221-222, *223t*
pig model
duration of protection, 101, 104
infarct size (IS) and area at risk (AR), 99-101, *100f*, *102f*
transmural distribution, 101, *103f*
partial coronary artery occlusion without intervening
reperfusion, 104-106, *106f*, *107f*
ventricular pacing, 108, *109f*, *110f*, 112
tolerance to A_1-adenosine agonists, 200
Myocardial protection
delayed, 236-239, *237f*, 251, 256-257
cardiac pacing, 254-257, *255f*, *256f*, *257f*
clinical implications, 244-245
drug-induced, 251-254
mediators, 242, 244
timecourse, 239-240, *240f*
triggers and signaling pathways, 240-242, *243f*
chronic high altitude hypoxia (HAH), 262-270
Myocyte. *See* Cardiomyocyte.
Myoglobin, 266
Myosin, 269

N

Na^+/Ca^{2+} exchange, 26, 169, 179
Na^+/H^+ exchange, 26, 169, 179
Na^+/K^+ ATPase, 179
Nattel S, 179
Nicorandil, 124, 132, 133
Nitric oxide (NO), 49, 50-51, 52, 214, *215t*, 216, 253, 256
Nitric oxide synthase, 254, 256
Noma A, 130
Noradrenaline, 54-55, 157, 168-169, 171, *172f*, 173, 174, *174f*, 254. *See also* Catecholamines.
"No-reflow" phenomenon, 23
5'-Nucleotidase, 89, 122, 173
memory of preconditioning, 196

O

Okadaic acid, 200
Okazaki Y, 121
Oldroyd KG, 120
Opiate receptors, 157
Opie LH, 53
Ottani F, 119
Ovize M, 98, 104, 105
7-Oxo-prostacyclin, 252

P

P-1075, 132
Parratt JR, 177, 198, 208
Pentobarbital, 8, *9t,* 42, 136
Pertussis toxin, 148, 152, *153f, 154t, 155f, 156f,* 179, 186, 220
Phenylephrine, 71, 171, 196, 197, 200
8-Phenyltheophylline (8PT), 5, *5f,* 70, 71
Phorbol 12-myristate 13-acetate (PMA), 188
Phosphatase inhibitors, 200
Phosphodiesterase, 6, 53, 54
Phosphoinositide metabolism
  G proteins, 157
Phospholipase A_2 (PLA_2), 149
Phospholipase C (PLC), 10-12, *11f,* 148, 149, 179, 186
PIA (phenylisopropyl adenosine), 200
Pinacidil, 132, 133, 135, 201
Piper HM, 21
Platelet thrombi, 118
Podzuweit T, 36
Polymyxin B, 10, 11, *11f,* 73, 175, 189
Prazosin, 174
Prostacyclin, 6, 50, 51, 252, 268
Prostaglandins, 268
Protein kinase A (PKA), 169, 186
Protein kinase C (PKC), 10-12, *11f,* 27, 50, 69-70, 72-73, 82-83, 139, 148, 169, 175, 180, 186
  bradykinin, 198
  delayed myocardial protection, 241-242, *243f*
  G proteins, 157, 186
  myocardial preconditioning, 188-189, *190t-191t,* 192-194, *192f, 195f,* 196-197, *197f,* 198-199, *199f,* 201, 220-221
Przyklenk K, 12, 69, 73, 108

R

R-PIA (R(-)N_6-2-phenylisopropyl adenosine), 5, 10, *11f,* 138, 212, *212f*
Rakusan K, 264
Renal ischemia, 91
Reperfusion injury, 68-69. *See also* Mechanical reperfusion injury.
Reserpine, 171, 174
Richard G, 178
Rigor bonds, 27
Rohmann S, 136
Rona G, 252
Ruiz-Meana M, 23

S

Sanz E, 28
Sarcolemma, 21
  ischemia/reperfusion injury, 26
Schaper (initial not given), 239
Schulz R, 136
Second window of protection. *See* Myocardial protection, delayed.
Selye H, 252
Shiki K, 40
Shizukuda Y, 69
Simpson PC, 186
Sleph PG, 10
Sodium 5-hydroxydecanoate, 133, 136, 138
8-SPT (8-sulfophenyltheophylline), 139, 188, 241
Staurosporine, 10, 11, *11f,* 188, 192-193
Steare SE, 235
Strasser RH, 194, 242
Stunning, 3-4, 62, 68, 69
Superoxide dismutase (SOD), 238, 242, 244, 269
SUR (sulfonylurea binding site), 140
Szekeres L, 252

T

Tajima M, 270
Tan HL, 137
Tanaka M, 239
Tani M, 73
Thornton JD, 8, 136, 153, 173, 188
Thrombolysis in Myocardiual Infarction 4 (TIMI 4) trial, 80, 119
Thyroid metabolism, 267
Tomai F, 137
Toombs CF, 8, 136, 174
Tricyclic antidepressants, 169
Tsuchida A, 6, 71, 138, 175
Turek Z, 264
Turrens JF, 69
Tyramine, 173

V

Van Winkle DM, 138
Van Wylen DGL, 7, 196
Vegh A, 108, 177, 238
Ventricular antiarrhythmics
  adrenoceptors, 170
  preconditioning, 35-36, *37f,* 38-44, *39f, 40f, 41f, 42f, 44f, 45f,* 55, *213f,* 252
    catecholamines, 175, *176f,* 177-179, *177f, 178f*
    endotoxin, 253
    mechanisms, 44, 46, *46t,* 198
      endogenous substances, 53-54
      endothelium-derived substances, 46-52, *47f, 49f, 50f*
      inhibition of cardiac sympathetic responses, 54-55
    pertussis toxin, 153-154, *154t, 155f, 156f*
Ventricular pacing, 108, *109f, 110f,* 112
Vogt A, 12, 189

W

Walker DM, 117
Weselcouch EO, 173
Williams DO, 122
Winter CB, 178

X

Xylazine, 8, 9, *9t*, 136

Y

Yamasaki K, 2
Yamashita N, 242
Yao Z, 72, 134, 135
Yellon DM, 122, 137, 139, 235, 251
Ytrehus K, 10, 99

Z

Z-1046, 54, 55